# NITROGEN AND ENERGY NUTRITION OF RUMINANTS

Ray L. Shirley
Department of Animal Science
University of Florida
Gainesville, Florida

1986

ACADEMIC PRESS, INC.
**Harcourt Brace Jovanovich, Publishers**
Orlando   San Diego   New York   Austin
London   Montreal   Sydney   Tokyo   Toronto

Copyright © 1986 by Academic Press, Inc.
ALL RIGHTS RESERVED.
NO PART OF THIS PUBLICATION MAY BE REPRODUCED OR
TRANSMITTED IN ANY FORM OR BY ANY MEANS, ELECTRONIC
OR MECHANICAL, INCLUDING PHOTOCOPY, RECORDING, OR
ANY INFORMATION STORAGE AND RETRIEVAL SYSTEM, WITHOUT
PERMISSION IN WRITING FROM THE PUBLISHER.

ACADEMIC PRESS, INC.
Orlando, Florida 32887

*United Kingdom Edition published by*
ACADEMIC PRESS INC. (LONDON) LTD.
24–28 Oval Road, London NW1 7DX

**Library of Congress Cataloging in Publication Data**

Shirley, Ray L.
  Nitrogen and energy nutrition of ruminants.

  Bibliography: p.
  Includes index.
  1. Nitrogen in animal nutrition.   2. Ruminants
—Feeding and feeds.   3. Bioenergetics.   I. Title.
II. Title: Energy nutrition of ruminants.
SF98.N5S45   1985       636.08'52       85-22984
ISBN 0–12–640260–4 (hardcover) (alk. paper)
ISBN 0–12–640261–2 (paperback) (alk. paper)

PRINTED IN THE UNITED STATES OF AMERICA

86 87 88 89     9 8 7 6 5 4 3 2 1

# Contents

Foreword     xi

Preface     xiii

**1 Introduction**     1

**2 Feed Intake**     5

**3 Activity of Rumen Microbes**     9

    I. Introduction     9
    II. Nutrition of Rumen Bacteria     12
    III. Methanogenic Bacteria     19
    IV. Proteolytic Bacteria     20
    V. Degradation of Amino Acids     24
    VI. Influence of Urea as a Soluble-Nitrogen Factor     25
    VII. Microbial Adaptation to Biuret     25
    VIII. Protein Synthesis by Rumen Microbes     26
    IX. Interactions Between Bacterial Species     33
    X. The Rumen Protozoa     34
    XI. Microbes in Relation to Rumen Functions     37
    XII. Efficiency of Energy Utilization by Microbes     42
    XIII. Lactic Acid and Other Nonvolatile Organic Acids     43
    XIV. Acidosis Syndrome and Performance of Ruminants     46
    XV. Bloat in Ruminants     53
    XVI. Dilution Rates in the Rumen     57
    XVII. Rumen Fermentation Manipulation     58
    XVIII. Oxygen in the Rumen     59
    XIX. Effect of Inorganic Ions     60
    XX. Temperature and Rumen Function     61
    XXI. Modeling of Nitrogen Metabolism in the Rumen     61
    XXII. Effect of Pesticides on Rumen Microbes     63
    XXIII. Markers in Ruminant Nutrition     64
    XXIV. Toxic Substances in the Rumen     67
    XXV. Enzyme Supplements and Digestibility     74
    XXVI. Gluconeogenesis in Cattle     75

## 4 Monensin and Other Antibiotics Fed to Ruminants — 79

    I. Introduction — 79
    II. Monensin — 79
    III. Lasalocid — 84
    IV. Amicloral — 86
    V. Avoparcin — 86
    VI. Tylosin — 86
    VII. Chlortetracycline — 87

## 5 Nutritional Energetics — 89

    I. Introduction — 89
    II. Nutritional and Energy Terms — 90
    III. Fermentation Pathways — 95
    IV. Rationale of Energy Systems for Ruminants — 97
    V. The California Net Energy System (CNES) — 100
    VI. Net Protein ($NP_m$ and $NP_g$) Requirements — 114
    VII. The Blaxter (ARC) ME System — 117
    VIII. Net Energy Value of Feeds for Lactation — 119
    IX. Lactation and Reproduction Studies — 121
    X. Conditions Affecting DE and ME of Diets — 123
    XI. Effect of VFA on ME — 125
    XII. Effect of Sex on Utilization of Energy — 125
    XIII. Relative Efficiency of Fat and Protein Synthesis — 126
    XIV. DE Cost of Protein Production — 126
    XV. Preformed Protein Sources in Finishing Diets — 127
    XVI. Effect of Withdrawal of Protein in Finishing Diets — 128
    XVII. Insolubility of Protein in Ruminant Feeds — 129
    XVIII. Starch Utilization by Ruminants — 130
    XIX. Dietary Fat for Ruminants — 135

## 6 Amino Acids — 141

    I. Introduction — 141
    II. Beef Cattle Fed Supplementary Amino Acids — 142
    III. Amino Acids for Lactating Cows — 143
    IV. Amino Acids for Sheep — 144
    V. Degradation of Amino Acids in the Rumen — 144
    VI. Upgrading of Plant Protein in the Rumen — 145
    VII. Methionine Hydroxy Analog — 145
    VIII. Labile Protein and Protein Turnover — 147

## 7 Nonprotein Nitrogen Utilization — 149

    I. Introduction — 149
    II. Nonprotein Nitrogen Adaptation and Utilization — 151
    III. Sulfur Requirements for NPN Utilization — 154

Contents            vii

|     |     |     |
| --- | --- | --- |
| IV. | Quality of Nitrogen Sources Fed Ruminants | 155 |
| V. | Ruminants Fed Low-Protein Forages with NPN | 156 |
| VI. | Ruminal Ammonia Concentration and NPN | 161 |
| VII. | Ammonia Levels in the Rumen of Feedlot Cattle | 164 |
| VIII. | Ammonia and Synthesis of Metabolizable Protein | 165 |
| IX. | Factors in MP Formation and Utilization | 165 |
| X. | Nitrogen Utilization for Dairy Cattle | 167 |
| XI. | Urea for MP in the Urea Fermentation Potential (UFP) System | 168 |

## 8  Energy and Nitrogen Utilization in Feedstuffs    173

|     |     |     |
| --- | --- | --- |
| I. | Introduction | 173 |
| II. | Net Energy | 173 |
| III. | Corn | 174 |
| IV. | Grains Treated with Alkali | 175 |
| V. | High-Moisture Corn | 175 |
| VI. | Corn Silage | 177 |
| VII. | Grass Silages | 183 |
| VIII. | Sorghum Grain | 185 |
| IX. | Citrus By-Products | 188 |
| X. | Sunflower Meal | 191 |
| XI. | Fermented Ammoniated Condensed Whey (FACW) | 192 |
| XII. | Feather Meal and Hair Meal | 193 |
| XIII. | Molasses and Other Liquid Feeds | 194 |
| XIV. | Alcohol Production By-Products | 198 |

## 9  Protein and Energy in Forages and Roughages    203

|     |     |     |
| --- | --- | --- |
| I. | Introduction | 203 |
| II. | Utilization of Corn and Sorghum Residues | 206 |
| III. | Grazing Intensity | 208 |
| IV. | Supplementation of Grazing Ruminants | 209 |
| V. | Cereal Residue | 212 |
| VI. | Mechanical Processing of Cereal Straws | 214 |
| VII. | Sugarcane and Bagasse | 214 |
| VIII. | Aquatic Plants | 215 |
| IX. | Pineapple Plant Forage | 217 |
| X. | Apple Pomace | 218 |
| XI. | Dried Celery Tops | 219 |
| XII. | Tomato Pulp | 219 |
| XIII. | Potato Processing Residue | 220 |
| XIV. | Coffee Grounds | 220 |
| XV. | Paper | 221 |
| XVI. | Pecan Hulls | 221 |
| XVII. | Oyster Shells and Plastic Polymers | 222 |
| XVIII. | Wood and Wood By-Products | 222 |
| XIX. | By-Products of the Essential Oil Industry | 226 |

viii  Contents

|  |  |
|---|---|
| XX. Lignocellulose Materials with Alkali | 228 |
| XXI. Enhancing Utilization of Concentrate Feeds | 231 |
| XXII. Infrared Reflectance Spectroscopy | 233 |

## 10  Effects of Processing Feedstuffs on Nitrogen and Energy    235

|  |  |
|---|---|
| I. Introduction | 235 |
| II. Corn-Processing Studies | 235 |
| III. Processing of Sorghum Grains | 237 |
| IV. Processing of Barley | 239 |
| V. Processing of Wheat | 240 |
| VI. Cottonseed Meal Processing | 240 |
| VII. Soybean Meal Processing | 240 |
| VIII. Processing of Forages | 241 |
| IX. Steam Treatment of Crop Residue | 241 |

## 11  Production Practices Affecting Nitrogen and Energy Nutrition    243

|  |  |
|---|---|
| I. Introduction | 243 |
| II. Spring versus Fall Calving | 243 |
| III. Growth Rate and Reproduction of Beef Heifers | 244 |
| IV. Forage Grazing and Concentrate Supplements | 245 |
| V. Caloric Efficiency of Cow–Calf Production | 246 |
| VI. Creep Feeding | 247 |
| VII. Feeding Frequency | 248 |
| VIII. Compensatory Growth | 248 |
| IX. Finishing Cattle in Drylot | 252 |

## 12  Nitrogen and Energy in Animal Excreta    257

|  |  |
|---|---|
| I. Introduction | 257 |
| II. Processing and Handling of Animal Excreta | 257 |
| III. Utilizing the Nutrients in Animal Excreta | 258 |
| IV. Poultry Litter | 258 |
| V. Swine Excreta | 261 |
| VI. Cattle Excreta | 261 |
| VII. Feeding Value of Methane Fermentation Residue | 263 |
| VIII. Sewage Sludge in Livestock Diets | 263 |

## 13  Minerals and Water in Nitrogen and Energy Nutrition    265

|  |  |
|---|---|
| I. Introduction | 265 |
| II. Nitrogen and Sulfur | 266 |
| III. Effect of Calcium | 267 |
| IV. Nitrogen and Trace Minerals | 268 |
| V. Water in Ruminant Nutrition | 268 |

|     |                                              |     |
| --- | -------------------------------------------- | --- |
|     | VI. Water Intake in Dairy Cows               | 272 |
|     | VII. Water Intake in Sheep                   | 272 |

## 14 Body Composition versus Nutritional and Other Factors — 275

    I. Introduction — 275
    II. Effect of Diet, Breed, and Sex — 276
    III. Realimentation of Cull Cows — 279
    IV. Dietary Fats — 280
    V. Implants — 281
    VI. pH of Muscle — 281
    VII. Predicting Composition of Beef Carcasses — 281
    VIII. Body Composition of Live Animals — 283

## 15 Endocrines and Nitrogen and Energy Nutrition — 285

    I. Introduction — 285
    II. Crossbreeding and Hormones — 285
    III. Reproduction and Nitrogen and Energy Intake — 286
    IV. Milk Production — 286

## 16 Effects of Ambient Temperature on Utilization of Nutrients — 289

    I. Introduction — 289
    II. Fasting Body Weight Loss — 289
    III. Metabolic Body Rate — 290
    IV. Calorigenic Effect or Heat Increment — 290
    V. Interactions with Environmental Temperature — 290
    VI. Heat Stress — 291
    VII. Effect of Cold — 297
    VIII. Housing and Management — 299

## 17 Composition of Feeds for Ruminants — 301

    I. Introduction — 301
    II. Dry Matter — 301
    III. Protein — 301
    IV. Crude Fiber and NDF — 302
    V. Energy — 302
    VI. Minerals — 303
    VII. Vitamin A — 303
    VIII. Tables on the Composition of Feedstuffs — 304

**Glossary** — 305

**Bibliography** — 309

**Index** — 349

# Foreword

This is the eighth in a series of books in animal feeding and nutrition. The books in this series are designed to keep the reader abreast of the rapid developments in this field that have occurred in recent years. As the volume of scientific literature expands, interpretation becomes more complex, and a continuing need exists for summation and for up-to-date books.

*Nitrogen and Energy Nutrition of Ruminants* is written by Dr. Ray L. Shirley, a distinguished scientist who is recognized worldwide for his outstanding work in animal nutrition and who has done an excellent job in assembling a large volume of information on the subject. He has brought together both basic and applied research information and indicated how it can be used in ruminant nutrition. The book is written to fit the needs of a course in nitrogen and energy nutrition of ruminants, but can also be used as a reference for other courses in nutrition and by college and university students and teachers. It is a valuable source of information for county agents, farm advisors, teachers of vocational agriculture, consultants, veterinarians, and livestock producers, and it will also be helpful to feed manufacturers, dealers, and others concerned with producing the many different supplements, feeds, and other ingredients used in ruminant feeding and nutrition.

Increasing attention is being paid to the seriousness of the world's food problem. Many third world countries have increased food production, but this increase is not keeping pace with rapid population growth. Over one billion people now suffer from chronic malnutrition. Every $2\frac{1}{2}$–3 years, the world's population increases by 220–240 million people, and about 87% of this population growth is taking place in countries which are the least able to feed themselves.

Many scientists feel there is a need to double animal protein production in the next 20 years in order to improve the protein status of the world's rapidly growing population. In addition to excellent quality protein, animals provide many important vitamins and minerals. The developing countries of the world have about 60% of the world's animals but produce only 19% of the world's meat, milk, and eggs. Better feeding and nutrition would increase their production of animal foods. One important problem is the development of feeding programs which provide the energy and nitrogen needed in animal diets. This book provides information on a wide range of feed resources, including range, grasslands, plant and animal by-products, cellulosic wastes, crop residues, roots, nuts, as well as other vegetable crops, fruit crops, and animal wastes, for animals that otherwise would contribute little in feeding mankind. This book can therefore be very helpful in increasing animal food products for human consumption throughout the world.

Tony J. Cunha

# Preface

This book was written as an expansion of notes utilized by the author in a course on nitrogen and energy nutrition of ruminants taught at the University of Florida. No available book has provided sufficient coverage of the many new developments and insights on nitrogen and energy nutrition for the course, and one summarizing the more recent publications in the field is especially needed for teachers, as well as for extension people, consultants, and producers of cattle and sheep.

This volume covers research on various nitrogen and energy feedstuffs and defines terminology commonly utilized in nitrogen and energy nutrition. The utilization of nitrogen and energy in oilseed meals, fish meals, cereal grains, distillers' residues, molasses, silages, grasses, hays, crop residues, animal waste, and nonprotein nitrogen sources is discussed. Details are given on development and utilization of net energy systems, systems for balancing total nitrogen, and nonprotein nitrogen with total digestible nutrients (TDN) or energy components of ruminant diets. Discussions are presented on metabolism, feedlot, milking, and grazing trials. Growth stimulants, processing of feedstuffs, type of animal, and environmental and management factors that affect feed intake, growth, feed efficiency, and quality of product are reviewed.

Emphasis is given to the contributions of ruminal microbes in upgrading forage and nonprotein nitrogen sources to higher-quality bacterial protein, as well as their ability to downgrade high-quality protein and waste nitrogen when protein is fed in excess of microbial needs. Research is presented on means to increase bypassing of the rumen to prevent nitrogen wastage when ruminants are fed concentrate diets. Contributions of ruminal microbes in utilizing cellulosic materials as lignocellulose and hemicellulose as well as starch and other carbohydrates are discussed.

The author wishes to acknowledge Professors C. B. Ammerman, D. B. Bates, J. H. Conrad, G. K. Davis, J. F. Easley, Don Hargrove, W. E. Kunkel, J. K. Loosli, Lee McDowell, A. Z. Palmer, R. S. Sand, and A. C. Warnick, who read various chapters and gave helpful suggestions. Also, thanks are due to Sarah McKee and Cheryl Coombs for typing the book and to my wife, Sarah, for encouragement and editorial help.

Ray L. Shirley

# 1

# Introduction

Nitrogen and energy are closely associated dietary factors in the nutrition of ruminants. Ruminal microbes require nitrogen for cellular protein synthesis and multiplication, since they utilize energy from lignocellulose and other cellulosic cell wall constituents as well as from starch and simpler metabolites. Certain phases of nitrogen and energy nutrition can be separated from other nutrients such as minerals and vitamins, but the many interrelationships between nitrogen and energy appear to warrant a book dealing primarily with these two dietary factors. Nitrogen and energy feedstuffs greatly exceed the other dietary factors, both in quantity and in cost, in commercial operations with ruminants.

Understanding of the nitrogen and energy nutrition of ruminants has been enhanced by investigations in many phases of basic chemistry, physics, microbiology, physiology, endocrinology, genetics, and environment, as well as general animal husbandry. Basic chemistry concepts apply to all nutrition, and concepts of physics apply especially to energy units and body functions. The many bacterial and protozoal species of the rumen allow ruminants to utilize grasses, roughages, and many waste products that have a high content of lignocellulose, hemicellulose, and other cell wall constituents that nonruminants cannot utilize. Meeting the dietary nitrogen requirements of the microbes results in a significant increase in their capacity to derive energy from such refractory dietary ingredients.

The term *dietary nitrogen* instead of *dietary protein* is commonly used with ruminants. This is because ruminal microbes can utilize nonprotein nitrogen (NPN) sources such as urea, biuret, and ammonium salts, as well as plant and animal proteins. Nonruminants have very limited, if any, capacity to utilize NPN. Some ruminal microbes prefer ammonia to amino acids for synthesis of bacterial cell proteins. The rumen's microbial population can function at above the maintenance level of the host with only NPN in the diet if sufficient dietary energy is provided. If enough nitrogen from either protein or NPN sources is present, ruminants can obtain energy from grasses and other roughages for maintenance plus the production of meat, milk, and wool.

## 1. Introduction

Ruminal microbes degrade dietary protein to amino acids, and then to ammonia and various non-nitrogen-containing fragments. The microbes then resynthesize microbial protein from these substances. The process results in an upgrading of forage protein to microbial protein that has higher levels of essential amino acids and greater biological value than plant proteins. However, with the high-concentrate diets fed ruminants for increased meat and milk production, the capacity of ruminal microbes to degrade dietary protein may exceed their capacity to resynthesize ammonia into microbial protein, and ammonia nitrogen may be lost through urinary urea excretion. To alleviate this loss of nitrogen and obtain maximum growth or production by ruminants, many studies have been made on factors that allow the dietary protein to bypass the rumen when it exceeds the requirements of the microbes.

Normal bypassing of the rumen by dietary protein with high-concentrate diets is generally highest with grains, lowest with forages, and intermediate with oilseed meals. Heating of protein or treatment of proteinaceous materials with formaldehyde or tannic acid will increase bypassing and decrease the loss of dietary nitrogen due to excessive deamination in the rumen. Residues from alcohol fermentation of grain and fish meals are sources of protein that are resistant to degradation in the rumen.

Recent development of the Urea Fermentation Potential (UFP) system at Iowa State University, Ames and the total digestible nutrients (TDN)–crude protein system at the University of Wisconsin, Madison, as well as the crude protein and metabolizable energy requirement tables of the National Research Council (NRC) bulletins, emphasize the importance of properly balancing dietary nitrogen and energy for various classes of ruminants and various levels of production. Understanding these systems is essential to the economical utilization of urea as a source of dietary nitrogen.

Utilization of crop residues, alcohol and methane fermentation residues from grains, wood by-products, and animal wastes should provide energy and nitrogen sources for ruminants that are economical for certain levels of production. A proper balance of available nitrogen and energy will enhance the nutritional value of such feedstuffs. Treatment of lignocellulosic materials with alkali has been demonstrated to increase greatly the store of carbohydrates available for ruminal microbes.

The development of large feedlots has necessitated studies on the more precise energy requirements for various weights and types of cattle. The California Net Energy System, which recognizes the net energy for maintenance ($NE_m$) and net energy for gain ($NE_g$) requirements of various weights of cattle and rates of gain, has provided a guideline for optimum feeding programs for feedlot cattle. By utilizing the UFP system or the TDN–crude protein system for dietary urea when applicable, dietary nitrogen should be provided efficiently. For example, an understanding of the protein and energy requirements of lactating cows has greatly enhanced dairy milk production, as well as meeting the dietary requirements of lactating beef cattle and sheep.

Maximum production of meat and milk by ruminants has been increased greatly by research on dietary requirements for common salt, phosphorus, calcium, cobalt, copper, iron, sulfur, iodine, and other minerals. Ruminant production has been enhanced by monensin, lasalocid, Ralgro, Synovex, melangestrol acetate (MGA), and other growth factors. Antibiotics and various chemicals that control disease and parasites are also essential to the efficient utilization of nitrogen and energy in ruminant diets.

Good breeding and selection of animals, and proper supplies of supplemental minerals, nitrogen, and energy feeds to ruminants on pasture during drought and nongrowing seasons, result in more healthy and productive livestock. Of course, management of ruminants exposed to extreme cold, heat, wind, rain, mud, and other stress factors is essential to prevent maintenance requirements of nutrients from becoming excessive. Finally, healthy, productive ruminants need a well-balanced diet that promotes sufficient feed intake to realize the potential of the animals and the goals of the producer.

# 2
# Feed Intake

The only input to the animal's energy reservoir is feed. Some outputs are quite constant, such as basal metabolic rate, while others, such as lactation, are dependent on energy balance. A single unit of energy input is a meal; however, a small but consistent delay in satiety could cause energy to accumulate in the animal's body. Energy balance is not maintained in ruminants when poor-quality roughage is fed, and insufficient amounts may be eaten due to bulkiness or unpalatability. Palatable concentrate diets readily lead to energy accumulation in feedlot cattle and sheep.

If energy balance is considered the regulated system and feed intake is an important regulator of that system, a controller system may be described, such as that proposed by Baile (1968). Baile's scheme summarizes the components thought to control the feed intake of ruminants. The input to the system is feed and the outputs include heat loss, work, milk, reproduction, and urine and fecal energy losses. At least three areas were identified by Baile (1968) as part of the controller system:

1. One component of the feedback system is related to the state of lipid depots, i.e., maintenance of lipostasis. A second component is involved in the control of meal size by hunger and activity. It appears to include neither temperature changes nor gastrodistension, although either may function under some conditions as a safety valve. The volatile fatty acids, especially acetate, may also be a component of the feedback system. Feed intake is depressed following intraruminal injections of acetate, propionate, or butyrate.

2. It is likely that there are stretch receptors in the rumenoreticulum that limit feeding. The rumenoreticulum may be an important receptor or detector site in the ruminant for control of feed intake.

3. The ventromedial area of the hypothalamus has at least some important characteristics of a memory or reference element. Temporary or permanent loss of part of this area causes an increase in the body's desired energy level. The lateral area of the hypothalamus has contradictory characteristics. When it is stimulated sufficiently, it mobilizes feeding; when it has been lesioned, feeding is not mobilized, even though severe depletion may exist.

It is not likely that glucose provides feedback for meal-to-meal control of feeding in ruminants, as it does in nonruminants, since neither blood glucose nor insulin levels change much with feeding; intravenous, intraperitoneal, or intracerebral–ventricular injections of glucose and/or intraabomasal injections of a liquid diet high in short-chain carbohydrates cause little or no decrease in feed intake, and glucose is not usually an important end product of digestion in ruminants (Baile, 1968).

More recently, Della-Fera and Baile (1984) pointed out that while the central nervous system and certain areas of the brain have primary roles in controlling hunger and satiety, peptides of the central nervous system play important roles in feed intake. There is evidence that opiate peptides are involved in hunger and that cholecystokinin peptides are involved in satiety. Dinius and Baile (1977) fed elfazepam at levels of 0.5, 1.0 or 2.0 ppm in a 90% ground orchard grass hay diet to beef steers for 117 days. Animal responses to the three levels of elfazepam were similar. Average daily gain, daily dry feed intake, and feed–gain ratios were 0.41, 8.2, and 20.7 kg, respectively, for the control steers and 0.52, 8.8, and 17.3 kg for the treated steers. In a similar trial with steers with 1.0 ppm elfazepam and 30 ppm monensin sodium added to the feed and Synovex-S implantation, feed intake was increased by elfazepam and decreased by monensin. The feed–gain ratio was improved by the implant.

Balch and Campling (1962) suggested that voluntary intake of roughages is related to the amount of digesta in the reticulorumen, which, in turn, is dependent on the rate of digestion of feed particles and their rate of passage out of the rumen. Montgomery and Baumgardt (1965) found that as the nutritive value of the diet increased, factors regulating intake changed. They suggested that when diets are high in nutritive value, food intake may be regulated to keep energy intake constant. McCullough (1969) fed six diets *ad libitum* to steers, providing ratios of concentrate to hay of 100 : 0, 95 : 5, 90 : 10, 80 : 20, 70 :30, and 60 : 40. As the proportion of hay in the diet increased, the daily dry matter intake decreased. The net energy intake tended to increase as the proportion of concentrate rose from 60 to 90% but then remained fairly constant. The addition of small amounts of hay to the concentrate diet caused an increase in intake.

When Bae *et al.* (1981) fed dry Holstein cows four hay levels (50, 75, 100 and 125% of the NRC recommended dry matter intake), they found that the greater levels resulted in increased eating and rumination times and in a greater number of rechewings and boluses. Eating time rose with increases in cell wall constituents. When Holstein steers were fed a complete mixed ration by Chace *et al.* (1976), diurnal feeding patterns were observed with 60.9% of the meals occurring between 0600 and 1800 hours. For individual meals, body weight accounted for less than 30% of the variation. Meal size was relatively constant in grams per kilogram or grams per kilogram of body weight raised to the three-fourths power.

Bond *et al.* (1976) fed twin beef steers *ad libitum* water and diets that contained 0, 30, or 80% forage and found that the feeding and drinking patterns were similar before, during, and after deprivation of feed, water, or both. Feed intake was reduced about 50% regardless of the type of diet when water was withheld from steers. Fenderson and Bergen (1976) fed four diets containing 10.7, 20.2, 32.5, and 40.0% protein to steers and observed that dry matter intake for the two high-protein diets was depressed primarily during days 2 and 3, with a subsequent recovery to initial levels between days 5 and 10.

Other factors affecting feed intake of ruminants include (1) nutritional deficiencies, (2) metabolic disorders, (3) environmental temperature, and (4) drugs. These conditions are discussed at appropriate places in the text.

# 3

# Activity of Rumen Microbes

## I. INTRODUCTION

There is a great variety of microbes in the rumen. The microbial ecosystem allows the rumen to control the energy and nitrogen nutrition of cattle and other ruminants. Through enzymatic processes, microbes break down cellulosic, amylolosic, and nitrogenous compounds in the diet and convert them into utilizable by-products and bacterial cells (Table 3.1, Figs. 3.1 and 3.2). Some of the fermentation products may be absorbed directly across the ruminal wall or may pass into the lower intestinal tract, where they are absorbed directly or further digested prior to absorption.

Disappearance of dry matter from the rumen is a complex process and depends on a number of factors, including chemical and physical properties of feeds, specific fermentation rates of components of feedstuffs, microbial growth rates, and passage rates to the omasum (Baldwin, 1970; Beever et al., 1972; Hungate et al., 1970; Maeng et al., 1976; Bergen and Yokoyama, 1977; Maeng and Baldwin, 1976a).

Intensive degradation of ingesta takes place in the rumen as carbohydrate polymers are hydrolyzed to saccharides; these saccharides, in turn, are fermented to numerous products, as shown in Fig. 3.1 (Russell and Hespell, 1981). While many intermediates are formed, those that accumulate are acetate, propionate, butyrate, carbon dioxide, and methane. Diet composition and frequency of feeding affect the ratios of these products. Unusual feeding conditions may cause other products such as lactate, ethanol, and formate to accumulate. Rumen microbial cellulases and glucanases described by King and Vessal (1969) have a principal role in the early stages of roughage breakdown.

Proteins are degraded by microbes to amino acids, ammonia, carbon dioxide, and short-chain, branched-chain, or aromatic fatty acids, as shown in Fig. 3.2 (Russell and Hespell, 1981). While ammonia is a major source of nitrogen for bacterial growth, peptides and amino acids may provide approximately 40% of bacterial nitrogen with high-fiber/low-protein diets (Nolan and Stachiw, 1979).

TABLE 3.1.
Ruminal Bacteria and Their Fermentative Properties[a]

| Species | Animal diets[b] | Functionality[c] | Fermentation products[d] |
|---|---|---|---|
| *Bacteroides succinogenes* | Many | C, A | F, A, S, –C |
| *Ruminococcus albus* | Many | C, X | F, A, E, H, C |
| *Ruminococcus flavefaciens* | Many | C, X | F, A, S, H, –C |
| *Butyrivibrio fibrisolvens* | Many | C, X, PR | F, A, L, B, E, H, C |
| *Clostridium lockheadii* | Coarse hay | C, PR | F, A, B, E, H, C |
| *Streptococcus bovis* | High grain | A, S, SS, PR | L, A, F (?) |
| *Bacteroides amylophilus* | High grain | A, P, PR | F, A, S, –C |
| *Bacteroides ruminicola* | Many | A, X, P, PR | F, A, P, S, –C |
| *Succinimonas amylolytica* | Forage-grain | A, D | A, S, –C |
| *Selenomonas ruminantium* | Many; grain | A, SS, GU, LU, PR | A, L, P, H, C |
| *Lachnospira multiparus* | Legume pasture | P, PR, A | F, A, E, L, H, C |
| *Succinivibrio dextrinosolvens* | High grain | P, D | F, A, L, S, –C |
| *Methanobrevibacter ruminantium* | Many | M, H | M |
| *Methanosarcina barkeri* | Many; molasses | M, H | M, C |
| *Spirochete species* | Many | P, SS | F, A, L, S |
| *Megasphaera elsdenii* | High grain | SS, LU | A, P, B, V, CP, H, C |
| *Lactobacillus vitulinus* | Lush pasture; high grain | SS | L |
| *Anaerovibrio lipolytica* | Forage; high lipids | L | A, P, S, –C |
| *Eubacterium ruminantium* | Forage | SS | F, A, B, C |
| *Vibrio succinogenes* | | H | S |

[a]From Russel and Hespell (1981).
[b]It is doubtful that any one species is completely absent from any rumen, but given diets indicate where the organism is more numerous.
[c]C = cellulolytic; X = xylanolytic; A = amylolytic; D = dextrinolytic; P = pectinolytic; PR = proteolytic; L = lipolytic; M = methanogenic; GU = glycerol-utilizing; LU = lactate-utilizing; SS = major soluble sugar fermentor; H = hydrogen utilizer.
[d]F = formate; A = acetate; E = ethanol; P = propionate; L = lactate; B = butyrate; S = succinate; V = valerate; CP = caproate; H = hydrogen; C = carbon dioxide.

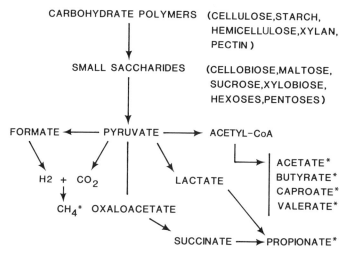

**Fig. 3.1.** Generalized scheme for ruminal degradation and fermentation of carbohydrates. The products marked with an asterisk represent terminal products and accumulate in the rumen (From Russell and Hespell, 1981).

Peptides and amino acids need to be deaminated to produce the branched-chain fatty acids that are growth factors for several species of bacteria, especially the cellulolytics (Allison *et al.*, 1958). Table 3.1 lists many of the species of rumen microbes involved in the breakdown of common ruminant diets (Russell and Hespell, 1981).

Ruminal microbes provide the enzymes necessary to break down cellulose,

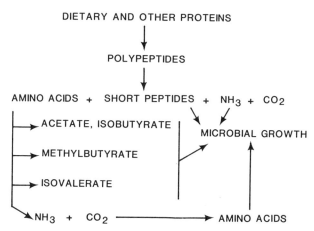

**Fig. 3.2.** Generalized scheme for ruminal degradation of proteins (From Russell and Hespell, 1981).

hemicellulose, starch, dextrin, pectin, protein, and lipids, as well as those that produce methane or utilize glycerol, lactate, hydrogen, and other intermediate substrates. The end products of some species of bacteria such as formate, ethanol, lactate, succinate, hydrogen, and carbon dioxide become the substrates of other species. The major pathway employed by rumen microbes for hexose breakdown is that of Embden–Meyerhof–Parnas (Joyner and Baldwin, 1966; Wallnofer et al., 1966). A likely pathway for degradation of pentoses and desoxyhexoses is a combination of a pentose cycle plus glycolysis (Turner and Robertson, 1979). While pyruvate and phosphenolypyruvate are major products formed from hexose and pentose degradation, they are further fermented to an array of products (Table 3.1). Rumen bacteria may have various morphological shapes such as rods, cocci, and vibrios. Bacteria vary considerably in physical dimension, motility, sporulation, capsulation, and substrate niche.

## II. NUTRITION OF RUMEN BACTERIA

All bacteria need a source of energy and nitrogen, but they are quite variable in their requirement for vitamins and minerals. It is difficult to give precise requirements. Some strains require $C_4$ to $C_6$ branched- and straight-chain fatty acids (Bryant and Doetsch, 1954; Bryant and Robinson, 1961b). Ammonia is a primary intermediate in the degradation and assimilation of dietary nitrogen and is required by many rumen bacteria (Bryant and Robinson, 1962). Amino acids and peptides are required by several bacterial species (Abou Akkada and Blackburn, 1963; Baldwin, 1970). Bryant and Robinson (1962) found that 82% of rumen bacteria grown on a relatively nonselective medium grew with ammonia as the sole source of nitrogen, 25% did not grow unless ammonia was present, and 56% used either ammonia or amino acids. Using labeled nitrogen, Al-Rabbat et al. (1971) reported that 61% of microbial nitrogen was derived from ammonia and 39% from amino acids and peptide nitrogen. These values correspond closely to those of Pilgrim et al. (1970), who reported that 78 and 64% of bacterial nitrogen was derived from ammonia in animals fed wheaten hay and alfalfa hay, respectively. The greatest percentage of microbial protein derived from urea nitrogen was with a diet that contained low protein.

*In vitro* trials with rumen bacteria utilizing urea as a source of nitrogen demonstrated that when glucose or cellulose was used as the substrate, methionine hydroxy analog (MHA) or DL-methionine increased bacterial nitrogen incorporation and the substrate digestion rate (Gil et al., 1973a), and showed that inorganic sulfate was as effective as MHA or DL-methionine in increasing bacterial protein production only when fermentation was continued beyond 18 hr with starch and 24 hr with cellulose. Glucose was fermented by rumen bacteria utilizing urea with or without MHA *in vitro* (Gil et al., 1973c), and the analog had no effect on the amino acid composition of bacterial protein. *In vitro* fermentation of glucose and urea by rumen bacteria produced an increase of about 2.5 times in the logarithmic growth rate due to the addition of MHA to the medium; and cystine, cysteine, and methionine each stimulated bacterial growth similarly to MHA (Gil *et al.*, 1973b).

Maeng *et al.* (1976) observed that the optimum ratio of nonprotein to amino acid nitrogen for rumen microbial growth was 75% urea nitrogen and 25% amino acid nitrogen. Washed rumen cell suspensions had more microbial dry matter, nitrogen, ribonucleic acid, and deoxyribonucleic acid in media containing urea plus amino acids compared to urea only (Maeng and Baldwin, 1976b).

Veira *et al.* (1980) fed to calves corn-basal diets that contained 10.2, 12.2, 14.1, and 16.1% crude protein and observed no apparent effect of the rumen ammonia concentration on microbial synthesis in spite of levels of less than 5 mg of N per 100 ml of rumen fluid with the low-protein diets. However, it may be that with low-protein diets, NPN utilization could be increased if the rate of ammonia formation from the natural protein was slowed. To test this possibility, Stern *et al.* (1978) determined rumen microbial protein synthesis with urea and heated and unheated peanut meal diets but found no difference in the amount of microbial protein synthesized per gram of dry matter digested due to the solubility of the peanut meal; however, the amount of microbial protein synthesized was lower with the urea-containing diets.

Magnesium, calcium, potassium, sodium, and phosphate are required by *Bacteroides succinogenes* and probably are required for many rumen bacteria (Bryant *et al.*, 1959). *In vivo* and *in vitro* studies by Evans and Davis (1966) showed that sulfur and phosphorus increased microbial activity for cellulose digestion. Vitamins such as biotin, *p*-aminobenzoic acid, folacin, thiamine, pyridoxine, and pantothenic acid are required by some rumen bacterial species (Hungate, 1966).

## A. Variation in the Number of Microbes

Kistner *et al.* (1962) found that short- and long-term fluctuations in bacterial counts of cellulose-, starch-, glucose-, xylose-, and lactate-fermenting organisms in individual animals were greater than corresponding diurnal changes. Similar or greater changes in the concentration of starch-digesting bacteria were reported by Hamlin and Hungate (1956) in two steers on grass–alfalfa hay, and by Hungate (1957) for different types of cellulolytic bacteria in 25 cows fed various combinations of timothy hay, concentrates, and salt. Bryant and Robinson (1961c) observed several fluctuations in bacterial concentration in a cow on a grain diet.

Figure 3.3 shows the effect of feed and water intake on total rumen bacterial counts in cattle fed hay compared to grain diets. There was a drop in count

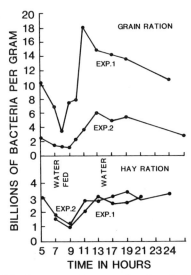

**Fig. 3.3.** Total bacterial colony counts from samples of ruminal contents collected at various times of day from a cow fed alfalfa hay or grain once a day (From Bryant and Robinson, 1961c).

between 5 and 7 hr corresponding to the first water intake of the day. Then there was a further drop 1 to 2 hr after feeding, possibly due to dilution of rumen fluid with feed and saliva. The bacterial concentration then increased for 2–4 hr and subsequently declined with the grain diet or plateaued with the hay diet to levels approaching those in the early morning before water and feed were consumed. The differences in counts in the two experiments when the animal was fed grain are difficult to explain, but the rumen pH was low during the second experiment.

Galyean et al. (1981a) studied the rumen fluid of steers deprived of feed and water for 32 hr, with and without transportation, and found that rumen fluid pH increased and bacterial and protozoal counts decreased in both groups compared to controls. Recovery of microbe counts was slower with the transported steers.

## B. Cellulolytic Microbes

Among the rod-shaped organisms are the cellulolytic bacteria known as *Bacteroides succinogenes, Ruminococcus flavefaciens, Ruminococcus albus, Clostridium longisporium, Butyrivibrio fibrisolvens,* and *Clostridium lochheadii. Bacteroides succinogenes* is a primary organism in the breakdown of cellulose in hay. Some strains ferment glucose, cellobiose, pectin, and starch, but not xylans (hemicellulose) (Hungate, 1966).

The cellulolytic cocci comprise up to 84% of the cellulolytic bacteria that develop in a cellulose–agar medium (Kistner et al., 1962). Most strains of cellulolytic cocci ferment cellulose and xylan and utilize cellobiose but not glucose or most other mono- and disaccharides (Bryant *et al.*, 1958b). Strains of rumen cellulolytic cocci commonly require short, branched-chain fatty acids, biotin, *p*-aminobenzoic acid, pyridoxamine, and some thiamin, riboflavin, folacin, pantothenic acid, methionine, or vitamin $B_{12}$. Ammonia is essential for most strains, and some prefer it to amino acids (Bryant, 1963).

King and Vessal (1969) pointed out that the cellulase enzyme complex has the following components: $C_1$ is an enzyme whose action is unspecified; it is required for the hydrolysis of highly oriented cellulose by $\beta 1 \rightarrow 4$ glucanases. The $\beta$-1→4 glucanases ($C_x$) are the hydrolytic enzymes, frequently referred to by the synonym *cellulase*. The $x$ in $C_x$ emphasizes the multicomponent nature of this enzyme. The $\beta$-1→4 glucanases consist of two types: (1) *exo*-$\beta$-1→4 glucanases, which successively remove single glucose units from nonreducing ends of cellulose chains, and (2) *endo*-$\beta$-1→4 glucanases, which attack internal linkages at random. The glucanases are highly active on the $\beta$-dimers of glucose, including cellobiose, and are called $\beta$-*glucodimerases*.

Cellulose in plant cell walls varies in purity from about 90% in cotton fibers to about 45% in wood, although it has a very similar molecular structure in cotton and wood (King and Vessal, 1969). Cotton fibers contain no intercellular substance, while wood fibers form a cohesive three-dimensional structure whose integrity is dependent on large amounts of intercellular substance. Both types of fibers have primary and secondary walls. Within each layer of the secondary wall, cellulose and other constituents occur in long, slender bundles called *microfibrils*. Within each microfibril, linear molecules of cellulose are bound laterally by hydrogen bonds and associated in various degrees of parallelism. Regions that contain highly ordered (parallel) molecules are called *crystalline regions*, and those of lesser order are called *amorphous regions*. The crystalline form increases as fermentation proceeds and the molecules reorient themselves into crystalline cellulose. This process accounts in part for the increased difficulty of hydrolysis of cellulosic materials as fermentation proceeds. The crystalline alignment of cellulose molecules provides less room for bacteria and their cellulases to contact.

The enzymatic degradation of cellulosic materials is dependent on a number of factors: (1) the moisture content of fiber; (2) the size and diffusibility of enzyme molecules in relation to the size and surface properties of the gross capillaries; (3) the degree of crystallinity; (4) the degree of polymerization; (5) the nature of extraneous substances; and (6) the nature, concentration, and distribution of substituent groups (King and Vessal, 1969). Moisture has three major roles in the fermentation of cellulose. Water swells fibers so that substrate is more available to enzymes; it provides a medium of diffusion for extracellular enzymes, as well as for nutrients liberated to supply microbes for their growth; and it is necessary for hydrolysis (King and Vessal, 1969). Many of the cellulases of rumen bacteria are extracellular rather than surface enzymes (King, 1956; Gil and King, 1957). Diffusibility of enzymes is affected by the capillary structure of cellulose. The length of the cellulose polymer varies from approximately 15 glucose units with γ-cellulose to 10,000–14,000 glucose units with α-cellulose. Thus, the degree of polymerization is important for exoglucanases. Fortunately, most cellulases are endogluconases and hydrolyze cellulose at random along the chain (King and Vessal, 1969).

Cellulose and hemicellulose of feedstuffs are relatively very insoluble and are slowly degraded by rumen microbes (Dehority, 1973; van Soest, 1973). In addition to solubility factors, structural factors such as lignification in association with forage fibers influence the fermentation rate of cellulose (van Soest, 1975), and delignification by chemical treatment increases cellulose digestion by rumen microbes both *in vitro* and *in vivo*. Hemicellulose degradation involves a relationship between cellulolytic bacteria that degrade the cellulose molecule and noncellulolytic strains of bacteria that utilize breakdown products.

Cellulose fibers usually contain about 1% ash, which includes essential elements for bacterial growth. Manganese, magnesium, calcium, and cobalt are stimulatory to cellulase production but mercury, silver, copper, chromium, and zinc salts are inhibitory (King and Vessal, 1969). Extraneous materials influence the susceptibility of fibers to enzymatic breakdown in several ways: as growth promoters such as thiamin; as poisonous substances such as phenolic materials that kill microbes; and as water-insoluble materials that physically reduce the accessibility of fibers to enzymes. In addition, there are certain specific enzyme inhibitors, and small amounts of nitrogen and phosphorus make many cellulosic materials too deficient to meet cellulolytic microbe requirements for these elements (King and Vessal, 1969).

Digestion of lignified fiber proceeds as if the cellulose consists of two definable components, one potentially digestible and the other indigestible (Waldo *et al.*, 1972). Passage of indigestible cellulose is proportional to the amount present, while the potentially digestible cellulose disappears from the rumen by both digestion and passage. Lignin, which is a complex two-dimensional polymer of *p*-hydroxycinnamyl alcohol and coniferyl units bound to cellulosic materials by physical or chemical bonds, needs to be dissociated before cellulase can be effective (King and Vessal, 1969).

Bryant and Burkey (1953) obtained total rumen microbial counts of $0.55 \times 10^8$ per milliliter of rumen fluid, with cellulytics constituting 15.6% of the total in forage-fed cattle.

## C. Hemicellulose Digesters

Hemicellulose consists largely of polymeric forms of xylose, arabans, hexosans, and uronic acids, and has a close association with cellulose and lignin in the cell walls of plants. It undergoes digestion in the rumen to about the same extent as cellulose (Heald and Oxford, 1953). Bacteria from the rumen of animals fed hay and grass produced 114 moles of acetic acid and 38 moles of propionic acid, respectively, per 100 moles of pectin (as galacturonic acid) (Howard, 1961).

## D. Starch Digesters

The monocellulolytic species such as *Streptococcus bovis, Bacteroides amylophilus, B. ruminicola, Succinimonas amylolytica,* and *Selenomonas ruminantium* include many starch-fermenting strains. *Streptococcus bovis* is one of the most commonly isolated starch digesters in the rumen (Perry and Briggs, 1955), and under certain conditions of changed feed, as when alfalfa hay–fed animals are suddenly given excess grain, their numbers increase almost explosively (Hungate *et al.*, 1952) and can cause fatal indigestion due to the high concentration of lactic acid produced.

Since starch is fermented, the numbers of *S. bovis* might be expected to be high in ruminants fed high-grain diets. Occasionally this is true (Gall and Huhtanen, 1951), but in most cattle well adapted to high-grain diets, the concentration of the microbes differs little from that in hay-fed animals (Hungate, 1957). *Streptococcus bovis* bacteria can be ingested by *Entodinium* protozoa (Gutierrez and Davis, 1959) and presumably serve as food. This may result in a rapid turnover of *S. bovis* organisms, and their concentration may not reflect their importance. Animals fed a high-grain diet had a higher turnover rate of lactate in the rumen than hay-fed ones (Jayasuriya and Hungate, 1959), suggesting that *S. bovis* was more active with concentrate diets. Ammonia can serve as the sole source of nitrogen for some strains of *S. bovis* (Wolin *et al.*, 1959; Prescott *et al.*, 1959), but in other strains, amino acids are assimilated in preference to ammonia (Bryant and Robinson, 1961c).

*Bacteroides ruminicola* has strains that digest starch (Bryant *et al.*, 1958a) and more types of sugars than *B. amylophilus*. These strains are found in the rumens of cattle fed almost any diet. Their fermentation products are largely succinic, acetic, and formic acids that are similar to those of other rumen *Bacteroides*. *Bacteroides ruminicola* strains appear to be relatively more important in animals fed low-starch diets, making up 64% of the cultivable starch digesters in an animal fed wheat straw but only 10% of the starch digesters in an animal fed only a grain mixture (Bryant *et al.*, 1958a).

*Succinimonas amylolytica* is a starch-fermenting microbe found in the rumen whose products resemble those of *Bacteroides* except that only a trace of propionic acid and no formic acid is formed (Bryant *et al.*, 1958a). It ferments only starch and hydrolysis products of starch. Carbon dioxide is required for growth. *Selenomonas ruminantium* has amylolytic strains (Bryant, 1956) and will ferment many sugars. Fermentation products vary among strains of *S. ruminantium* (Hobson and Mann, 1961), but all strains produce acetate and propionate; some carbon dioxide is formed, but no hydrogen or volatile alcohols. Butyric, succinic, lactic, and formic acids have also been identified as products of some strains. *Streptococcus bovis* produces α-amylase (Hobson and MacPherson, 1952).

## E. Fermenters of Sugars

The polysaccharide-fermenting bacteria also utilize mono- or disaccharides. Hungate (1966) stated that it would be interesting to know whether the activity of rumen polysaccharidases is inhibited when sugars are present, since this could be the mechanism for decreasing fiber utilization when feed contains a great deal of sugar. Xylose, glucose, fructose, sucrose, and raffinose occur in red clover (Bailey, 1958a,b), and ryegrass contains 1% glucose, 1% fructose, 9% sucrose, and 19% fructan (Thomas, 1960). Some bacteria isolated in pure cultures cannot digest polysaccharides. In the rumen, they presumably depend on mono- and disaccharides. Since these simple sugars are available only during short periods after feeding, the bacteria depending on them would be handicapped during nonfeeding periods unless sugars were maintained by hydrolysis of polysaccharides by other microorganisms (Hungate, 1966).

## F. Lipolytic Bacteria

Mixed rumen bacteria hydrolyze fats into glycerol and fatty acids (Garton *et al.*, 1958; Wright, 1961). *Veillonella alcalescens* ferments glycerol, which suggests that this organism is involved with fat utilization in the rumen (Johns, 1953). Motile rod bacteria were isolated by Hobson and Mann (1961) from the rumen on a medium containing linseed oil; their fermentation products included acetate, propionate, butyrate, and hydrogen sulfide, but no formic or lactic acids. Oltjen and Williams (1974) fed cattle and sheep purified diets with and without lipids and, on the basis of rumen microbial concentrations and patterns as well as daily body weight gains, suggested that dietary lipids are not required for ruminal function or for the performance of the adult ruminant.

## G. Acid-Utilizing Bacteria

*Selenomonas* microbes utilize lactic acid and others decompose succinate, since it does not accumulate in the rumen even though it is produced by many bacterial types (Hungate, 1966). Some bacteria in the rumen attack formate. In feedlot cattle, *Pegaspaera elsdenii* occurs in conjunction with *S. bovis* and other bacteria that ferment starch to lactate. The lactate is attacked by *P. elsdenii*, producing acetic, propionic, butyric, valeric, and caproic acids, as well as carbon dioxide and traces of hydrogen. *Pegaspaera elsdenii* may be an important fermenter of lactate in cattle during adaptation to high-grain diets (Gutierrez *et al.*, 1959). It utilizes the acrylic acid pathway for propionate production (Ladd, 1959; Ladd and Walker, 1959; Baldwin and Milligan, 1964; Baldwin *et al.*, 1965). Schulman and Valentino (1976) found that *in vitro* fermentation of lactate by rumen microbes under hydrogen produced 4–8 times more propoionate and 2.5–3.2 times as much volatile fatty acids (VFA) as obtained under a nitrogen atmosphere. Propionate formation was proportional to the hydrogen concentration.

*Pegaspaera elsdenii* and *Veillonella alcalescens* contain ferredoxin, which serves in both species as a hydrogen acceptor in the breakdown of pyruvate. Hydrogenase catalyzes the oxidation of reduced ferredoxin, forming hydrogen (Valentine and Wolfe, 1963). Hobson *et al.* (1958) reported that *P. elsdenii* was abundant in the rumen of young calves but decreased as the calves grew older. *Vibrio succinogenes* oxidizes hydrogen or formate, with the reduction of malate or fumarate to succinate or of nitrate to ammonia (Wolin *et al.*, 1961; Jacobs and Wolin, 1963). Cytochromes *b* and *c* are present and presumably act during the electron transfer from the reducing substrate to the one being oxidized. Growth with fumarate, malate, or nitrate occurs in the complete absence of oxygen in the rumen.

## III. METHANOGENIC BACTERIA

The principal microbe producing methane in the rumen is *Methano-bacterium ruminantium* (Smith and Hungate, 1958). This organism requires rumen fluid as a source of branched- and straight-chain volatile fatty acids and acetate, heme, and other growth factors (Bryant, 1965). Hydrogen and carbon dioxide provide a source of energy and are converted to methane and water.

Methane production and dry matter intake by cattle have a linear relationship (Kriss, 1930). The amount of carbohydrate digested has been related to the amount of methane produced in the rumen by cattle (Bratzler and Forbes, 1940) and sheep (Swift *et al.*, 1948). The relationship with cattle was expressed by the equation $E = 4.012X + 17.68$, where $X$ is hundreds of grams of carbohydrate digested; that for sheep was $E = 2.41X + 9.80$. Blaxter and Clapperton (1965) observed methane production in cattle and sheep at a maintenance dietary intake (percentage of gross energy intake) to be $3.67 + 0.62$ digestible energy (DE) (%). The rate of change in the percentage of gross energy lost as methane by an increase in intake equal to one times maintenance was given as: $b = 0.054\text{DE}$ (%) $- 2.26$ for mixed diets. Moe and Tyrrell (1979), working with Holstein cows, determined the amount of methane using a regression equation relating methane production to digestible nutrients: $CH_4$ (Mcal) = 0.439 + 0.273 ± 0.015 kg digestible soluble residue + 0.512 ± 0.078 kg digestible hemicellulose + 1.393 ± 0.097 kg digestible cellulose.

## IV. PROTEOLYTIC BACTERIA

Hungate (1966) reported that proteolytic non-spore-forming bacteria in rumen contents ranged from $10^4$ to $2 \times 10^7$. Some strains digest bacterial cells and become relatively more abundant when substrates of other bacteria are in short supply, i.e., just prior to feeding. Increased ammonia concentration in the rumen contents before feeding may be due to such action (Moir and Somers, 1956). J. B. Russell et al. (1981) worked with cultures of rumen bacteria containing varying levels of carbohydrates, casein, and ammonia, and found active proteolytic strains of S. bovis that showed a ratio of protein degraded to bacterial protein synthesized of approximately 1.5.

Whetstone et al. (1981) studied the effect of monensin on nitrogen metabolism by rumen microbes in a semicontinuous culture system. Casein degradation ammonia and microbial protein decreased linearly with increases in the level of monensin (0, 1, and 4 mg/kg incubation mixture), while increases occurred in nonammonia, nonmicrobial nitrogen, α-amino N, nitrogen, and total peptides. Several workers (Warner, 1956; Blackburn and Hobson, 1960; Fulghum and Moore, 1963; Borchers, 1965) have isolated and identified proteolytic bacteria. Considerable research has been done on protein metabolism in the rumen, and a number of reviews have been published (Chalupa, 1975; Hogan, 1975; Smith, 1973; Tamminga, 1979; Orskov, 1982).

Dietary protein is extensively broken down in the rumen. Initially, protein chains are broken by hydrolytic degradation of peptide bonds or proteolysis, giving peptides and amino acids. This is followed by deamination. The optimum pH for both proteolysis and deamination is between 6 and 7 (Blackburn and Hobson, 1960; Lewis and Emery, 1962). Hendrickx and Demeyer (1967) found that the pH maxima for ammonia production in vitro were 4.5, 5.6, 6.7, and 7.7. Under most nutritional conditions in the rumen, the pH will allow extensive breakdown of dietary protein. Amino acids are either incorporated into protein within the bacterial cell or degraded to VFA, ammonia, carbon dioxide, methane, and some heat of fermentation. The end products are excreted back into the rumen medium. Protozoa engulf small feed particles and bacteria (Coleman, 1975), and proteolysis of dietary protein occurs inside the protozoal cell.

Degradation of dietary protein is believed necessary to provide microbes with required precursors for their own protein synthesis as either ammonia, alpha-keto acids, or intact amino acids. Since degraded products often occur in excess of bacterial requirements, it may be that the microbes benefit from the high-energy adenosine triphosphate (ATP) yielded by degraded amino acids. Formation of 1 mole of terminal pyrophosphate bonds in ATP require at least 12 kcal (Armstrong, 1969), but hydrolysis of peptide bonds yields only 3 kcal of free energy per mole (Baldwin, 1968), which indicates that proteolysis does not yield ATP. In addition to deamination, microbes degrade amino acids by transamination and decarboxylation (Prins, 1977). Under usual conditions, decarboxylation to yield amines does not seem important, but it may become significant at a low pH in the rumen. The most important degradation pathway is thought to be deamination, followed by decarboxylation of the α-keto acid formed, giving 1 ATP per decarboxylation as well as some additional ATP with oxidation of some cofactors formed.

*In vitro* methods of measuring degradation are usually based on the release of ammonia ($NH_3$) after incubation with fluid from the rumen or by estimating the proportion of nitrogen that goes into solution after incubation for a definite period. Studying protein degradation by $NH_3$ release after incubation is difficult because of the simultaneously occurring problem of microbial growth, in which part of the released $NH_3$ becomes incorporated into microbial protein. Measurements *in vivo* are generally carried out with animals cannulated in the abomasum or small intestine. In such animals, undegraded dietary protein from the rumen can be estimated as the difference between total and microbial protein entering the cannulated section. Microbial protein can be determined using markers such as nucleic acids, diaminopimelic acid, aminoethylenephosphoric acid, or isotopes such as $^{35}S$, $^{32}P$, or $^{15}N$ (Clarke, 1977). Different results are often obtained by these different methods (Tamminga, 1979).

## A. Factors Influencing Protein Degradation

Solubility of dietary protein, as well as structural differences caused by disulfide bridges and crosslinking of protein, appear to be important factors in ruminal degradation of protein (Nugent and Mangan, 1978). When the principal protein fractions are albumins and globulins, the solubility is higher than when feeds contain mainly prolamins and glutelins (Wohlt *et al.*, 1976). Solubility of protein in feedstuffs is affected by pH (Isaacs and Owens, 1972). Drying of forages in the field allows proteases to become active, and protein solubility increases (Sullivan, 1973). Part of the carbohydrates and proteins will be degraded during silage making due to fermentation, resulting in the nitrogen-containing end products of the fermented protein occurring in the soluble fraction (McDonald and Whittenbury, 1973).

Essentially all of the soluble nitrogen and 40–50% of the insoluble nitrogen are degraded in the rumen (Mertens, 1977). The extent of protein breakdown is a function of the rate of proteolysis and the retention time in the rumen. Retention time is influenced by the particle size of diet ingredients and the level of feed intake (Balch and Campling, 1965; Church, 1970).

Tamminga *et al.* (1979) studied the effect of level of feed intake on protein breakdown in the rumen of dairy cows equipped with reentrant cannulae in the small intestine. The cows were fed mixed diets of hay and concentrates with three levels of protein and two levels of feed intake. The six diets had 30% soluble protein. At the low level of feed intake, 26% of the total dietary nitrogen and 37% of the insoluble dietary nitrogen escaped degradation in the rumen. At the high level of feed intake, the corresponding values were 42 and 60%, respectively. Retention time in the rumen varies from one diet to another and between animals (Balch and Campling, 1965) as well as between species (Church, 1970). Hungate (1966) gave ruminal retention times for cattle ranging from 1.3 to 3.7 days and from 0.8 to 2.2 days for sheep. Fluid retention time is usually much less, but it also probably affects the rate at which food particles pass through the rumen.

## B. Methods for Decreasing Degradation of Protein

Considerable research has been done seeking practical ways to decrease degradation of dietary protein in the rumen in order to obtain maximum absorption of utilizable amino acids (Tamminga *et al.*, 1979). Such efforts would not result in a decrease of microbial protein synthesis in the rumen or or make dietary protein so undegradable that it cannot be hydrolyzed in the lower alimentary tract and utilized. Protein degradation in the rumen can be decreased to a limited extent by diets with ingredients containing protein with natural resistance to microbial breakdown. However, most research has involved some form of feed processing such as heat treatments and grinding or the use of chemical agents such as aldehydes or tannins.

Most attention has been given to formaldehyde. Increases in postruminal protein flow after treatment with formaldehyde have generally been significant, but the best production responses have usually been reported for wool growth (Fergusan, 1975). Barry (1976b) fed formaldehyde-treated alfalfa hay to sheep and observed increased wool production with increasing amounts of sulfur-containing amino acids in the diet. Wool proteins are high in sulfur-containing amino acids. Meat production responses measured as the rate of growth or nitrogen retention with dietary formaldehyde treatments have been variable and usually much smaller than wool growth responses (Chalupa, 1975; Clark, 1975). The small change in meat production with formaldehyde treatment of diets is probably due to a less important role of sulfur-containing amino acids with this product than in wool production. Generally, responses to formaldehyde-treated diets fed to milk cows have been small except in the experiments by Verite and

Journet (1977), which gave statistically improved milk production. Apparently, no single amino acid is clearly limiting for milk production, and Tamminga and van Hellemond (1977) concluded that under most feeding conditions, the protein supply in the small intestine of dairy cows is sufficient for milk production of at least 25 kg/day.

Tamminga *et al.* (1979) reviewed various diets treated with different levels of formaldehyde fed to milk cows and concluded that formaldehyde levels above 10 g per kilogram of protein has a negative effect on production. Apparently, at such high levels, formaldehyde not only protects protein from degradation in the rumen but also decreases enzymatic hydrolysis in the abomasum and small intestine. Recommended levels are 0.8–1.2% formaldehyde per 100 grams of protein (weight/weight, w/w) for protection of casein, 2% for oilseed meals, and 3% for legume grass silage (Broderick, 1975). Barry (1976a) suggested that application rates should be expressed in grams of formaldehyde per kilograms of degradable true protein.

Formaldehyde serves two purposes if applied to silages. It prevents excessive degradation of protein during fermentation of the silage, and by reducing the solubility of the protein, it gives protection against microbial degradation in the rumen when the silage is fed (Tamminga, 1979). Beever *et al.* (1977) found that pepsin-soluble nitrogen in silage decreased from 82% in the untreated material to 78% in 6% formaldehyde-treated material, and obtained a further reduction by drying the silage at high temperature. Degradation of protein in the rumen of sheep measured by reentrant cannulae was reduced from 85% for untreated silage to 22% for formaldehyde-treated silage and to 16% for formaldehyde-treated silage after heat drying. More than half of the protein that bypassed the rumen also bypassed the abomasum and ileum; this reaction may have been due to the high level (6%) of formaldehyde utilized. Wilkinson and Penning (1976) suggested 3–5% formaldehyde in silage treatment. Losses of as much as 35% of the formaldehyde applied may occur in silage making (Barry, 1976a).

## C. Effect of Protein Solubility

Several workers have shown that protein degradation in the rumen is proportional to protein solubility (McDonald, 1952; Ely *et al.*, 1967; Mangan, 1972; Hume, 1974). The rapid degradation of soluble protein will result in an increased rate of amino acid hydrolysis and ammonia release, with loss of ammonia across the wall of the rumen.

Denaturation of protein by heat causes unfolding of hydrophobic tertiary-structure amino acids, thereby decreasing protein solubility and reducing *in vivo* protein fermentation (Clark, 1975). Disulfide linkages in protein have been found to make proteins more resistant to hydrolysis by rumen proteinases (Mahadevan *et al.*, 1980). Treatment of serum albumin, rapeseed meal, and fish meal with mercaptoethanol or performic acid, which are known to disrupt disulfide linkages specifically, increased protease liberation of amino acids. This clearly indicates that factors other than solubility affect fermentation of proteins.

## V. DEGRADATION OF AMINO ACIDS

Scheifinger *et al.* (1976) studied the degradation of amino acids by five major genera of rumen bacteria under *in vitro* conditions. The results indicated that not all amino acids are degraded by all strains of bacteria and that degradation occurs at different rates. Chalupa (1976) studied the degradation of amino acids by rumen microbes of cattle under *in vitro* and *in vivo* conditions. He found that *in vitro* rate constants for essential amino acids (EAA) indicated that arginine and threonine were rapidly degraded (0.5–0.9, m$M$/hr), lysine, phenylalanine, leucine, and isoleucine formed an intermediate group (0.2–0.3 m$M$/hr), and valine and methionine were less rapidly degraded (0.1–0.14 m$M$/hr). The *in vivo* EAA rates of degradation were about 1.5 times greater than those obtained *in vitro,* but a significant linear relationship between the two conditions indicated that similar degradative pathways predominated. When threonine, arginine, lysine, phenylalanine, leucine, or isoleucine were incubated alone or in combination with a mixture of EAA, similar amounts of EAA were degraded. Methionine and valine degradations were approximately twice as great when these nutrients were fermented alone compared to fermentation with other EAA. The data demonstrated that specific amino acids were degraded at different rates and that there were interactions between certain ones. Chalupa (1976) concluded that, with the possible exception of methionine, free amino acids cannot survive degradation in the rumen.

Burris *et al.* (1974a) studied the effects of different protein supplements (soybean meal, fish protein, and linseed protein) fed to steers on the amino acid composition and amino acid availability for rumen bacteria. The bulk amino acid composition of rumen bacteria was not altered by the protein source. The release of threonine, serine, glutamic acid, valine, methionine, phenylalanine, lysine, and total amino acids by pepsin–pancreatin digestion was not altered by protein treatments of the steer diets. Lysine was released at a higher rate from microbes obtained from steers fed soybean meal than from those fed fish meal. However, acetate, propionate, and ammonia levels in rumen fluid varied significantly among protein dietary treatments.

Working with an *in vitro* system, Broderick and Balthrop (1979) tested many compounds for their capacity to inhibit deamination of amino acids from a casein hydrolsate by rumen microbes. They found that hydrazine (and three of its derivatives), diphenyliodonium chloride, thymol, and sodium arsenite inhibited deamination of approximately 100% of total amino acids. Penicillin G, oxytetracycline, and chlorotetracycline caused less inhibition of deamination. Thiosemicarbazide, *p*-hydrazineobenzenesulfonic acid, and iproniazid gave no protection. Recovery of individual amino acids generally followed the trends observed for total amino acids.

## VI. INFLUENCE OF UREA AS A SOLUBLE-NITROGEN FACTOR

Aitchison *et al.* (1976) carried out trials with high-producing dairy cows fed diets supplied with 0, 8, 15, or 24% of the total dietary nitrogen as urea with three levels of dietary crude protein (12, 13, and 15%). The measured solubility of nitrogen in the diets ranged from 31.5 to 48.7%. They observed mean coefficients of utilization of 0.82 and 0.39 for soluble and nonsoluble nitrogen, respectively. Formulating mixed concentrates with different nitrogen solubilities may alter the composition and properties of nitrogen-free ingredients which may, in turn, affect nutrient utilization by microbes in the rumen or by ruminant tissues (Tamminga, 1979). Satter and Roffler (1975) reported that ammonia levels in the rumen greater than 5 mg/dl were ineffective in generating greater microbial protein synthesis and stated that the amount of urea that could be utilized for various levels of natural protein was dependent on the total digestible nutrients (TDN) in the diets. With dairy cows, levels of natural protein greater than 12% ruled out effective utilization of urea in their diets.

## VII. MICROBIAL ADAPTATION TO BIURET

Ingesta from the rumen of animals not fed biuret are generally devoid of biuretolytic activity. A number of workers have demonstrated adaptation periods of rumen microbes that vary from a few days to 8 weeks on biuret-containing diets (McLaren *et al.*, 1959; MacKenzie and Altona, 1964; Schroder and Gilchrist, 1969; Clemens and Johnson, 1973a). Schroder and Gilchrist (1969) observed that the capacity for biuret hydrolysis was lost within 1–5 days following biuret deletion from the diet, and Clemens and Johnson (1973b) found that animals fed biuret at 4-day intervals were unable to develop or maintain biuretolytic activity. However, animals fed biuret daily or at 2-day intervals developed and maintained biuretolytic activity.

A biuret-degrading bacterium from sheep rumen was isolated by Bellingham and Bernstein (1973). Wyatt *et al.* (1975) conducted experiments with varying dietary levels of biuret, dehydrated alfalfa, and molasses, and observed that when biuret supplied 50 and 80% of the total dietary nitrogen, biuretolytic activity was greater at day 14 than with 20% biuret. This difference was maintained for the remainder of the 35-day trial. Five percent molasses caused a slight improvement in biuretolytic activity with low levels of alfalfa. When biuret supplied 65% of the digestible protein requirement of sheep and alfalfa supplied 0, 3, or 10% of the diet, adaptation to biuret was complete by day 4.

## VIII. PROTEIN SYNTHESIS BY RUMEN MICROBES

Various markers characterizing microbial components have been used to estimate the quantity of microbial protein synthesized with different diets and leaving the rumen. Among these markers are diaminopimelic acid (DAP), aminoethylphosphoric acid (AEP), ribonucleic acid (RNA), and the isotopes $^{15}N$, $^{35}S$, and $^{15}P$. Adenosine triphosphate (ATP) has been employed as a marker of microbial activity, and differences in the amino acid profile of components entering the duodenum have been utilized. Weller *et al.* (1958) used DAP to estimate the synthesis of bacterial protein, since DAP is present in the cell membrane of many types of rumen bacteria but is absent in plant material. Protozoa contain traces of DAP as bacteria have been observed in entodiniomorphs and holotrichs by electron microscopy (Stern *et al.*, 1977a,b). DAP measures bacterial protein synthesis only. Czerkawski (1974) proposed methods for utilizing DAP and AEP, and these can be used together to estimate total microbial protein synthesis by rumen microbes.

RNA and DAP markers were compared by McAllan and Smith (1974) in determining the contribution of microbial nitrogen in the duodenal digesta of a protozoa-free calf and a faunated cow. Good agreement was found by the two methods in the calf but not with the cow. With the RNA and DAP methods, microbial nitrogen to total nonammonia nitrogen ratios in the cow varied from 0.78 to 0.40, respectively. This variation was believed to be due to the fact that the DAP method does not include measurement of the protozoa.

Evans *et al.* (1975) based a method for measuring microbial protein synthesis on differences in the constant amino acid profile of microbial protein and the protein of each dietary component behaving as a single entity. They compared this technique with the DAP method and found an $r$ value of 0.92 based on 21 samples. The amino acid method is limited by a lack of knowledge of the degradation rates of different dietary proteins (Nikolic and Jovanovic, 1973).

The use of ATP as a marker for microbial protein synthesis in the rumen was recommended by Forsberg and Lam (1977) based on the following assumptions: (1) ATP is present in all living cells and is absent from dead cells; (2) ATP is similar in concentration in all microbes; and (3) extraction and assay of ATP are relatively simple. Wolstrup and Jensen (1978) found the concentration of DNA in rumen biomass to vary from 0.07 to 0.25% and to be dependent upon the nitrogen source in the ruminant diet.

## VIII. Protein Synthesis by Rumen Microbes

Hendrickx (1961) first used $^{35}$S as a marker for microbial protein synthesis in the rumen. A method for *in vivo* measurement of protein synthesis by ruminal microbes was based upon the incorporation in microbial protein of sulfur derived from $^{35}$S-labeled inorganic sulfate infused continuously into the rumen. Walker and Nader (1975) and compared this technique to that of the DAP method in 27 trials and found a correlation coefficient of $r = .68$, with DAP estimates averaging about 30% less than $^{35}$S estimations. The difference was thought to be due to the fact that the DAP procedure does not account for protozoal protein.

Synthesis of microbial protein has been determined by quantitating $^{15}$N incorporation into microbes from ($^{15}$NH$_4$)$_2$SO$_4$ (Pilgrim *et al.*, 1970) or $^{15}$NH$_4$Cl (Mathison and Milligan, 1971). These procedures depend on incorporation of nitrogen from ammonia and do not account for protein synthesized by microbes from amino acids or peptides. However, when Kennedy and Milligan (1978) compared $^{15}$N with $^{35}$S as markers for ruminal protein synthesis in sheep, they found that the values were only 12% less with $^{35}$S than with $^{15}$N.

Phosphorus uptake and incorporation into microbial phospholipids was related ($r = .98$) to ruminal protein synthesis. Bucholtz and Bergen (1973) proposed an *in vitro* method for determining microbial protein based on this observation. Van Nevel and Demeyer (1977) pointed out that this technique depends on the nondegradation of nonlabeled cells and a constant cell composition during growth, and stated that neither of these assumptions is valid. Harmeyer *et al.* (1976) made comparative studies of microbial protein synthesis using $^{15}$N and $^{35}$S as markers and demonstrated that exchange and degradation processes occurred in rumen contents. The isotope techniques did not distinguish between no-growth and negative-growth conditions. Naga and Harmeyer (1975) pointed out that negative growth values occur in situations where synthesis is less than degradation of microbial matter.

In Table 3.2, data on rumen microbial protein synthesis obtained by various techniques are summarized for representative studies (Stern and Hoover, 1979). Data are expressed as grams of crude protein (CP) synthesized per 100 g organic matter digested (OMD) in the rumen of sheep and cattle. The 68 observations in Table 3.2 give a mean of $16.9 \pm 6.2$ g CP synthesized per 100 g OMD, with values ranging from 6.3 to 30.7. Variations in values may be due to limitations of the methods applied or to sources and ratios of nitrogen and carbohydrates, dietary sulfur, frequency of feeding, or rumen dilution rate. When sheep were fed semipurified protein-free diets and urea was added to increase the dietary nitrogen from 0.95 to 1.82%, increased microbial protein production occurred (Hume *et al.*, 1970), indicating that sheep would have depressed protein synthesis if dietary CP were below 11%.

Cattle require approximately 2 percentage units more of CP for maximum microbial growth than sheep fed similar diets (Satter and Roffler, 1975). Microbial protein synthesis peaks at approximately 12–13% dietary CP with dairy

TABLE 3.2.
Efficiency of Rumen Microbial Protein Synthesis[a]

| Reference | Animal | Method | Major dietary constituents | Microbial synthesis (g CP/100 g OM digested) |
|---|---|---|---|---|
| Hume et al. (1970) | Sheep | Protein-free[b] | Semipurified diet + urea | 13.3 |
| Hume (1970) | Sheep | Protein-free[b] | Semipurified diet supplemented with: | |
| | | | Urea | 17.1 |
| | | | Gelatin | 19.8 |
| | | | Casein | 23.3 |
| | | | Zein | 22.5 |
| Hogan and Weston (1970) | Sheep | Substantial degradation[c] | Dried forages | 23.1 |
| Hogan and Weston (1971) | | Substantial degradation[c] | Straw + urea | 27.5 |
| Mathison and Milligan (1971) | | Direct zein determination[d] | Barley | 16.3[e] |
| | | $^{15}$N | Hay | 12.5[e] |
| Lindsay and Hogan (1972) | Sheep | DAP | Lucerne hay | 22.0 |
| Liebholz (1972) | Sheep | $^{35}$S | Dried red clover | 25.0 |
| | | | Straw supplemented with: | |
| | | | Lucerne meal | 6.3 |
| | | | Starch | 9.5 |
| | | | Lucerne meal + casein | 6.2 |
| | | | Starch + casein | 10.5 |
| | | | Wheat gluten | 6.9 |
| | | | Starch + wheat gluten | 11.6 |
| Pitzen (1974) | Sheep | DAP | Corn + urea | 20.1 |
| | | | Corn + soybean meal | 16.5 |
| Walker et al. (1975) | Sheep | $^{35}$S | Dried forages | 15.1 |
| | | | Fresh forages | 24.6 |
| Ulyatt et al. (1975) | Sheep | DAP | Fuanui ryegrass | 16.2 |
| | | | Manawa ryegrass | 30.7 |
| | | | White clover | 19.8 |

| Reference | Animal | Marker | Diet | Value |
|---|---|---|---|---|
| Chamberlain et al. (1976) | Sheep | DAP | Sugar beet pulp + barley | 14.3 |
| Hagemeister et al. (1976) | Cattle | DAP | Forages: | |
| | | | ryegrass, 200 kg N/ha | 16.3 |
| | | | ryegrass, 400 kg N/ha | 14.0 |
| | | | clover grass | 15.9 |
| | | | Mixed rations (45% N supplied by:) | |
| | | | Casein, untreated | 21.0 |
| | | | Casein, treated with formaldehyde | 19.4 |
| | | | Coconut pellets | 12.4 |
| | | | Soybean meal | 20.6 |
| | | | Fish meal | 22.3 |
| | | | Yeast | 22.0 |
| | | | Rapeseed meal | 20.6 |
| | | | Peanut meal | 19.5 |
| | | | Horsebean meal | 19.0 |
| | | | Urea | 20.8 |
| Kropp et al. (1977b) | Cattle | RNA | Grass + soybean meal | 9.9[e] |
| | | | Grass + soybean meal + urea | 11.6[e] |
| Kropp et al. (1977a) | Cattle | RNA | Cottonseed hulls + soybean meal | 23.0 |
| | | | Cottonseed hulls + urea | 23.5 |
| Beever et al. (1977) | Sheep | $^{35}$S | Silage | 16.7 |
| | | | Formaldehyde-treated silage | 6.6 |
| | | | Dried formaldehyde-treated silage | 7.7 |
| Crooker (1978) | Sheep | $^{35}$S | Semipurified diet supplemented with: | |
| | | | Peanut meal | 17.4 |
| | | | Peanut meal + urea | 19.7 |
| | | | Heated peanut meal | 24.3 |
| | | | Heated peanut meal + urea | 31.6 |
| Kelly et al. (1978) | Sheep | DAP | Ryegrass silage + formic acid (cutting): | |
| | | | Wilted spring silage | 15.6 |
| | | | Early direct-cut autumn silage | 7.5 |
| | | | Late direct-cut autumn silage | 10.0 |

(*continued*)

TABLE 3.2. (Continued)

| Reference | Animal | Method | Major dietary constituents | Microbial synthesis (g CP/100 g OM digested) |
|---|---|---|---|---|
| Prigge et al. (1978) | Cattle | RNA | Dry, rolled corn | 7.5[e] |
| | | | Steamed, flaked corn | 9.4[e] |
| | | | High-moisture corn, whole shelled and treated with propionic acid | 14.8[e] |
| | | | High-moisture corn, ground prior to ensiling | 9.5[e] |
| R. H. Smith et al. (1978) | Cattle | DAP | Flaked maize and straw supplemented with: | |
| | | | Peanut meal | 10.6 |
| | | | Fish meal | 18.1 |
| | | | Soybean meal | 17.5 |
| | | | Heated soybean meal | 16.3 |
| | | RNA | Peanut meal | 11.9 |
| | | | Fish meal | 16.9 |
| | | | Soybean meal | 15.6 |
| | | | Heated soybean meal | 18.1 |
| Allen and Harrison 1979 | Sheep | 35S | Dried grass, nuts, and ground maize | 15.3 |
| | | | Dried grass, nuts, ground maize, and monensin | 12.6 |
| M. D. Stern (unpublished data) | Cattle | DAP | Corn silage, hay, and corn gluten meal | 21.8 |
| | | | Corn silage, oat straw, and ground corn | 27.7 |

[a]From Stern and Hoover (1979).
[b]It is assumed that with a protein-free diet, the amount of protein nitrogen flowing from the rumen may be equated with the daily production of microbial protein in the rumen.
[c]It is assumed that protein was substantially degraded in the rumen; therefore, nearly all of the protein leaving the rumen was of microbial origin.
[d]Zein was separated from microbial protein by precipitating the latter with 80% ethanol (McDonald, 1954).
[e]Efficiency expressed as grams of CP per 100 g of DM digested.

cows (Burroughs et al., 1975b; Satter and Roffler, 1975). Ammonia concentrations in the rumen will increase if CP is added above these levels but without an increase in microbial protein production. The 12–13% dietary CP level is not fixed. Actually, the amount of protein synthesized by rumen bacteria at a given dietary level of CP is dependent on the TDN or fermentable energy of the diet up to about 12–13% CP (Satter and Roffler, 1975). Microbial protein synthesis also depends on the amount of dietary NPN, the extent of dietary protein degradation, salivary nitrogen input, and efficiency of microbial growth (Satter et al., 1977).

*In vitro* studies have shown maximal microbial growth when the ammonia-nitrogen concentration was 5–8 mg per 100 ml of the fermentation medium (Allison, 1970; Satter and Slyter, 1974). *In vivo* observations by Hume et al. (1970) showed maximal microbial growth when ammonia-nitrogen in the rumen reached approximately 9 mg/100 ml. Okorie et al. (1977), in *in vivo* studies, obtained maximal protein synthesis with ammonia-nitrogen levels of 7 mg per 100 ml of rumen fluid, which is quite close to the 5 mg ammonia-nitrogen level observed *in vitro* for maximal production of protein by Satter and Slyter (1974).

The source of dietary protein and its susceptibility to rumen degradation may influence the amount of nitrogen available for protein synthesis. Low-roughage diets were found to require a minimum of 2 g of available nitrogen per 100 g OMD for efficient microbial production of protein (McMeniman and Armstrong, 1977). Thomas (1977) pointed out that starchy cereal diets, especially corn, are likely to be inadequate in nitrogen. Forages can also be deficient in available nitrogen for rumen microbial protein synthesis depending on species, maturity, and drying (Ulyatt et al., 1975; Walker et al., 1975). Kropp et al. (1977a) fed steers hourly a low-quality roughage with soybean meal or soybean meal plus urea and found the microbial protein synthesized to be relatively constant regardless of the source of nitrogen. Chen et al. (1976) observed that both processed distillers solubles (SDS) and centrifuged processed distillers solubles (CDS) increased the synthesis of microbial protein *in vitro*. They demonstrated that the stimulatory effect was not due to valeric acid, a vitamin mixture, or an amino acid mixture.

Efficiency of rumen microbial growth when urea is the sole source of nitrogen may be limited to lack of preformed amino acids. Nitrogen provided from urea, gelatin, casein, and zein resulted in synthesis of 17.1, 19.8, 23.3, and 22.5 g CP per 100 g OMD by rumen microbes in sheep (Hume, 1970). Hume (1970) suggested that since gelatin was deficient in several amino acids including methionine, this may explain why gelatin produced lower microbial synthesis of protein than casein. Satter et al. (1979), working with steers fed protein and urea-containing diets, found that when an adequate dietary supply of preformed amino acids was available, proline, arginine, histidine, methionine, and phenylalanine were derived from the medium to a greater extent than other amino acids, and that synthesis of proline, arginine, and histidine increased with the urea diet, while methionine and phenylalanine failed to do so. This suggests that methionine and phenylalanine may be limiting for bacteria when diets are low in protein and high in NPN.

Substituting sunflower protein in a diet containing low-quality coastal bermuda grass increased microbial protein synthesis, but no benefit occurred when urea was supplemented (Amos and Evans, 1976). However, Kropp et al. (1977b) observed increased microbial protein synthesis in steers fed a low-quality roughage when urea replaced equivalent amounts of soybean meal nitrogen. Despite changes in the dietary nitrogen source, the bulk amino acid composition of rumen bacteria remained quite similar (Weller, 1957; Bergen et al., 1968a,b; Burris et al., 1974a).

Starch and sugars are more effective than other carbohydrates in increasing the utilization of degraded dietary nitrogen for microbial growth both *in vitro* and *in vivo* (McDonald, 1952; Chambers and Synge, 1954; Lewis and McDonald, 1958; Phillipson et al., 1962; Offer et al., 1978; Stern et al., 1978). McAllan and Smith (1976) compared starch to other carbohydrates and found that it provided the greatest amount of energy for rumen bacteria. Differences in digestion vary between the starch in corn and barley (Orskov et al., 1971a; Durand et al., 1976). In continuous cultures, an increase in microbial growth was observed with increased levels of nonstructural carbohydrates by Stern et al. (1978). They concluded that major factors affecting the utilization of degraded dietary nitrogen were the type and rate of carbohydrate availability.

Burroughs et al. (1975a) developed a Urea Fermentation Potential (UFP) system in which special attention was paid to balancing nitrogen and energy in ruminant diets. They estimated that rumen microbial synthesis was 10.4% of the TDN of the diet. Satter and Roffler (1975) demonstrated that for given dietary concentrations of TDN, certain levels of crude protein or NPN were required for maximum bacterial protein synthesis. Their observation showed that levels of ammonia above 5 mg per 100 ml of rumen fluid had no benefit in increasing bacterial protein synthesis for a given TDN level in diets. Stern et al. (1978) used three isonitrogenous diets with differing starch-to-cellulose ratios in fermentors and observed that substituting starch for cellulose increased the synthesis of microbial protein. Utilizing four isonitrogenous diets with differing levels of urea and protein solubility (heated versus unheated peanut meal), Stern et al. (1978) observed no effects of urea or protein solubility on microbial growth when expressed as grams of CP synthesized per 100 g of dry matter digested; however, the microbial protein synthesized daily (in gram) was lower for the urea diets.

Gil et al. (1973a), in *in vitro* fermentation trials with rumen bacteria and utilizing urea as the sole source of nitrogen, observed that when glucose or cellulose was used as substrate, methionine hydroxy analog (MHA) or DL-methionine increased bacterial protein synthesis more than inorganic sulfate. The maximum effect on growth with MHA occurred at a level of 0.2 mg per milliliter of medium, and the free acid and calcium salt of the analog and DL-methionine gave equivalent stimulation for bacterial protein synthesis (Gil *et al.*, 1973b). When 18 different amino acids and MHA were utilized *in vitro* with glucose and urea, only those that contained sulfur significantly increased bacterial protein synthesis (Gil *et al.*, 1973c).

Freitag *et al.* (1970) fed steers two concentrations of protein (7 and 11%) and two sources of nitrogen (urea and cottonseed meal pellets) and observed no difference in bacterial protein in rumen fluid. When they fed these sources of protein with low- and high-energy diets, they found that the cottonseed meal resulted in more bacterial protein with the low-energy diets. With the high-energy diets, slightly more bacterial protein was present 7 hr after feeding the urea compared to the cottonseed meal diets.

## Starea and Microbial Protein Synthesis

Starea is a product produced by passing finely ground corn, barley, wheat, or sorghum grains with urea through a cooker-extruder under moisture, temperature, and pressure conditions that gelatinize the starch. Helmer *et al.* (1970b) fed grain diets to cows supplemented with either soybean meal or starea and found that more grain was consumed and more milk was produced than with the same diet supplemented with urea. Starea improved the utilization of urea *in vitro*, since the concentration of ammonia was lower with starea than with unprocessed corn and urea, and more ammonia was converted to microbial protein (Helmer *et al.*, 1970a). These studies demonstrated decreased ammonia concentration and increased microbial protein synthesis when sorghum or corn grain starch was 100% gelatinized compared to 10% gelatinization. Stiles *et al.* (1970) fed starea and a cracked corn plus urea diet to two pairs of identical twin cattle and found the total quantity and the concentration of bacterial and protozoal nitrogen to be greater in the rumen contents of the starea-fed animals. More microbial protein was synthesized *in vitro* from corn starea (44% crude protein) or expanded corn plus urea (44% crude protein) substrates than from unprocessed corn grain plus urea (44% crude protein).

## IX. INTERACTIONS BETWEEN BACTERIAL SPECIES

In order to maintain a steady state of characteristic rumen bacteria, there must be a continuous replication of diverse indigenous species. Their energy is derived from plant polymers, principally cellulose and starch, which are fermented to acetate, propionate, butyrate, and other VFA, methane, and carbon dioxide by activities of the major bacterial species. None of the carbohydrate-fermenting species produce methane. The methane-producing species utilize hydrogen, formate, and carbon dioxide products of carbohydrate fermentation of other species for growth and methanogenesis. Concurrent with the activities of carbohydrate fermentation, interactions occur that provide nutrients necessary for growth of the major species. Formation of ammonia and branched-chain VFA are examples of how some species fill the requirements of other species.

Limited urease activity of a few bacteria provides essential ammonia for urease-negative microbes supplied with urea diets. Synthesis of B vitamins by some microbes provides essential nutrients for other species. Understanding the qualitative and quantitative aspects of species interactions is necessary in describing the contributions of individual species to overall rumen fermentation (Wolin, 1975). *Selenomonas ruminantium* does not ferment cellulose by itself, but depends on *B. succinogenes* to provide it with energy from cellulose. Then *S. ruminantium* decarboxylates succinate produced from cellulose by *B. succinogenes,* and the combined fermentation results in the formation of acetate, propionate, and carbon dioxide from cellulose (Wolin, 1975).

## X. THE RUMEN PROTOZOA

According to Hungate (1966), Gruby and Delafond (1843) first saw rumen protozoa and observed that some of them ingested parts of plant cells. The protozoa in the rumen and reticulum were viable and active, while those in the omasum and abomasum were immobile and disintegrating. No traces of their bodies were found in the duodenum. The protozoa were considered an unexpected source of animal food in the herbivorous ruminant. Rumen ciliates have evolved into a very specialized group fitted to survive only in the rumen or closely related habitats. Rumen protozoa are anaerobic, can ferment plant materials for energy, ingest bacteria, and grow in the presence of billions of accompanying bacteria. Hungate (1966) presented a description of the morphology and a classification of common rumen protozoa. Transfaunation may occur by dams licking and grooming their calves and the calves later licking off saliva and digesta (Becker and Husing, 1929).

Protozoa are sensitive to acidity, and their numbers in the rumen decrease as the pH is lowered (Purser and Moir, 1959). For instance, at pH 5.9, 5.6, and 5.3, the protozoa counts were $6.2 \times 10^5$, $4.7 \times 10^5$, and $3 \times 10^5$/ml, respectively, and chiefly entodinium. The entodinium are more resistant to acid than holotrichs (Abou Akkada *et al.,* 1959). All protozoa are killed rapidly at very low pH levels. Quinn *et al.* (1962) reported that protozoa did not survive extended exposure to acidities outside the pH range of 5.5–8.0. Various investigators have reported concentrations of protozoa in rumen of cattle fed different diets. Gutierrez (1955) reported $2.7 \times 10^3$ dasytrichs and $0.3–0.4 \times 10^3$ isotrichs per milliliter of rumen fluid in cattle fed hay and concentrate. Hungate (1957) found

2.7 × $10^4$ diplodinia, $10^4$ isotrichs, and 0.9 × $10^4$ dasytrichs per milliliter in cattle fed timothy hay and concentrates.

Loss of nitrogen from holotrichs fermenting glucose may indicate active metabolism of nitrogen, or it may be caused by protozoa cell contents being extruded due to the pressure of deposition of large amounts of amylopectin (Heald and Oxford, 1953). The cells will burst if too much sugar is available; this has been observed with normal feeding (Warner, 1964). Starch in holotrichs digested in the abomasum and intestine may constitute about 1% of carbohydrate requirements of the ruminant host (Heald, 1951).

Starch is actively ingested and utilized by all entodiniomorphs (Sugden, 1953). The ingestion of available starch continues until the protozoa are packed and swollen. *Epidinium* was abundant in cattle fed a high-starch diet (Bond *et al.*, 1962). Cellulose was demonstrated to be digested by protozoa (Hungate, 1943). Individual *E. maggii* were starved until starch resources were exhausted and were then fed cellulose, which they rapidly ingested and digested. Sugden (1953) reported that ground cotton cellulose in preference to finely ground hay was digested by polyplastron protozoa obtained directly from the rumen. There is cellobiase activity in *Epidinium,* and various hemicelluloses are digested by it (Bailey *et al.*, 1962).

Rumen protozoa are actively proteolytic (Warner, 1955), and it has long been assumed that protozoa utilize bacteria as a source of nitrogenous feed (Margolin, 1930). Conclusive evidence for this was obtained when Gutierrez and Hungate (1957) demonstrated ingestion of bacteria by *Isotricha prostoma,* with formation of vacuoles similar to those of *Paramecium*. Protozoa may be able to obtain energy by fermentation of protein, since ammonia is released when concentrates are fed to protozoa *in vitro* (Williams *et al.*, 1961). Less direct evidence was found when protozoa remained active for a considerable period after the disappearance of all iodophilic substances, which suggested that their source of energy could be protein. Coleman (1972) did much to quantitate the engulfment of bacteria by rumen protozoa. Nour *et al.* (1979) substituted urea for 0, 36, 55, or 100% cottonseed cake protein and determined the total count, pool size, and composition of protozoa in the rumen of cattle, sheep, buffalo, and goats. The 0 and 100% urea-nitrogen diets resulted in the highest protozoal counts, and sheep and goats had more protozoa than cattle and buffalo. Increasing the urea-nitrogen in the diet gave more *Entodinia* but reduced *Isotricha, Diplodinium,* and *Polyplastron* to negligible numbers with the 100% urea-nitrogen diet. It appears that protozoa digest bacteria and assimilate bacterial nitrogen as well as other nitrogenous substances. Proteins may be a source of protozoal energy during fermentation, and they may be important in the nitrogen economy of the host.

Osman *et al.* (1970a) inoculated Zebu calves with whole-rumen contents from a mature cow, and a thriving mixed population of ciliate protozoa became established 2 weeks later. Ruminal volatile fatty acids and ammonia concentrations in the inoculated calves showed a pronounced increase above the values observed in uninoculated calves. The inoculated calves grew faster and had greater feed efficiency than the controls.

Williams and Dinusson (1972) maintained calves of various breeds and both sexes for 833 days of age free of ciliated protozoa. Some of these calves were faunated with single inoculations of *I. prostoma, I. intestinalis, Entodinium simplex, E. caudatum,* and *E. bursa* protozoa. These single species were maintained in the rumen of the isolated calves for 370–420 days, at which time the calves were mixed with mature cattle. Williams and Dinusson (1973) reported on the weight gains and volatile fatty acid levels in the rumen of calves in isolation stalls with and without ciliated protozoa. Samples from the rumen of calves with established populations of *Entodinium* sp. or *Isotricha* sp. had higher levels of propionic acid than samples from calves with a mixture of *Entodinium* sp. and *Isotricha* sp. or samples from fauna-free calves. There were no significant differences in the total volatile fatty acid concentrations of rumen fluid from inoculated and protozoa-free calves. Daily weight gains were equivalent in inoculated and protozoa-free calves.

## Interrelation between Protozoa and Bacteria

Microflora inhabiting the rumen are dense and include approximately $10^{10}$ bacterial and $10^6$ protozoal cells per milliliter. There are approximately 200 species of bacteria and 20 species of protozoa (Bryant, 1959). Coleman and co-workers (Coleman, 1960, 1969; Coleman *et al.*, 1972) incubated $^{14}$C-labeled bacteria (*Butyrivibrio fibrisolvens* and *S. bovis*) isolated from the rumen and washed suspensions of *Entodinia* protozoa. The bacteria were taken up by the protozoa at a steadily decreasing rate with time, and the protozoa contained all the bacterial $^{14}$C that had disappeared. The number of bacteria engulfed by *E. caudatum* was similar for all small bacteria tested (*Proteus vulgaris, Escherichia coli, S. faecalis,* and *B. fibrisolvens*) at 3000–4000 bacteria per protozoon per hour, but *Entodinia* engulfed only 630 *B. megaterium*. However, since *B. megaterium* has approximately 6.5 times the volume of *E. coli*, the volume of *B. megaterium* engulfed was about 24% higher than that of *E. coli* (Coleman, 1975).

Generally, the interference of a bacterial species with engulfment of another species by protozoa was proportionate to their size when studying *Escherichia coli, Clostridium welchii* and *Lactobacillus casei* (Coleman, 1964). In the presence of 10, 52, 260, and 1670 rice starch grains per *E. caudatum* protozoon, bacteria were engulfed progressively for 4 hr at all starch concentrations. However, with the two greatest concentrations of starch, bacteria engulfment was depressed by 18 and 64%, respectively, during the first hour compared to controls without any starch (Coleman, 1975). Williams and Dinusson (1973) found rumen bacteria and protozoa of cattle to have comparable ranges of 17 amino acids, with aspartic and glutamic acids, leucine, lysine, and isoleucine representing over 50% of the total protein.

Eadie et al. (1970) observed $2 \times 10^6$ Entodinia, $10^5$ Epidinia, $5 \times 10^4$ Isotricha spp., $2 \times 10^5$ medium-sized protozoa (e.g., Eremoplastron spp.), and $10^{10}$ bacteria per milliliter of rumen fluid of a heifer on a restricted high-grain diet. The approximate volume occupied by these various organisms per milliliter of rumen contents were calculated to be: Entodinia, 0.07 ml; Epidinia, 0.03 ml; Isotricha, 0.05 ml; medium-sized protozoa, 0.03 ml; and bacteria, 0.03 ml. Under these conditions, bacteria occupied less than 20% of the total volume of the protozoa. However, when the surface area of the microorganisms was calculated, the bacteria had four times the area of the protozoa. Metabolic activity is closely related to surface area.

## XI. MICROBES IN RELATION TO RUMEN FUNCTIONS

Rumen microbes depend on their host for intake of food, its mixing and propulsion, saliva secretion, and passage of substances through the rumen wall. Much has been learned about ruminal microbes and their functions by utilizing fistulas of the rumen. Numerous types of fistula plugs for cattle have been described (Yarns and Putnam, 1962; Bowen, 1962).

The ruminant mouth is adapted for effective grazing by having a tough pad on the upper jaw that replaces the incisor teeth. This gives a firm base against which the lower teeth press the forage and hold it while a movement of the head breaks the attached plants. After several bites have been taken, the material is swallowed en masse or as a bolus. The average weight of a bovine bolus is about 100 g with constituent saliva (Hungate, 1966). Very little chewing occurs during feeding, just enough to form a bolus with saliva prior to swallowing. This allows maximum grazing of available forage in a short time. While the chewing action of the ruminant during feeding does not thoroughly grind fibrous materials, it liberates approximately 65% of cell contents from fresh forages (Mangan, 1959). The bolus, upon being swallowed, is propelled into the rumen, and muscular concentrations of the rumen tends to mix it with previously ingested feed and saliva.

The rumen and reticulum of forage-fed cattle hold an amount of digesta roughly equal to one-seventh of the total body weight. They may have a smaller percentage, i.e., one-tenth of the body weight when animals are fed more easily digested diets (Yadawa and Bartley, 1964). The reticulum is small but has a large opening into the rumen and looks somewhat like a pouch of the rumen. Together the rumen and reticulum form one large sac known as the *rumen-reticulum,* which is normally filled with digesta and microorganisms. Hungate (1966) described the morphology and function of the rumen-reticulum as well as other sections of the alimentary tract of ruminants, namely, the omasum, abomasum (true stomach), and intestines. Flatt et al. (1958) showed that the rumen of a young animal can be distended by inserting nylon sponges composed of indigestible material. The walls were very thin and weighed no more than the rumen walls of an undistended control. Blaxter et al. (1952) observed similar results when feeding hay to calves during initial stages, but after continued feeding on roughage, the weight of the rumen tissue became greater than that of controls on a nonroughage diet.

VFA and other products of rumen fermentation are necessary for the increased tissue weight of the rumen wall and for the development of the papillae and muscle tissue (Sander et al., 1959). Sudweeks (1977) varied citrus pulp, corn, and soybean meal to give three levels of concentrate (10, 40, and 70%) in diets fed steers and observed them for chewing time and VFA production. Chewing time was reduced with each increase in concentrate but was not affected by the type of feedstuff. Linear equations indicated a relationship between chewing time and VFA levels with 40 and 70% concentrate, but no difference was observed at 10%.

Rumen fluid pH gradually rises as roughage is consumed and the concentration of acids diminishes (Godfrey, 1961). Acid and ammonia are formed by fermentation of concentrates that enter the rumen. Markoff (1913) recognized the importance of saliva in neutralizing acids formed in rumen fermentation.

## A. Mixing in the Rumen-Reticulum

Muscular movement of the walls of the rumen-reticulum causes mixing of fresh ingesta with the microbes, spreads saliva throughout the contents, enhances absorption by replenishing fermentation acids absorbed by the epithelium, counteracts flotation of materials during fermentation, and assists comminution and passage of digesta to the lower alimentary tract (Hungate, 1966). Rumen contents may vary between 6 and 18% in dry matter. Coarse hay diets may result in particles of ingesta that float due to air caught in hollow stems. The particles become heavier than water when the gas ultimately dissolves. The specific gravity of the total contents, exclusive of the gas phase, does not become much greater than that of water. Specific gravity values of approximately 1.03 were reported by Balch and Kelley (1950). Specific gravity would be sufficient to cause ingesta to settle in the rumen if it were not for muscular mixing movements that disperse the material. Actually, stratification occurs due to copious carbon dioxide and methane gases from fermentation, which tend to stratify ingesta in the opposite direction (Balch and Kelley, 1950). Mixing of digesta by rumen contractions is insufficient to counteract the factors causing stratification, and measurable differences in rumen contents at different locations have been shown

(Smith *et al.,* 1956). Animals fed ground, pelleted diets are more uniform, though they are still not homogeneous. Composition varies with time due to ingestion of the diet, entrance of saliva, and passage of materials to the lower part of the alimentary tract.

The contents of the rumen can be obtained from a slaughtered animal, a stomach tube, or a fistula. The advantages of using a slaughtered animal is that one can measure the total volume of the digesta and mix it somewhat uniformly. However, in commercial abattoirs, feed is usually withheld for 16–24 hr prior to slaughter, and data may not be applicable to the usual feeding conditions. Samples obtained by stomach tube tend to be more liquid than the average for the entire contents of the rumen and, of course, the tube will only enter the center regions of the rumen. A tube with a lumen of 1.25 inches, such as a garden hose, can be used with large cattle with or without a suction pump (Hungate, 1966). The fistulated animal is best for most intensive investigations, since it does not cause undue disturbances in rumen function if samples of limited size are obtained. While the fistula makes the contents of the rumen more accessible, in practice it may not result in homogeneous sampling of the contents. For sampling the ingesta of grazing cattle, Weir and Torell (1959) recommended an esophageal fistula that collects the ingesta directly in a bag.

## B. Rumination

Rechewing of food, or chewing the cud, is a characteristic of ruminants and is intimately associated with feeding of herbages. Comminution of resistant plant parts by rumination has been postulated as essential for complete action of microbial enzymes in reducing particles. However, this factor is probably minor compared to the importance of rumination in reducing particles to a size that can proceed through the lower alimentary tract. Demonstration that very find grinding increases the digestibility of cellulose in forage indicates that rumination may increase fermentation to some extent (Dehority and Johnson, 1961). Observations of grazing cattle showed that the maximum time spent in rumination is about 8 hours per day (Gordon, 1958). Since there is not complete separation of ruminated from nonruminated material, some material goes through more than one cycle of rumination. Schalk and Amadon (1928) found that the ruminated bolus contained very little recently ingested food, probably because the ruminated bolus was collected in the dorsal sac of the rumen and was slow to mix. Bae *et al.* (1979), using an automatic jaw-motion–recording apparatus with rams fed 400, 800, 1200, and 1600 g hay at four daily feedings, found that the rumination time, number of chews, and number of boluses increased with increasing levels of hay intake.

The partial separation of fine and coarse particles between the rumen and reticulum depends on the spatial configuration of these compartments and on the mixing contractions. Reticulum contents are more liquid than those of the rumen. This is due in part to saliva that is secreted into this compartment during nonfeeding periods and in part to the rapid reticulum contractions that expel into the rumen the coarse materials that float. Slow contractions of the ventral sac of the rumen cause liquid to well up around the mass of digested solids, spilling over into the reticulum. Some particles are carried with the liquid, but the main mass stays in the rumen. The reticulo-omasal orifice is the site of separation of coarse and fine particles.

## C. Composition of Saliva

The total secretion of saliva involves contributions from the parotid, submaxillar, and sublingual salivary glands. Saliva is a source of recycled nitrogen both as urea and proteinaceous substances and as various nutrient and buffering ions such as carbonates, phosphates, chloride, sodium, potassium, and calcium (Hungate, 1966). The relative contributions of saliva and food in providing calcium, potassium, phosphorus, and chloride were evaluated by Bailey (1961). Sodium was added principally by saliva, but the other elements were supplied about equally by food and saliva. Submaxillary saliva excretion of nonurea proteinaceous material was markedly stimulated by inflation of the rumen, but inflation of the rumen or the use of carbachol as a stimulant had little effect on parotid or sublingual secretion of protein nitrogen (Phillipson and Mangan, 1959).

McManus (1961) reported that the rate of nitrogen secretion in saliva is fairly constant and does not increase proportionally with increased saliva secretion. Sialic acid ($N$-acetylneuraminic acid) constitutes 22.4% of bovine salivary mucoprotein and is responsible for the viscous nature of total salivary secretion (Gottschalk and Graham, 1958), along with 15.7% $N$-acetyl galactosamine. Nisizawa and Pigman (1960) found saliva to contain 38% protein, 32.3% sialic acid, and 27% hexosamine. Rumination stimulates salivary secretion, but not to the same extent as feeding. Coarse, dry feeds cause more secretion than lush green forages (Weiss, 1953; Denton, 1957). A new feeding regimen must continue for several days before salivary flow becomes relatively constant at the new level.

## D. Volatile Fatty Acids (VFA) and Their Absorption

Accurate quantitative determination of individual VFA in microamounts was not possible until liquid-liquid partition chromatography procedures were adapted by Elsden (1946). With the development of gas-liquid chromatography, James and Martin (1952) and El-Shazly (1952) demonstrated that small amounts of higher straight- and branched-chain fatty acids such as isobutyric, valeric, isovaleric, and 2-methylbutyric acids were present in the rumen. Gray *et al.* (1952) found that a maximum of 5% of the VFA was formic acid. When rumen contents are sampled several hours after feeding, only traces of formic acid are found. This acid is metabolized rapidly in the rumen (Carroll and Hungate, 1955), and its concentration is generally less than 0.02 μmole/ml. Matsumoto (1961) reported 0.0049% formic acid in hay; this would contribute to the acid concentration in the rumen of hay-fed animals.

Barcroft *et al.* (1944) demonstrated that fermentation acids were absorbed from the rumen when they showed that portal blood leaving the rumen contained a greater concentration of VFA than did entering arterial blood. The VFA are metabolized to a limited extent while passing through the rumen wall (Pennington, 1954). Very little butyric acid reaches the blood at low levels of nutrition (Kiddle *et al.*, 1957). Oxygen consumption was found to be greater in tissue slices of rumen epithelium with butyric acid as substrate than with other VFA (Pennington and Pfander, 1957). Butyrate is converted mostly to β-hydroxybutyrate by rumen epithelium (Hird and Weidemann, 1964), and a similar conversion occurs on passage through the wall of the omasum (Joyner *et al.*, 1963). To some extent, propionic acid is utilized by the rumen wall as propionyl-coenzyme A (CoA) is formed and carboxylated to methylmalonyl-CoA. The methylmalonyl-CoA is rearranged to succinyl-CoA, which, in turn, reacts with propionic acid to regenerate propionyl-CoA, and releases succinic acid, which is utilized by the tissue (Pennington and Sutherland, 1956).

In well-fed ruminants, absorption of VFA never decreases the rumen contents below 40 μmoles/ml, or about 0.3% (weight/volume, w/v), and the usual concentration is between 0.6 and 0.9% (w/v) (Hungate, 1966). The distances required for the VFA to diffuse through the digesta in order to reach the rumen wall prevent them from reaching a low concentration except during starvation. After the food is ingested, fermentation increases the concentration of the acids until the gradient across the rumen epithelium into the blood causes the rumen-to-blood diffusion rate to equal the average rate of production minus the rate of passage to the omasum. The average steady state of distribution of the three principal VFA in sheep rumen was found to be 62% acetic acid, 21% propionic acid, and 15% butyric acid (Pfander and Phillipson, 1953). Similar data have been obtained in varying feeding programs and geographic localities.

When VFA production diminishes, absorption and passage to the omasum exceed production, and the concentration in the rumen decreases. In the omasum, conditions for absorption are more favorable than those in the rumen, and very little VFA are in the digesta passing on to the abomasum. Any acids that do pass on to the abomasum are absorbed (Ash, 1961; Johnston *et al.*, 1961). There is somewhat faster absorption of VFA at pH 5.7 than at pH 7.5 (Danielli *et al.*, 1945; Parthasarathy and Phillipson, 1953).

Since the undissociated acids penetrate the epithelium more rapidly than anions, absorption of VFA are more rapid whenever a high production rate causes the pH to drop. The VFA are generally dissociated as anions in normal rumen contents in the pH range of 5.7–6.7. Acetic acid dissociates slightly more to anions than propionic and butyric acids, and this may account for its slower absorption. Acetic acid may also be less absorbed due to less solubility in the lipoidal substances of the cell membranes. Conversion of butyrate and propionate in the epithelium tissues of the rumen makes the diffusion gradient greater for these acids than for acetic acid, which may account for their greater specific absorption rates. The rate of ammonia transport to the blood is increased with decreased absorption of VFA across the rumen (Hogan, 1961).

## XII. EFFICIENCY OF ENERGY UTILIZATION BY MICROBES

The amount of growth of microorganisms was found to be directly proportional to the amount of ATP generated by the catabolism of the energy source (Bauchop and Elsden, 1960). Reviews of this subject (Forrest and Walker, 1965; Payne, 1970; Stouthamer, 1969) indicated that $Y_{ATP}$ (dry weight of cells produced per mole of ATP in grams) was a constant for various microbes and had a magnitude of 10.5. Bauchop and Elsden (1960) observed that the $Y_{ATP}$ and growth rate of microorganisms were maximized at a specific growth rate. They suggested that $Y_{ATP}$ may not be constant even at optimum growth rates and that it will vary with the growth rate.

Isaacson et al. (1975) studied $Y_{ATP}$ production by bacteria in rumen fluid at three dilution rates (0.02, 0.06 and 0.12 per hour), each at four growth-limiting glucose concentrations (5.8, 9.9, 12.7, and 25.0 m$M$), and observed no changes in the number of viable cells per mole of glucose fermented, ATP production, or $Y_{ATP}$ with changes in glucose concentration. However, the increased growth rate at high glucose concentrations tended to reduce ATP production from glucose fermentation, since a dilution of 0.02–0.12 gave a reduction of approximately 10% in yield. This gave an increase of 120% in $Y_{ATP}$ with rising growth rate when coupled with the large increase in viable cells per mole of glucose fermented. Enhanced production of bacteria increased the amount of energy used for maintenance and thereby prevented any change in the efficiency of glucose use. However, the efficiency of bacterial yield from hexose increases with the dilution rate, and the abomasal supply of bacterial protein will increase by almost twofold as the dilution rate increases from 2 to 12% per hour (Owens and Isaacson, 1977).

Energy for maintenance as well as for growth is required by bacteria (Stouthamer and Bettenhaussen, 1973). For this reason, new terms have been developed as follows (Owens and Isaacson, 1977):

Hexose (1 mole) → acetate, butyrate, propionate ($CO_2$, $CH_4$) + ATP (3.6–5.6 moles)

$Y_{ATP}$ = 10.5 g of dry cells per mole of ATP
1 mole of ATP = 10.5 g of dry cells
$m_e$ = moles of ATP per gram of dry cells per hour
$Y_{ATP}^{max}$ = yield in grams of dry cells per mole of ATP above maintenance

Thomas (1973), using the value of 10.5 g/mole for $Y_{ATP}$ as the yield of microbial matter per mole of adenosyl triphosphate, calculated a value of 27 g bacterial protein nitrogen per kilogram of organic matter truly digested and about 36 g nitrogen per kilogram of organic matter apparently digested. The organic matter apparently digested was considered to be that which disappeared between the food as eaten and the duodenum contents, while the organic matter truly digested was the sum of that apparently digested plus that of the microbial mass. According to Czerkawski (1978), an analysis of considerable data found the average efficiency of synthesis of microbial matter in the rumen to be 19.3 g of nitrogen per kilogram of organic matter truly digested.

## XIII. LACTIC ACID AND OTHER NONVOLATILE ORGANIC ACIDS

Many organic acids that are not classified as VFA occur in the rumen, usually in low concentrations. Sometimes they occur in abundance and are absorbed. Lactic acid is usually present in low concentrations in the rumen, but when concentrate feedstuffs replace high-forage diets, it may accumulate. In most cases of fast change from roughage to concentrate diets, there is a rapid blooming of *S. bovis* in the rumen, with accompanying production of lactic acid (Hungate *et al.*, 1952; Briggs *et al.*, 1957; Ryan, 1964). The acidity may drop to pH 4.0–4.5 and may remain low for 7–10 days, inhibiting activity of the rumen wall (Ash, 1956); and death may result if sufficient carbohydrate is available. Lactic acid is almost all L (+) in the rumen prior to feeding of grain diets and changes to approximately equal concentrations of D- and L-forms during adaptation to a high-grain diet (Ryan, 1964).

*Streptococcus bovis* may reach a level of several billion per milliliter when high-concentrate diets are suddenly introduced to the animal. It may then rapidly diminish in number, to be followed by a predominance of lactobacilli (Hungate *et al.*, 1952). Although *S. bovis* outgrow the lactobacilli initially, they are inhibited by the high acidity they cause, whereas the lactobacilli are not sensitive to pH in this low range. The time taken to shift from forage to high-grain diets influences the symptoms caused by ulcerative rumen lesions and liver abscesses in cattle (Jensen *et al.*, 1954). Adaptation over 12 days caused significantly more damage than adaptation over 30 days. Calcium carbonate dosing is ineffective in decreasing the rumen and liver damage. However, feeding 3% bicarbonate can increase the acetate : propionate ratio and maintain the milk fat content of cows fed high-grain diets (Davis et al., 1964; Emery et al., 1964).

Other organic acids reported to occur in the rumen include oxalic, glyceric, malic, glycolic, malonic, succinic, fumaric, and adipic acids (Bastie, 1957). Some of these acids are constituents of forage, others are common intermediates in metabolic reactions of cells, and still others are waste products of microbe metabolism. Higher unsaturated fatty acids are hydrogenated in the rumen.

## A. Effect of Rumen Acidity on Fermentation

Slyter (1981), working with rumen microbes from a steer fed an orchardgrass diet in culture fermenters adjusted to various levels of pH, determined the effect of pH on microbial activity for utilization of orchardgrass. It was found that the percentage of neutral detergent fiber (NDF) digestibilities at initial pH 7.5, 7.0, 6.5, 6.0, 5.5, and 5.0 were 51.4, 50.7, 44.0, 41.2, 25.2, and 11.5, respectively. Corresponding values for micromoles of VFA produced per gram of culture fluid were 45.2, 47.6, 42.9, 34.8, 23.4, and 9.6 at decreasing pH levels. Both NDF digestion and VFA production were reduced significantly at pH 6.0 or less.

In the pH range 4.5–7.0, the lower the pH the more rapid the absorption of VFA. Undissociated VFA are absorbed more rapidly than ionized forms that occur at higher pH values. Salivary secretion during feed ingestion assists in controlling pH by supplying sodium bicarbonate and phosphate during most rapid fermentation and production of acids. The pH in hungry ruminants moves toward alkalinity, and a diet high in protein and low in carbohydrates will tend to change in this direction. Smith (1941) reported that cows fed alfalfa hay had a pH of 6.27, but when fed alfalfa hay plus beet pulp, the pH was 6.0.

## B. Acidosis and Rumen Function

Acidosis is the result of excessive consumption of fermentable carbohydrates, reducing pH in the rumen as well as producing toxic factor(s) (Slyter, 1976). Concentrate diets cause a variety of effects, ranging from excellent productivity at one extreme (Wise et al., 1965; Tremere et al., 1968; Oltjen et al., 1966) to death of the ruminant at the other (Hungate et al., 1952). Many interrelated factors determine how a particular ruminant responds to the change from a roughage to a concentrate diet. Generally, what is most important is to control the animal's feed intake until it is completely adapted to the diet. However, adapted ruminants may overreact and develop acidosis and die or develop milder forms of acidosis (Dirksen, 1970). In the growing ruminant, feed intake and the growth response to a concentrate diet sometimes declines over time (Oltjen et al., 1966; Koers et al., 1976).

Ruminants have developed acidosis after eating diets containing wheat, barley, corn, unripe corn standing in the field, oats, sorghum, molasses, brewer's grains, potatoes, peaches, pears, and apples, as well as compounds such as lactic acid, sulfuric acid, VFA, butryic acid, starch, glucose, and lactose (Slyter, 1976). Silages and other fermented feeds that contain lactic acid could intensify the acidosis caused by concentrate diets. Sheep can consume more feed without acidosis problems than cattle on a body weight basis. Dinius and Williams (1975) observed that cattle in poor condition had more frequent and more severe disorders such as diarrhea than well-nourished cattle when a change was made from forage to concentrate diets.

Healthy ruminants fed concentrate diets at about two-thirds to three-fourths of the *ad libitum* intake have greater numbers of protozoa in the rumen than animals fed *ad libitum* (Christiansen et al., 1964; Eadie et al., 1970; Slyter et al., 1970). Frequent feeding at restricted intakes increases the number of protozoa in sheep (Moir and Somers, 1956) and cattle (Putnam et al., 1966). Healthy cattle fed concentrate diets *ad libitum* had an increase in rumen bacteria about ten-fold greater than that of roughage-fed animals; few, if any, protozoa were present (Slyter et al., 1965, 1970; Eadie and Mann, 1970).

Ghorban et al. (1966) observed that diets containing dried beet pulp or cracked corn and corn silage resulted in the most ruminal lactate when they evaluated a number of diets for production of lactate and VFA. *Selenomonas ruminatium* strains that are known to produce large quantities of lactic acid sometimes occur in high numbers in cattle fed urea–corn diets *ad libitum* (Caldwell and Bryant, 1966). The proportion of this organism increased from 5.2 to 16.8% when the amount of corn–urea diet was increased from 2.8 to 5.6 g/4 hr/500 ml in an artificial continuous culture maintained at a pH of 6.1 (Slyter and Weaver, 1972).

Jayasuriya and Hungate (1959) studied the turnover rate of lactic acid by incubating lactate-2-$^{14}$C with rumen fluid from animals fed roughate or concentrate diets. Their results indicated that less than 1% of the ingesta was fermented by lactate in the rumen of animals fed forage diets and less than 17% was fermented in those fed concentrates. Whanger and Matrone (1967) estimated that as much as 30% of lactate was converted to propionate through the acrylate pathway by rumen microbes in sheep fed an adequate sulfur diet but little by this pathway with a sulfur-deficient diet.

Baldwin et al. (1962) concluded that lactate was a minor contributor to the formation of succinate when they used labeled lactate to determine whether lactate was converted to propionate via the acrylate or randomizing pathways. Working with labeled glucose, Baldwin et al. (1963) concluded that propionate was labeled with the isotope as though 70–100% was formed through the succinate route and 0–30% via the acrylate pathway.

The percentage of butyrate and higher acids generally rose with increased dietary protein (Balch and Rowland, 1957), increased sugar (Orskov and Oltjen, 1967), and increased grain (Oltjen *et al.,* 1966). Greater numbers of protozoa in the rumen may increase butyric acid (Eadie and Mann, 1970). Hay-grain diets produced more butyrate and less acetate than hay diets (Waldo and Schultz, 1956). Acetate appears to decrease and propionate to increase in silage diets compared to hay diets. The primary end product of lactate metabolism is acetate, but oxidation of lactate to pyruvate requires the synthesis of butyrate from acetate to maintain an oxidation–reduction balance (Satter and Esdale, 1968). Adequacy of dietary sulfur also increases the synthesis of butyrate (Whanger and Matrone, 1967) as well as higher fatty acids from acetate (Whanger and Matrone, 1967; Slyter *et al.,* 1971). Beede and Farlin (1977a) evaluated the effect of 16 antibiotics on lactic acid production in an *in vitro* rumen with a mixed microbial population and found that it was effectively decreased by bacitracin methylene disalicylate, capreomycin disulfate, novobiocin, and oxamycin.

## XIV. ACIDOSIS SYNDROME AND PERFORMANCE OF RUMINANTS

Excessive intake of readily fermentable carbohydrates by ruminants often causes the rate of lactic acid production to exceed the rate of its conversion to VFA in the rumen. The practice in feedlots of rapidly increasing the concentrate portion of the diet frequently results in a problem called *acidosis*. Ruminal accumulation and subsequent absorption of lactic acid may result in rumen stasis, diarrhea, body dehydration, systemic acidosis, and death of animals (Huber, 1976). The rate of lactic acid absorption, and not the total quantity absorbed, appears to determine whether compensatory mechanisms such as buffering capacity are sufficient to maintain the hydrogen ion concentration of body fluids in a satisfactory range (Juhasz and Szegedi, 1968). The unabsorbed lactic acid in the alimentary tract also contributes to the lactic acid syndrome by disrupting the microbial population in the rumen as well as decreasing rumen motility.

Experimental acidosis has been produced by manually engorging rumen-fistulated animals with grain or by voluntarily overfeeding starved ruminants. Acidosis in steers was produced by abruptly changing them from an all-alfalfa diet to one containing 90% grain (Uhart and Carroll, 1967). The steers had a marked increase in lactic acid in the rumen and decreases in rumen and urine pH.

Elam (1976) discussed many management problems of large cattle feedlots that contribute to the development of acidosis. These problems are usually encountered when starting cattle on feed, when graduating cattle to higher-concentrate rations, when feeding a high-energy finishing ration for long periods, after weather changes, or after periods in which some problems such as a mill breakdown has resulted in animals being without feed and therefore hungry.

## A. Starting Cattle on Concentrate Diets

The usual problem with new cattle that have never eaten mixed feed is getting them to consume enough rather than too much. However, cattle that have previously eaten out of feed bunks usually start on concentrate feed rapidly and sometimes develop digestive disturbances. Elam (1976) suggested starting cattle on feedlot rations by beginning with a diet that contains 40–60% roughage plus long hay *ad libitum* for the first few days. When the animals appear to be eating near capacity, the roughage level in the mixed feed can then be gradually decreased. The level of $NE_g$ should be increased about 10% at a time; that is, if the cattle were started on diets that contained 40 Mcal $NE_g$/45.5 kg, they should be moved up to 44 Mcal, then to 48 Mcal, and increased at similar increments until maximum intake levels of energy are provided. It is difficult to feed a high-grain, low-roughage diet without eventually having some cattle develop acidosis or bloat. Replacing some of, the grain in a high-energy diet with roughage and fat on an isocaloric basis, or replacing some grain with other high-energy ingredients such as molasses or dried beet pulp, will tend to decrease acidosis (Elam, 1976).

Generally, more acidosis occurs in summer and may be due to more fluctuations in feed intake. High temperatures will cause a decrease in feed intake, and overeating may occur when the weather cools. Rains may increase molding of feeds in bunks, and feed intake will decrease until it is removed; this may be followed by overeating.

## B. Dietary Wheat and Lactic Acid

High levels of wheat in cattle diets were associated with high concentrations of lactic acid in the rumen (Ahrens, 1967). Varner and Woods (1975) conducted *in vitro* and *in vivo* studies with Hereford steers to compare the fermentation characteristics of various wheat varieties to those of corn. Wheat in the diets reduced feed intake and increased lactate in the rumen fluid compared to corn. *In vitro* fermentation indicated that wheat was more digestible and supported more lactate concentration than corn. Generally, Scout 66, Turkey, and Warrior wheats supported higher *in vitro* lactate levels than Gage wheat. Elam (1976) stated that wheat generally produces the highest levels of acidosis with high-grain diets. Corn and sorghum grain diet are also highly conducive to acidosis, while barley causes the smallest problem.

## C. Histamine and Other Amines in Acidosis

Several investigators have suggested a relationship between histamine in the rumen and acidosis of cattle (Ahrens, 1967) and sheep (Dain *et al.*, 1955; Irwin *et al.*, 1972). Koers *et al.* (1976) conducted trials with cattle and sheep to determine the relationship of lactic acid to pH and histamine in subacute and acute acidosis. Histamine and pH of rumen fluid were equivalent in wheat- and corn-containing diets. In lambs engorged with a corn diet, rumen pH declined from 6.77 to 4.74 after 4 hr and remained low for 60 hr. There were ruminal peaks of lactate at 4–6 hr postengorgement and at 60 hr, while histamine peaks occurred at 2, 10, 24, and 60 hr postengorgement. Infusion of histamine and lactic acid into the rumen of sheep 1 hr postfeeding showed that histamine disappearance was slower with added lactic acid than when infused alone. In another trial, infusing histamine into the rumen of sheep had no effect on feed intake. It was concluded that histamine did not play a major role in ruminant acidosis.

Irwin *et al.* (1979) conducted *in vitro* and *in vivo* trials to study ruminal lactic acid, histamine, tyramine, and tryptamine. Thirty hours after glucose was added to rumen fluid, *in vitro* pH decreased and lactic acid increased, as would be expected in acidosis. Tyramine increased dramatically, and when graded levels of histidine were added with glucose, histamine levels increased accordingly. When ewes were dosed with a mixture of 90% glucose and 10% casein, the rumen pH decreased but the histamine concentration was unchanged.

J. R. Wilson *et al.* (1975) compared rumen fluid pH, histamine, total lactate, L (+) lactate, and VFA in sudden death syndrome cattle, cattle with lactic acidosis produced through grain engorgement, and healthy feedlot cattle maintained on a high-energy finishing diet for more than 100 days. The rumen fluid from lactic acidotic cattle was high in histamine and lactic acid and low in pH and total VFA. Concentrations of histamine, lactic acid, and total VFA varied in the rumen fluid of sudden death syndrome cattle but showed no significant differences from those in healthy feedlot cattle.

## D. Antibiotics and Lactic Acidosis

The proliferation of *S. bovis* and *Lactobacillus* species, major lactic acid–producing rumen bacteria, initiates acidosis (Hungate *et al.*, 1952; Krogh, 1961; Mann, 1970; Slyter, 1976). Carbohydrate is fermented to lactic acid by *S. bovis*, which lowers rumen pH and facilitates proliferation of lactobacilli. Numerous research efforts have been made to control acidosis by preventing the proliferation of *S. bovis* with selective antibiotics.

Among the antibiotics evaluated in the control of acidosis are tetracycline hydrochloride and penicillin G (Nakamura *et al.*, 1971), oxamycin (Beede and

Farlin, 1977b), and thiopeptin (Muir and Barreto, 1979; Muir *et al.*, 1981; Kezar and Church, 1979). Dennis *et al.* (1981a) reported that the two polyether antibiotics (ionophores), lasalocid and monensin, inhibited major lactate-producing rumen bacteria except *Selenomonas* species *in vitro*. Monensin inhibited *Lactobacillus* and some strains of *S. bovis,* and lasalocid inhibited both *Lactobacillus* and *S. bovis.*

Nagaraja *et al.* (1981) investigated the effectiveness of monensin and lasalocid in preventing induced acidosis in fistulated steers and cows. Administering the antibiotics (1.3 mg per kilogram of body weight) for 7 days before experimentally inducing acidosis with corn (27.5 g per kilogram of body weight) effectively prevented acidosis. Two days were sufficient to prevent glucose-induced (12.5 g per kilogram of body weight) acidosis. The different responses probably were due to differences in the amount of carbohydrate used to induce acidosis. Cattle treated with the antibiotics had higher rumen pH and lower L(+) and D(−) lactate concentrations than cattle that received no antibiotics. The control cattle exhibited classic signs of acidosis such as lowered blood pH, increased blood lactate, and depleted alkali reserves with a pronounced base deficit. Cattle treated with the ionophores exhibited no signs of systemic acidosis.

Muir *et al.* (1981) tested the effectiveness of thiopeptin, a sulfur-containing peptide antibiotic, in preventing acidosis in fistulated bull calves and Angus feedlot steers. Fistulated bull calves that were given 40 g ground wheat per kilogram of body weight developed both ruminal and systemic acidosis. Single doses of 0.75–1.50 mg thiopeptin per kilogram of body weight added to a wheat slurry resulted in 80% less lactic acid in the rumen fluid. Steers adapted to an alfalfa diet and abruptly changed to a micronized milo diet containing 0, 11, or 22 ppm thiopeptin produced 70% less rumen lactate with both the 11- and 22-ppm levels of the antibiotic than did controls. In both studies, the thiopeptin allowed normal fermentation to proceed, as indicated by the levels of rumen VFA.

## E. Endotoxin and Lactic Acidosis

A toxic substance was isolated from the rumen ingesta of cattle and sheep suffering from acidosis that depressed blood pressure, rumen, and isolated intestine motility and caused leukopenia (Dougherty and Cello, 1949). Mullenax *et al.* (1966) observed similar responses in animals given endotoxin intravenously and those with acidosis. It was suggested that endotoxin of gram-negative rumen bacteria, when released and absorbed, could play a role in the lactic acidosis syndrome. The presence of endotoxin in the rumen of cattle with lactic acidosis was demonstrated by Huber (1976). Nagaraja *et al.* (1978a) observed that rumen bacteria are a source of endotoxins and that cell-free rumen fluid of grain- or hay-fed cattle contains free endotoxins.

Nagaraja *et al.* (1978b) induced lactic acidosis in fistulated cows by engorging them with a 1 : 1 mixture of ground corn and wheat. The endotoxin level in the rumen increased 15–18 times within 12 hr after lactic acidosis occurred. This increase was accompanied by a change from predominately gram-negative to gram-positive bacteria. The presence of significant granulocytosis that occurs with acidosis was suggestive of systemic action of rumen bacterial endotoxin.

## F. Rumen Buffering

Ruminants evolved with forage-type diets that were fibrous, low in energy, and generally available in limited quantities. Such low-energy diets favored a rumen fermentation that produced a limited amount of acid and may have contributed to a lower buffering capacity in the rumen and small intestine. Ruminant animals sometimes encounter difficulty in handling high-energy diets that produce high levels of acid in the rumen, where it adversely affects microbes, or passes on to the intestines, where it interferes with digestion of starch and other nutrients. Noller *et al.* (1980) reported that cattle fed large quantities of poorly buffered high-energy feeds show (1) reduced feed intake, (2) lower production, (3) lower pH in the small intestine, large intestine, and feces, (4) lower digestibility of starch, (5) lower digestibility of dietary energy, (6) lower fiber digestion, (7) possible rumen parakeratosis, and (8) inflammation of the small intestine mucosa. Effective dietary buffers cause many of these problems to be prevented or reduced in severity.

Different materials may be used to buffer acidity, such as dietary ingredients or chemical buffers such as sodium bicarbonate. For instance, 0.75% of sodium bicarbonate in ration dry matter is frequently fed dairy and beef cattle. However, it is most effective in the rumen and is of little value farther down the alimentary tract (Noller *et al.*, 1980). The effectiveness of many mineral buffers depends on their source, particle size, and chemical and physical properties. Substances such as limestone and protein also help control pH in the lower gastrointestinal tract. Rations that contain alfalfa may not benefit from added chemical buffers, since alfalfa itself is an excellent buffer.

## G. Feed Utilization with Buffering Substances

Calcium carbonate (0.74%) in high-concentrate diets fed to beef steers improved the digestibility of crude protein, cellulose, and energy, with about half of the total improvement in digestibility due to increased cellulose digestion (Varner and Woods, 1972). Wheeler and Noller (1977) found that the addition of limestone or dolomitic limestone to high-moisture corn diets increased fecal pH while decreasing the concentration of fecal starch. When the animals were slaughtered, digesta in all segments of the gastrointestinal tract had elevated pH levels. The investigators theorized that the increase in intestinal pH produced a more favorable environment for amylase activity, resulting in more starch being digested in the small intestine.

## H. Beef Cattle Fed Buffers

Greater use of high-concentrate finishing rations has stimulated studies of buffering agents due to improved performance when they are present in diets. A mixture of sodium bicarbonate, potassium bicarbonate, and magnesium carbonate at 3.02% of a high-concentrate diet fed cattle improved both feed intake and daily weight gains (Nicholson *et al.*, 1960). Wise *et al.* (1965) observed that 5% sodium bicarbonate, potassium bicarbonate, or a mixture of the two buffers in an all-concentrate diet fed steers improved feedlot performance. Cullison *et al.* (1976) added 0.5% sodium bicarbonate to the drinking water of steers fed a pelleted forage-concentrate diet and observed increases in feed intake and daily weight gain. Addition of sodium bicarbonate to drinking water increased the number of microbes in the rumen but had no effect on pH or VFA production there (Lassiter *et al.*, 1963).

In some studies, beef cattle performance failed to improve when buffers were added to the diet. For instance, when Nicholson *et al.* (1962) fed 5.7% sodium bicarbonate in a high-concentrate diet to steers, they found no benefits in feed intake, weight gains, or nutrient utilization. Wise *et al.* (1961) reported marked decreases in feed intake and weight gains in steers fed a pelleted low-fiber diet that contained 11% sodium bicarbonate.

Russell *et al.* (1980) fed feedlot steers a high-corn diet with 0.9% $NaHCO_3$, 1.8% limestone, or a combination of both buffers at these levels, and took rumen samples via stomach tube and fecal samples by rectal grab at 56 and 112 days of feeding. Steers fed the buffered diets had slightly smaller weight gains and feed efficiencies than those fed the control diet, but steers fed the 1.8% limestone had higher carcass marbling scores than those in the other groups. Carcass fat thickness was less for the limestone and buffer combination groups and contributed a lower yield grade. Buffer supplementation increased fecal pH but had little effect on starch level in the feces.

## I. Dairy Cattle Fed Buffers

Oltjen and Davis (1965) observed that buffers added to high-grain rations increased the acetic acid in the reticulorumen. Depressions in milk fat from dairy cows fed high-grain diets have been associated with low concentrations of acetic acid and high concentrations of propionic acid in the rumen (Armstrong and Blaxter, 1957; Ensor *et al.*, 1959). However, Emery and Brown (1961) observed an increase in milk fat from milk cows fed 454 g of either sodium bicarbonate or potassium bicarbonate per day without any change in acetic acid levels in the rumen.

Lactating dairy cows fed limited quantities of hay plus *ad libitum* amounts of a grain mixture supplemented with 454 g of sodium bicarbonate daily increased their milk fat test by an average of 0.83% in two trials (Emery *et al.*, 1964). Addition of 272 g calcium carbonate daily in the above diets had no influence on milk fat, and total milk production was not affected by the addition of either sodium bicarbonate or calcium carbonate to the diets. However, 10–20% less grain was consumed by cows that were fed sodium bicarbonate. Davis *et al.* (1964) fed cows a high-grain diet containing 3% of a sodium bicarbonate and potassium bicarbonate mixture and obtained a 0.57% improvement in the milk fat test. Milk fat depression in cows fed large quantities of grain was reversed when a buffer mixture containing 454 g potassium bicarbonate, 381 g sodium bicarbonate, or 191 g magnesium carbonate per day was incorporated in the diet (Miller *et al.*, 1965).

Wheeler and Noller (1976) added limestone to high-grain diets fed dairy cows in order to raise the pH of the intestines because sodium and potassium are more readily absorbed from the upper part of the tract than is calcium. While cows fed 2.71% limestone in 55% grain diets had no change in feed intake, milk production, or milk fat percentage, they gained 0.66 kg daily compared to a daily body weight loss of 0.27 kg for the controls.

## J. Cement Kiln Dust and Ruminant Response

Inclusion of cement kiln dust in diets fed finishing steers and lambs has been reported to increase gains, to improve feed efficiency, and to have favorable effects on ruminal and carcass characteristics (Wheeler and Oltjen, 1977). Noller *et al.* (1980) conducted a series of tests to characterize the reactivity of cement kiln dusts. There was little difference among various sources of kiln dust in terms of their capacity to neutralize acids. All kiln dusts were less reactive than calcium oxide but were similar to feed grade calcium carbonates. Kiln dust was found by x-ray analysis to contain calcium primarily as calcium carbonate, and varied from 49.5 to 85.7%. All kiln dusts were more reactive at pH 3.0 than at 6.0, and those that did not show a favorable response when fed to ruminants had less reactivity at pH 6.0 than dusts that gave a favorable response. The active agent appeared to be finely divided calcium carbonate.

Galyean and Chabot (1981) compared the effects of sodium bentonite, McDougall's buffer salts, clinoptilolite, and cement kiln dust on the rumen characteristics of beef steers fed a high-roughage diet. Intake of roughage tended to be higher with diets containing buffer salts than with the other treatments (7.19 versus 6.72 kg per head daily). In some cases, supplements containing the buffer salts and kiln dust were rejected. Seven hours after supplements were offered, the rumen pH of steers consuming buffer salts was higher (6.74) than with the other treatments (6.57). Fecal pH was higher in steers fed buffer salts, clinoptilolite, and kiln dust than in those consuming sodium bentonite and in the controls. VFA, ammonia, protein, ash, osmolality of a rumen bacterial fraction, and rumen liquid volume were equivalent with all treatments.

Rumsey (1982) determined the effect of dietary kiln dust on the composition of tissue gain of Hereford steers in a slaughter balance trial. Half of the steers were implanted with Synovex-S. Steers implanted consumed more feed dry matter, gained faster, and had better feed efficiency than unimplanted steers. The Synovex-S implants also resulted in steers retaining 1.21 times more nitrogen, 1.23 times more ether extract, and 1.24 times more energy than controls. The kiln dust in diets had no effect on steer performance or composition of gain, although it did tend to have a synergistic effect on protein and ash retention in tissues.

## XV. BLOAT IN RUMINANTS

Gas is produced continuously and in large amounts in the rumen. Normally, the gas is eliminated without difficulty, but sometimes the rate of elimination falls behind the rate of production and gas accumulates, causing bloat. Gross distention of the rumen occurs if the disparity continues. Intraruminal pressure may rise as high as 70 mm Hg (Boda *et al.*, 1956). The inflated rumen interferes with respiration; carbon dioxide at excessive levels may be absorbed into the bloodstream, and a susceptible animal may die within an hour of commencing a feeding period (Dougherty and Habel, 1955; Davis *et al.*, 1965).

Beef cattle that have been fed primarily pasture forage are placed in feedlots and changed gradually from a roughage-starter ration to a high-grain concentrate ration over a period of 3–4 weeks. The final finishing diet usually has about 80% grain, 15% roughage, and 5% protein-mineral-vitamin supplement. The cattle are fed for about 120–180 days and gain 1.0–1.5 kg per head daily. They sometimes develop "grain bloat" slowly over 50–100 days of concentrate feeding. High-producing dairy cows fed 12–22 kg grain daily may also exhibit grain bloat. Excessive frothing of ingesta in the rumen occurs in both grain bloat and pasture bloat. In pasture bloat the primary foaming agents are derived from plants, while in grain bloat the principal source appears to be of microbial origin.

### A. Microbial Changes

Gutierrez *et al.* (1959) reported that *S. bovis* and *M. elsdenii* increased as cattle started to bloat with feedlot diets. They believed that the lactic acid streptococci contributed to frothing by producing gas from the breakdown of lactic acid and forming a filamentous mat and a stable foam. *Lactobacillus* counts were equivalent before and during grain bloat.

Bartley and co-workers (Bartley and Yadava, 1961; van Horn and Bartley, 1961) found that saliva inhibited and dispersed legume bloat. They postulated that bloat occurs when mucin of the rumen is lowered by reduced salivation when ruminants eat succulent feeds containing foaming agents. However, they were unable to show that saliva plays the same role in grain bloat. Several rumen microbes are mucinolytic and have been isolated (Mishra et al., 1967, 1968). Encapsulated rumen microbes have been found in great numbers in cattle fed feedlot diets, and it has been postulated that such capsules or slime may trap fermentation gases and cause bloat.

Meyer and Bartley (1971) attempted to correlate the incidence and severity of grain bloat with the relative viscosity and concentration of polysaccharides of bacterial origin. Rhamnose was analyzed, since it represented polysaccharides of microbial origin. Since hexose was part of the rhamnose analysis, it was called *apparent glucose*. The relative viscosity and concentration of cell-free apparent glucose and rhamnose were higher in cattle fed an all-concentrate diet than in those fed an all-alfalfa ration. The bloat score (scale of 0, no bloat, to 5, severe bloating resulting in death) was correlated with the relative viscosity and cell-free polysaccharide level in the rumen fluid. It was concluded that polysaccharides of bacterial origin could contribute to the viscosity of rumen fluid and thereby stabilize the froth in the rumen. Gutierrez et al. (1961) precipitated the rumen slime fraction of cattle fed a high-grain diet and found that it contained 33–35% crude protein, acid-hydrolyzable polysaccharides, and nucleic acids.

The low rumen pH associated with concentrate diets generally reduces the number of protozoa in ruminants. *Entodinia* protozoa ingest starch (Abou Akkada and Howard, 1960) and streptoccoci (Hungate, 1966). They can reduce bloat by preventing bacteria from utilizing starch and by reducing the number of bacteria associated with bloat.

## B. Feed Factors

Mead et al. (1944) observed that mild bloat occurred in ruminants fed finely ground hay and grain but not grain plus long alfalfa hay. Smith et al. (1953) produced frothy bloat in dairy cattle fed a diet of 61% barley, 16% soybean meal, 22% alfalfa meal, and 1% salt. Others have produced bloat on such high-protein rations (Lindahl et al., 1957; Lyttleton, 1960). Such diets are higher in protein than typical feedlot rations, but the frothy type of bloat is similar to that observed with feedlot cattle, and it can be produced more quickly on high-protein diets.

When cattle were fed 6.4 kg concentrate (78.3% corn, 20.7% soybean meal, 1% salt) and 1.8 kg alfalfa hay per animal per day, the bloat index increased over time; the maximum incidence and severity of bloat occurred 120–150 days after the cattle were fed the diet (Lindahl et al., 1957). Increased bloating occurred when alfalfa meal was substituted for alfalfa hay in the ration. Bloat was not affected by substituting barley for the corn or by pelleting the ration. Adding 1.4 kg ground corn cobs to the ration after bloat was initiated did not reduce the problem. Emery et al. (1960) reported that once feedlot bloat developed, feeding of equal parts of hay and concentrate failed to decrease its severity.

Bartley et al. (1975b) concluded that bloat can be eliminated only by removing more than 50% of the grain for 4–6 weeks. Further, if animals are fed a high-roughage ration for only a week and then a high-grain diet, they will start bloating again in 1–2 days. It takes several days for roughage to remove slime from the rumen wall. If the slime is not completely removed, it acts as a source of slime-producing microbes that multiply rapidly when a high-grain ration is fed again.

Hironaka et al. (1973) fed five sets of identical twin cows diets of coarse and fine particles (715 versus 388 μm mean particle size) and observed more bloat with the fine particles. They reported that when pelleted all-concentrate diets were fed at several commercial feedlots, a minimum of bloat occurred, while corresponding finely ground, nonpelleted rations resulted in severe bloat losses.

Elam et al. (1960), working with steers susceptible to bloat on a bloat-producing ration, found that a highly significant increase in feedlot bloat occurred when soybean oil was added to the diet. A greater incidence of bloat occurred in steers fed twice a day than in those allowed to eat *ad libitum*. However, Lindahl et al. (1957) did not observe any correlation of bloating with eating habits in steers consuming a bloat-producing diet. Elam and Davis (1962a) reported that adding fat to feedlot diets had no effect on the incidence of bloat, but either 4 or 8% mineral oil reduced bloat by 40%.

Bartley et al. (1975b) stated that many feedlot operators have observed bloat due to changes in weather, feeding times, missed feedings, under- or overfeeding, and other forms of improper management. Cattle may bloat when they are not greedy feeders.

## C. Animal Factors

Cattle have been shown to inherit the tendency to bloat (Reid et al., 1975). Heritable characteristics related to bloat include eructation proficiency, salivary capacity, secretion rate of saliva, and rate of eating. Dougherty and Habel (1955) observed that foam covering the cardia inhibited eructation of gas. Johnson and Dyer (1966) reported that bloaters appear to secrete less epinephrine than cattle that do not bloat. Nonbloating cattle appear to secrete less saliva than bloaters when consuming legumes (Lyttleton, 1960; Mendel and Boda, 1961). Decreased saliva secretion could promote bloating, since saliva can act as an antifoaming agent (Bartley and Yadava, 1961).

Histamine has been found to impair motility of the rumen (Hungate, 1966), and its concentration increases in the rumen of cattle fed high-concentrate diets due to decarboxylation of histidine (Dunlop, 1972). Steers bloating on an all-concentrate ration were treated with antihistamine, poloxalene plus antihistamine, or poloxalene alone by Meyer et al. (1973). Poloxalene alone did not cure the bloat, but when 50 g per head was mixed in the daily rations, intramuscular injection of antihistamine increased rumen contractions and reduced the bloat index.

## D. Poloxalene

Poloxalene is a polymer of polyoxypropylene-polyoxyethylene. It has been established as an effective legume bloat preventive for cattle (Bartley, 1965; Bartley et al., 1965; Stiles et al., 1967) but has less efficiency in preventing bloat in sheep (Lippke et al., 1969). Bartley and Meyer (1967) reported that poloxalene reduced bloat by about 50% when mixed in rations at levels of 10, 15, or 22.5 g per head daily. Essig et al. (1972) found that administration of poloxalene and penicillin appeared to reduce VFA in bloaters (steers) and nonbloaters grazing Ladino clover and reduced the severity of bloating.

Several reports have indicated the absence of undesirable side effects as a result of feeding the antibloat polymer (Helmer et al., 1965; Lippke et al., 1970). R. M. Meyer et al. (1965) reported that approximately 4% of dietary poloxalene intake is absorbed from the alimentary tract and appears in the urine, but none appears in the milk or tissues.

Administration of poloxalene is generally done through feed or feed supplements. Some researchers have incorporated the surfactant in the ration or sprinkled it over grain (Bartley et al., 1965; Foote et al., 1968). Grazing cattle may be given poloxalene by incorporating it in molasses salt blocks (Essig and Shawver, 1968; Foote et al., 1968; Stiles et al., 1967) or provided with molasses in tanks with lick wheels (Scheidy et al., 1972). Bartley et al. (1975a) reported that a molasses liquid supplement containing poloxalene and a molasses salt block containing poloxalene reduced the incidence and severity of bloat in cattle grazing wheat pastures.

## E. Feedlot Bloat

Excessive foaming of rumen contents occurs in feedlot bloat, just as it does in pasture bloat. However, feedlot bloat is considered to be of bacterial origin, while pasture bloat is largely due to substances of plant origin. Feedlot bloat usually occurs over a period of several weeks when ruminants are fed high-concentrate diets. While roughage is generally accepted as a dietary ingredient that decreases feedlot bloat, alfalfa is considered to cause bloat. High protein level and low cell wall content may be factors in enhancing alfalfa bloating, but no actual factor has been determined.

Although poloxalene is effective in dispersing foam in pasture bloat, it is less useful for prevention and treatment of feedlot bloat. A dimethyl dialkyl quaternary ammonium compound has been shown to be quite effective in preventing feedlot bloating in beef cattle, and sheep (Meyer and Bartley, 1972a). Mineral oil (4–8%) in the diet has been reported to reduce bloat by 40% (Elam and Davis, 1962b). Animal fats are not as effective as mineral oil, and soybean oil (8%) in the ration increases bloat (Elam and Davis, 1962b). The difference in action between mineral oil and other oils suggests that lipids have a role in feedlot bloat. Meyer and Bartley (1972b) reported that various surfactants, fatty acids, silicones, antibiotics, and dyes have not been demonstrated to be effective in feedlot bloat control.

## XVI. DILUTION RATES IN THE RUMEN

The rumen operates as a continuous culture in many ways, since feed enters the rumen at more or less regular intervals and digesta and microbes are similarly passed to the omasum. The diet consists of both insoluble feed particles and soluble fractions, which prevents the rumen from operating as a homogeneous system. This results in at least two major dilution rates, one of solids and one of liquids. Liquids turn over two to four times faster than solid feed particles, subjecting rumen microbes to different conditions for growth and washout, depending on the phase in which they exist (Russell and Hespell, 1981). Mineral salts added to diets increase rumen liquid dilution rates, thus altering the products of fermentation (Rodgers *et al.*, 1979). A decrease in the molar quantity of propionic acid is the most marked observation in this change in fermentation.

An increased rumen dilution rate may improve the efficiency of rumination by increasing the efficiency of bacterial growth and the flow of $\alpha$-linked glucose polymers, amino acids, and microbes to the abomasum (Stouthamer and Bettenhaussen, 1973; Harrison *et al.*, 1975). Lemenager *et al.* (1978a–c) reported that feeding monensin to beef cattle reduced the dilution rate, decreased the acetic acid, and increased the propionic acid in rumen contents. In contrast, sodium bicarbonate increased the dilution rate (Harrison *et al.*, 1975; Rodgers *et al.*, 1979). Adams *et al.* (1981) found that steers fed monensin had lower liquid dilution rates than those fed yeast culture or 5% sodium bicarbonate, but the rates were similar to those of steers fed 2.5% sodium carbonate and to those of the controls.

A number of studies have demonstrated a positive correlation between increased dilution rate and greater growth of bacteria (Hobson, 1965; Stouthamer and Bettenhaussen, 1973; Isaacson *et al.*, 1975; Harrison *et al.*, 1976; Kennedy *et al.*, 1976). Cole *et al.* (1976a–c) reported an increase in the dilution rate of 2.8–5.0%/hr along with increased protein synthesis in the rumen of 7.5–11.8 g per 100 g of digestible dry matter (DMD) when steers were changed from an all-concentrate diet to one that contained 14% roughate. Feeding monensin to sheep decreased the rumen dilution rate from 7.0 to 4.0%/hr, with a concomitant decrease in microbial synthesis from 24.5 to 20.2 g nitrogen per kilogram of organic matter digested (OMD) (Allen and Harrison, 1979).

## XVII. RUMEN FERMENTATION MANIPULATION

Dietary components of high nutritional value may be degraded to ones of less value in the rumen. It would be especially beneficial to the host animals if high-quality protein bypassed the rumen microbes and their enzymes. Very young ruminants are protected from rumen fermentation by a closed esophageal groove, or they may be fed diets that do not require pregastric digestion. The closed esophageal groove provides an extension of the esophagus from the cardia to the reticulo-omasal orifice. Bypassing the rumen in older animals with certain diets has improved growth and feed efficiency (Orskov, 1972). Degradation of proteins and amino acids can be minimized by heat and chemical treatments of dietary proteins, encapsulation of amino acids, and the use of amino acid analogs rather than free amino acids (Chalupa, 1975; Fergusan, 1975). Rumen lipolysis and hydrogenation of lipids can be prevented by coating minute oil droplets with formaldehyde-treated protein (Scott and Cook, 1975). Amino acid nitrogen and energy in feedstuffs that require digestion in the rumen cannot be improved by protective measures.

Control of rumen microbial activities by chemical agents may be an alternative to nutrient protection. Efficiencies of fermenting hexose to acetate, propionate, and butyrate are 62, 109, and 78%, respectively, due to differences in the production and utilization of metabolic hydrogen (Chalupa, 1977). To obtain maximum energy yields, electron acceptors need to be available to dispose of hydrogen. Some hydrogen is used in bacterial growth and for the hydrogenation of unsaturated fatty acids, but most hydrogen is utilized in the formation of propionate and butyrate from pyruvate and in reducing carbon dioxide to methane (Demeyer and van Nevel, 1975; Leng, 1970a,b).

Raun *et al.* (1976) found that monensin increased propionate production, with decreases in acetate and butyrate but with little effect on total VFA production. The antibiotic has improved the feed efficiency of cattle fed concentrate diets or on grazing pasture. Thornton and Owens (1981) fed monensin to growing steers in diets containing 12, 27, or 40% acid detergent fiber; they found that it decreased methane production by 16% at the two lower roughage levels and by

24% at the higher roughage level. Monensin increased propionate and decreased acetate in the rumen at the low and high roughage levels, respectively.

Halomethane analogs are among the more potent methane inhibitors (Czerkawski and Breckenridge, 1975; Demeyer and van Nevel, 1975; Trei et al., 1970a,b). Trei and Scott (1971) reported that methane production was reduced with amicloral, with an accompanying improved growth rate and feed efficiency in steers. Feeding amicloral to feedlot steers had no effect on feed intake or weight gains, but feed efficiency was increased by 9% (Horton, 1980b). Amicloral feed either alone or with monensin to growing steers decreased feed intake by 11%, had no effect on weight gains, but improved feed efficiency by about 12% (Horton, 1980a).

Efficiency of fermentation was improved as greater amounts of hydrogen were recovered in VFA, especially in propionate and butyrate (Chalupa, 1977). Concentrate and feed intake levels appear to affect energetic changes caused by analogs of halomethane. Diphenyl iodonium chloride has been reported to decrease deamination of amino acids (Chalupa, 1976) and to improve the efficiency of gains in cattle (Chalupa, 1977; Horton, 1980b). Chalupa et al. (1983) reported that 40 ppm of several diphenyliodonium chemicals in diets of growing steers increased nitrogen (N) absorption, utilization of absorbed N, and N retention. Acetohydroxamic acid is an inhibitor of ruminal urease (Brent et al., 1971), and was found to decrease ammonia concentration in the rumen and to increase nitrogen retention in sheep (Streeter et al., 1969).

## XVIII. OXYGEN IN THE RUMEN

Oxygen has been shown to be present in the dorsal sac of the rumen (Kandatsu et al., 1955). The low oxygen tension in mammalian tissues and the thick layer of metabolically active epithelial cells through which oxygen would have to diffuse in going to the rumen from the blood makes it doubtful whether such diffusion can account for the oxygen in rumen gas (Hungate, 1966). Swallowed air is more likely to be its source. The small amounts of oxygen mixed with the rumen contents are rapidly reduced (Broberg, 1957). Inhibition of cellulose digestion in the rumen of fistulated animals without a plug may be due to oxygen, particularly with reduced feed intake (Hungate, 1966). Johnson et al. (1958) observed inhibition of in vitro cultures with washed cell suspensions of rumen microbes.

### A. Rumen Gas Composition

McArthur and Miltimore (1961), using the gas-solid chromatographic technique, collected the data on rumen gas composition presented in Table 3.3.

**TABLE 3.3.**

**Composition of Rumen Gas**[a]

| Gas | Mean percentage | Sample variance | Analysis variance |
|---|---|---|---|
| Hydrogen | 0.18 | 0.0073 | 0.0027 |
| Hydrogen sulfide | 0.01 | 0.000007 | 0.000005 |
| Oxygen | 0.56 | 0.0151 | 0.0004 |
| Nitrogen | 7.00 | 0.6922 | 0.1277 |
| Methane | 26.76 | 0.7475 | 0.0620 |
| Carbon dioxide | 65.35 | 0.9293 | 0.1170 |

[a]From McArthur and Miltimore (1961).

## XIX. EFFECT OF INORGANIC IONS

According to Dobson and Phillipson (1958), potassium and other cations are not actively absorbed through the rumen, nor are the anionic inorganic radicals under ordinary conditions. When the potassium ion concentration is very high, chloride ion may be actively transported into the rumen from the blood, and vice versa (Sperber and Heyden, 1952). Parthasarthy and Phillipson (1953) and Heyden (1961) reported that potassium is absorbed from the rumen, but at a slower rate than sodium. The potassium concentration in the fluid fraction of the gastrointestinal tract is consistently several times greater than that in blood plasma, even though the contents of the tract are generally electronegative. Sodium in the rumen fluid exceeds potassium by a factor of 1.5–3.0. Aldosterone production is stimulated by a dietary sodium deficiency; as a result, potassium largely replaces sodium in saliva (Blair-West et al., 1965). Bell (1984) reviewed the influence of dietary sodium on feed intake of cattle.

Animals consuming high-roughage forage diets may have a fourfold greater intake of potassium than when fed concentrate diets. An osmotic deficiency in rumen fluid may be responsible for the results attributed to potassium deficiency (Ward, 1966). DuToit et al. (1934) reported that a diet that provided 0.32% potassium in the dry matter was sufficient to maintain milk production of approximately 8 liters/day over two lactation periods. Finishing steers fed 0.5–0.6% potassium on a dry matter basis made rapid weight gains (Roberts and St. Omer, 1965). Excess sodium chloride intake may increase osmotic pressure in the rumen to the point where dry matter consumption is diminished (Galgan and Schneider, 1951). This effect can be produced with excess salt intake without affecting the activity of rumen bacteria (Cardon, 1953); however, all protozoa are diminished, and some kinds are completely eliminated (Koffman, 1937).

Russell et al. (1980), in a feedlot trial with steers fed concentrate diets with added limestone and sodium bicarbonate, observed that the buffer decreased ruminal pH and increased butyric acid at 56 days, but at 122 days there were no differences in VFA patterns. The added limestone increased ruminal pH and the bicarbonate increased fecal pH but had little effect on fecal starch concentration. Steers fed diets with 2% sodium bentonite plus 1% potassium bicarbonate had higher ruminal pH at 4 and 8 hr postfeeding. Lactate production was depressed with diets that contained 1% bentonite plus 1% dolomite limestone or 1% bentonite plus 1% potassium bicarbonate by 2 hr postfeeding, but ruminal glucose, VFA, and rumen fluid osmolalites were not affected by treatments (G. W. Horn *et al.*, 1979).

## Absorption of Water

Material entering the omasum contains 90–95% water (Balch *et al.*, 1951), and the percentage of dry matter in the reticulum is less than in the rumen. Cattle absorb 100 liters/day in the omasum (Sperber and Heydén, 1952). In addition to water, VFA and other dissolved constituents are absorbed by the omasum (Gray *et al.*, 1954; Johnston *et al.*, 1961). The omasum is more efficient in removing the VFA than the rumen, and the effluent from the abomasum has 1/7th to 1/20th of the VFA of that in the rumen (Masson and Phillipson, 1952).

## XX. TEMPERATURE AND RUMEN FERMENTATION

Constant temperature of the rumen is of great importance in fermentation and has been used as a measure of the fermentation rate (Walker and Forrest, 1964). Rumen temperature is usually above the 38°C of body temperature (Wren *et al.*, 1961) and decreases as fermentation drops after feeding (Dale *et al.*, 1954). Cunningham *et al.* (1964) reported that very cold drinking water can lower the temperature of rumen contents below normal but has little effect on feed utilization. The temperature may rise as high as 41°C during active fermentation of alfalfa (Trautmann and Hill, 1949; Gilchrist and Clark, 1957); it is usually between 39 and 40.5°C (Krzywaneck, 1929). Findley (1958) found that 42°C is the upper temperature limit at which cattle can survive, indicating that hot climates allow little temperature leeway with some diets and survival.

## XXI. MODELING OF NITROGEN METABOLISM IN THE RUMEN

Modeling of nitrogen and energy metabolism in the ruminant involves the interactions of microbes, chemical and physical properties of feeds, and the host (Baldwin and Denham, 1979). The principal force in rumen microbial activity is fermentation of feed components to obtain ATP for maintenance and growth of the microbes. Three classes of fermentable substrates need to be considered in order to describe the rate of fermentation: (1) soluble nutrients such as sugars and organic acids, (2) nutrients such as starch and pectin, with intermediate solubility characteristics, and (3) insoluble substrates such as hemicellulose and cellulose. Ion transport may be a major component of ATP utilization for bacterial maintenance (Baldwin et al., 1977; Stouthamer and Bettenhaussen, 1976).

Microbial maintenance might include resynthesis of components of turned-over cells or replacement of lysed cells. Two types of lyses could be factors; one is predation of bacteria by protozoa (Jarvis, 1968) and the other is lysis of both bacteria and protozoa. Microbial growth rate and maintenance requirements per unit of substrate fermented would be expected to vary if the availability of nutrients other than energy varied. Baldwin and Denham (1979) present a diagram of nitrogen metabolism in the rumen (Fig. 3.4) in which three classes of dietary substances are considered: (1) soluble true protein (ASP), (2) insoluble true protein (AIP), and (3) nonprotein nitrogen (ANPN). Soluble protein enters the fluid phase of the rumen readily after ingestion of food (R801 in Fig. 3.4) and are hydrolyzed by the microbes to form amino acids (RSSA, R802) or pass with water from the rumen. Insoluble protein enters the large or small particle pools of the rumen. Solubilization or hydrolysis (R902) of insoluble protein requires physical association with microbes or their enzymes. Small particles of insoluble protein pass from the rumen with other particulate organic matter (R904).

In a model changed from a dried forage diet with nucleic acids as the principal source of NPN to one containing urea in a concentrate diet, the rate constants and stoichiometric coefficients for solubilization (R1002) and degradation (R1003) reactions need to be adjusted. Some amino acids (RSAA) pass to the omasum (R804), and the remainder are utilized for growth of microbes (R805) or degraded to ammonia ($RNH_3N$) or deaminated to VFA (R806). Ammonia in the

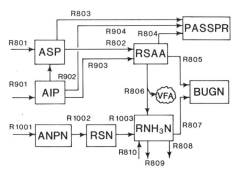

**Fig. 3.4.** Detail from a block diagram of nitrogen metabolism depicting rumen nitrogen pools and fluxes (From Baldwin and Denham, 1979).

rumen can be utilized for growth of microbes (R802), or may be absorbed directly through the wall of the rumen (R808), or may pass to the lower tract with water. Saliva may be a source of ammonia in the rumen. Critical features of a model include (1) proper description of feed nitrogen in terms of properties and amounts of ASP, AIP, and ANPN, (2) identification and quantitative characterization of factors that determine the amounts and proportions of ammonia and amino acids utilized for growth of microbes, and (3) computation of the host's contribution (Baldwin and Denham, 1979). Available data are generally inadequate and need to be investigated.

The problem of adequately describing the nitrogen portion of feeds was illustrated by Baldwin *et al.* (1977) in a model that failed to simulate adequately rumen ammonia concentrations when low-quality nitrogen diets were fed and/or with infrequent feeding. The authors pointed out three limitations that might have led to underestimates of ammonia concentrations. In Fig. 3.4 (1) reaction 902 was set at zero because of lack of data, (2) nitrogen entry from the animal (R810) was not adequately represented, and (3) possible effects of bacterial predation by protozoa and lysis of bacteria and protozoa were not included.

Baldwin and Denham (1979) point out an interesting modeling problem in regard to ammonia concentrations and enzyme activity. For instance, Satter and Roffler (1975) and Hogan (1975) observed that ammonia levels in the rumen did not limit the function of microbes until concentrations fell below 5 mg/100 ml, while Mehrez *et al.* (1977) observed that ammonia levels below 20–25 mg/100 ml did limit microbial function. Both of these observations may be correct, according to Baldwin and Denham (1979), since at 3–5 mg ammonia per 100 ml, glutamine synthetase is essentially saturated and its activity reaches a maximum, while glutamate dehydrogenase reaches maximum activity at ammonia concentrations of 20–25 mg/100 ml. The quantitative and relative contributions of the two enzymes in individual microbes and mixed populations in the rumen, as well as the effects of diet, have not been adequately observed. The effect of different levels of amino acids upon microbial growth can be expressed by equations based on data of Maeng and Baldwin (1976c).

## XXII. EFFECT OF PESTICIDES ON RUMEN MICROBES

Schwartz *et al.* (1973) found that less than 100 ppm of Bordeau mixture, toxaphene, dichlorodiphenyltrichloroethane (DDT), *O*-ethyl *O*-*p*-nitrophenyl phenylphosphonothionate (EPN), Zectran, dieldrin, parathion, Mobam, aldrin, Baygon, Black-leaf 40, malathion, 2,4,D acid, and Sevin had no effect on rumen microbe action in cell wall digestion of plant material *in vitro* using microbes from the rumen of goats. Levels of 1,000 ppm of the pesticides generally had inhibitory action on the microbes.

## XXIII. MARKERS IN RUMINANT NUTRITION

Dietary markers may be classified as internal (occurring naturally in the diet) or external (added to the diet). They have been utilized for determining absorption, passage rate, and retention time, as well as digestion of nutrients. Kotb and Luckey (1972) reviewed the use of natural and external markers in digestibility studies. They concluded that these markers offer the advantages of low cost and convenience, but suggested that they cannot be used without reservations.

Internal markers have the advantages of being already mixed in the diet. Substances such as lignin, indigestible organic matter, acid-insoluble ash, and plant chromogens have been utilized as internal markers. External markers such as chromic oxide, rare earths, and isotopes need to be mixed thoroughly with the feed or administered in boluses. This may not be possible with large numbers of ruminants or under grazing conditions. External markers may be excreted in a diurnal pattern, which increases the number of grab samples required or may preclude the use of such samples in digestion studies (Smith and Reid, 1954). Most markers have their limitations, but their usage eliminates the laborious task of conducting total fecal collections in digestibility trials.

Kleiber (1975) presented equations for utilizing a marker such as chromic oxide to evaluate the digestion of nutrients in animals. If the concentration of the marker in the diet is $C_i$ and $I$ is the daily dietary intake, then the amount of marker taken in per day is $I \times C_i$. If $F$ stands for the amount of feces excreted per day and $C_f$ for the concentration of marker in the feces, then the amount of marker excreted per day is $F \times C_f$. Thus, $I \times C_i = F \cdot C_f$ if the marker is indigestible, or $F/I = C_i/C_F$. Digestibility ($D$) is determined as $D = (I - F)/I = 1 - F/1$. By replacing $F/1$ by $C_i/C_f$, $D = 1 - C_i/C_f$, or 1 minus the ratio of the marker concentration in diet and in feces.

If one is interested in a component of the diet such as nitrogen, one can make similar calculations. That is, if the concentration of nitrogen in the diet is $n_i$ and that in the feces is $n_f$, then the intake of nitrogen amounts to $1 \times n_i$ and the excretion of nitrogen is $F \times n_f$. The digestibility is calculated as follows:

$$\frac{In_i - Fn_f}{In_i} = 1 - \frac{Fn_f}{In_i}.$$

In this equation, the ratio of $F/I$ may be replaced by the ratio of the marker concentrations, $C_i/C_f$, and the apparent digestibility is given by the following equation:

$$D_n = 1 - \frac{n_f}{n_i} \times \frac{C_i}{C_f}$$

where $D_n$ = apparent digestibility of nitrogen; $n_i$ and $n_f$ = nitrogen content per gram of diet intake ($i$) and feces ($f$), respectively; and $C_i C_f$ = concentration of indigestible marker in diet ($i$) and feces ($f$), respectively.

## A. External Markers

Chromic oxide ($Cr_2O_3$) has been widely investigated and used to estimate nutrient digestibility in diets (Barnicoat, 1945; Woolfolk *et al.*, 1950; McGuire *et al.*, 1966). Phar *et al.* (1970) utilized 453-kg Angus steers in the following treatments with $Cr_2O_3$: (1) 6.81 kg of a pelleted diet containing 0.5% $Cr_2O_3$ fed steers in crates; (2) 6.81 kg of the pelleted diet fed steers wearing fecal collection apparatus in individual pens; and (3) feed offered *ad libitum* to steers wearing fecal collection apparatus in individual pens. Excretions of $Cr_2O_3$, crude protein, and gross energy were determined from 2-hr grab samples over 48 consecutive hours. There were no significant differences in excretions due to time of sampling, but there were differences among steers for all treatments. The digestion coefficients calculated by the use of $Cr_2O_3$ were lower than the conventional coefficients for all treatments due to incomplete recovery of dietary $Cr_2O_3$. Crude protein digestibility coefficients calculated from the grab samples were significantly lower than conventional coefficients. Gross energy and $Cr_2O_3$ excretion patterns were not significantly correlated for any of the treatments.

Robles *et al.* (1981) separated alfalfa hay into leaves and stems and treated them in one of four ways: (1) unprocessed, control; (2) 1.0 g refluxed in 200 ml water for 20 min, water boiled; (3) 1.0 g refluxed in 200 ml concentrated $H_2SO_4$ for 1 hr, acid boiled, and (4) neutral detergent fiber (NDF). Preparations from each treatment were soaked in a gold (Au) solution for 72 hr (1 g forage + 12.3 mg $HAuCl_4 \cdot 3H_2O$ in 50 ml distilled water) and filtered. Another set of 1.0-g samples was refluxed for 3 hr with 1.0 g $Na_2Cr_2O_7 \cdot 2H_2O$ per 100 ml distilled water to mark the tissues with chromium (Cr), filtered, and dried. The concentration of the gold marker was determined by activation analysis, and that of chromium by atomic absorption spectrophotometry. The highest recovery of marker was found in the NDF, since some marker may have become attached to cell solubles in the unprocessed, water-boiled, and acid-boiled preparations. Lower *in vitro* digestibilities were observed in the marked leaves and stems than with those unmarked. Chromium reduced *in vitro* digestibilities of nutrients more than gold, except in NDF. Concentrations of markers appeared adequate for detection in ingesta or feces when added to the diet in small amounts.

Cross *et al.* (1973) reported the effect of water restriction at 60% of *ad libitum* intake using $Cr_2O_3$ as a marker with Angus steers, utilizing grab samples or a conventional digestion trial. There was an increase in dry matter digestibility but no effect on crude protein when water was restricted. It was concluded that a fluctuation in the ratio of nutrient to marker with time indicates a need for a grab sampling procedure that includes each 2-hr interval throughout the day.

Kyker (1962) reported that rare earths at low concentrations have an affinity for particulate matter as well as being unabsorbed from the digestive tract (Ellis, 1968). Various rare earth elements have been used as digestive markers with encouraging results (Ellis, 1968; Miller et al., 1967; Olbrich et al., 1971). Prigge et al. (1981) assessed the potential of ytterbium (Yb) and $Cr_2O_3$ either once (0800 hours) or twice (0800 and 1600 hours) daily. DMD and fecal output were estimated from fecal marker concentrations determined in grab samples collected at 0800, 1600 or 0800 and 1600 hours, and were compared with the values found by total collection of feces. Fecal output values estimated from ytterbium with single daily dosings were equivalent to those obtained by total collection, but $Cr_2O_3$ underestimated fecal output at 0800 hours and in the composited (0800 and 1600 hours) collections. Relative marker concentrations in feces obtained by grab sampling at 4-hr intervals indicated that twice daily dosing appeared to reduce diurnal variation. The diurnal variations observed with both indicators in the dosing schedule appeared to be about the same.

Teeter et al. (1981) studied the passage rates of liquids with ytterbium-labeled feedstuffs and ruminally dosed cobalt–ethylenediaminetetraacetic acid (Co-EDTA) in steers. Liquids had higher passage rates than ytterbium-labeled solids. Dilution rates of Co-EDTA and ytterbium-labeled feedstuffs were equivalent in ruminal, abomasal, and fecal samples. The time required for short and long ytterbium-labeled prairie hay to reach peak concentration was 14 and 18 hr for duodenal and 26 and 32 hr for fecal samples, respectively.

Young et al. (1976), in digestibility trials with steers, studied the reliability of dysprosium as a marker. Steers were fed alfalfa hay sprayed with dysprosium solution to give about 40 ppm in the forage. Mean fecal recovery of dysprosium was $90.6 \pm 0.2\%$ from the hay-treated samples compared to $80.4 \pm 0.3\%$ from bolus containing dysprosium. Grab sample data indicated that 2 days, at least, should be allowed for dosage before fecal collections are initiated to permit equilibrium in the digestive tract. It was concluded that grab sampling twice daily using dysprosium provided a reliable alternative to total fecal collections.

## B. Internal Markers

Shrivastava and Talapatra (1962) utilized acid-insoluble ash (AIA) of feed and feces as a digestibility marker for grazing sheep. Their estimated digestibility coefficients were equivalent to the values obtained by traditional total fecal collections. Van Keulen and Young (1977) utilized AIA using 2 normal, 4 normal, or concentrated hydrochloric acid as a marker in digestibility studies with sheep. Dry matter digestibility coefficients estimated by the AIA marker method, using all three acid treatments procedures, were equivalent to those determined by conventional total fecal collection.

Thonney *et al.* (1979) reported that AIA gave deviation from actual digestibility that ranged from $-3.62\%$ for 80% concentrate plus 20% late-cut hay to 1.40% for 40% concentrate plus 60% late-cut hay diets fed steers. Block *et al.* (1981) reported that the range of total recovery of AIA from wethers and lactating dairy cows was 98–102% when ort AIA was taken into account and 91–121% when ort AIA was not accounted for in *ad libitum* feeding of grain plus hay diets. Lignin, as determined by the 72% $H_2SO_4$ method of Ellis *et al.* (1946), was observed to give equivalent digestibilities with cattle and sheep in conventional trials when utilized as a marker (Ellis *et al.*, 1946; Forbes and Garrigus, 1950). Thonney *et al.* (1979) reported that permanganate lignin, when used as a marker with steers fed 0, 20, 40, 60, and 80% concentrate, with the remainder being early- or late-cut mixed grass hay, underestimated digestibility compared to the total collection method by an average of 23.9 percentage units. They concluded that permanganate lignin is an unreliable marker for measuring digestibility. Orskov (1982) presented an excellent critical review of markers utilized in ruminant nutrition.

## XXIV. TOXIC SUBSTANCES IN THE RUMEN

When toxic materials are in the diet, ruminal microbes may be considered a first line of defense. In some cases, ruminal microbes convert dietary substances into toxic compounds. Generally, production of more toxic substances by microbes in the rumen is less significant than their detoxification reactions. The relationship between ruminal microbes and their animal host is crucial in the health of ruminants. Particularly palatable feedstuffs may lead to problems due to excessive consumption of potentially toxic substances or their precursors. For instance, consumption of large amounts of palatable plants containing nitrates and/or cyanogenetic glycosides creates a potentially lethal situation due to the lag between initial intake of feed and the occurrence of adverse effects. High consumption of diets containing available carbohydrates or CP may lead to hazardous levels of lactic acid or ammonia in the rumen. Fasting, ruminal and abamasal pH changes, and antimicrobial substances may alter the number of microbes and reduce the detoxification capacity of the rumen.

Crane (1973) published a bibliography of toxic plant substances, a few of which will now be discussed.

### A. Cyanide (Prussic Acid)

This toxic substance occurs in many plants, such as sorghum, forages, and grasses of the *Cynodon* species (Schroder, 1976) as hydrocyanic acid (HCN) combined in glucosides. The glucosides are hydrolyzed to free acid in the rumen. The severity of toxicity upon consumption of cyanide-containing plants varies with the rate of hydrolysis, absorption of free HCN through the wall of the rumen, and the rate of HCN detoxification. The rate of cyanide ingestion is very important. Ruminants may tolerate considerable amounts of HCN if grazed forage does not exceed 15–50 mg HCN per kilogram of body weight per day (Rose, 1941; Franklin and Reid, 1944; Coop and Blakley, 1950), but 2–4 mg HCN per kilogram of body weight is a lethal single dose for sheep. The minimum lethal dose of the glucoside is 4.6 mg HCN per kilogram of body weight as the HCN is slowly released from glucosides.

Less cyanogenetic glucoside is required for a lethal dose with a full rumen than with one that is relatively empty, since the greater microbial populations in the full rumen will hydrolyze the glucoside more rapidly. However, if starved ruminants ingest large amounts of cyanogenetic plants that lack the relevant plant enzyme, a delayed but rapid hydrolysis may occur several hours later as the bacterial population and pH of the rumen change to a more normal level, and death may result (Coop and Blakley, 1950). Cyanide detoxification in the rumen appears to occur by oxidation of the hydrogen sulfide donor to yield thiocyanate (Coop and Blakley, 1949). Veterinarians may treat cyanide poisoning by intravenous administration of nitrite and thiosulfate salts.

## B. Oxalates

Many plants contain soluble oxalates, but only a few have levels that are toxic. The Rumex family of plants contains primarily potassium hydrogen oxalate, and the Chenopodiaceae family contains primarily sodium oxalate (James *et al.*, 1975). Oxalate consumed in plants may be degraded in the rumen to carbon dioxide and water, may be absorbed into the blood to react with various tissues, or may be excreted in the feces as insoluble oxalates. Ward *et al.* (1979) found that 25–33% of calcium in alfalfa was excreted as the oxalate salt. Stephens *et al.* (1972) fed bermuda grass, hydrilla, and hyacinth forages that contained 0.1, 0.3, and 0.8% oxalate (dry basis) in diets to steers in a metabolism trial and observed that 26, 21, and 20% of oxalate, respectively, was excreted in the feces. The oxalate level had no effect on the feed intake of steers.

Talapatra *et al.* (1948) concluded that since only a negligible part of dietary oxalate appeared in the feces and urine of cattle fed paddy straw, it was degraded in the rumen. Morris and Garcia-Rivera (1955) demonstrated by *in vitro* fermentation that oxalate was degraded by rumen microbes. Dobson (1959) found that the rumen contents of sheep accustomed to small intakes of oxalate adapted and were able to break down considerably greater amounts of oxalate than would have been possible with no adaptation period. If the rumen was bypassed, tolerance for oxalate was decreased.

## C. Pyrrolizidine Alkaloids

Alkaloids of the pyrrolizidine class are toxic and occur in plants of the genera *Heliotropium, Crotalaria,* and *Senecio.* Chronic liver disease occurs in livestock that consume these plants over extended periods of time, and death is not uncommon. The principal alkaloids of *Heliotropium europaeum* (heliotrine and lasiocarpine) were found to be metabolized in the rumen of sheep to nontoxic compounds (Dick *et al.,* 1963). Lanigan (1970) reported that the rate of *in vitro* degradation of the alkaloids occurred more rapidly in ruminal fluid of sheep grazing *H. europaeum* than in fluid from animals fed chaff. When the *H. europaeum* forage was added to the chaff diet of sheep, the ruminal fluid had a greater capacity to degrade the alkaloids *in vitro.*

## D. Nitrates and Nitrites

Nitrates are reduced in the rumen to nitrites, which cause toxicity if they accumulate. Microbes also reduce nitrite to hydroxylamine and finally to ammonia. Nitrites and amines may form hazardous nitrosamines in the alimentary tract. The rate of reduction of nitrate to ammonia is probably important to the degree of toxicity. Sensitivity to nitrate poisoning is greatest in pigs, followed by cattle, sheep, and horses. Absorbed nitrite forms methemoglobin, and animals die from lack of oxygen in their tissues. Nitrates may occur at toxic levels in many plants, depending on the amount of moisture in the soil, stage of plant growth, amount of sunshine, temperature, and other factors that affect the photosynthesis and conversion of nitrate nitrogen to protein in plant tissues. Sometimes water and fertilizer are sources of nitrates for livestock. Crops grown on fallowed soil generally contain higher nitrate levels than crops grown on land in continuous production (Kretschmer, 1958).

Cereal crops such as oats, barley, and wheat harvested for hay have been frequent sources of nitrate poisoning. Common weeds are often high in nitrate when growing on muck soils or marshland that have high nitrogen and relatively low phosphorus and potassium levels. Frequent abortions occur in cows consuming these plants. Diets fed ruminants that contain grains, molasses, or other high-energy feedstuffs allow greater utilization of nitrates than low-energy roughage diets. Shirley (1975b) and others (National Academy of Sciences [NAS], 1972; Emerick, 1974) reviewed the nutritional and physiological effects of nitrates, nitrites, and nitrosamines on animals.

### 1. Animal Adjustment to Nitrate

A few experiments have demonstrated how ruminants adapt to higher than usual intakes of nitrate. Jainudeen *et al.* (1964) fed heifers 440 and 660 mg nitrate per kilogram of body weight over 120–230 days of pregnancy and observed that they adapted to a deficiency of oxygen in their blood by increasing the number of circulating erythrocytes. Lichtenwalner *et al.* (1971) fed weanling steers and heifers enough potassium nitrate to maintain methemoglobin at approximately 20% of the total hemoglobin; the methemoglobin declined in the treated cattle with time.

The adaptability of sheep to nitrite was demonstrated by injecting them with 0.7 mmole of nitrite per kilogram of body weight for 35 days and then comparing them with control sheep injected with 0.9–1.2 mmoles of nitrite per kilogram of body weight on each of 3 subsequent days. The control sheep died within 2 hr with 87% methemoglobin after the first of these larges doses, whereas the pretreated sheep lived and had 68% methemoglobin (Diven *et al.*, 1964).

2. Dietary Nitrate and Utilization of Nutrients

When potassium nitrate was added to concentrate diets as levels such that the daily intake by wethers was 500 mg per kilogram of body weight, it increased the digestibility of dry matter from 2.6 to 7.6% and that of crude protein from 6.2 to 9.6% (Miyazaki *et al.*, 1974). Apparently, the nitrate served as a source of nitrogen for the rumen microbes and aided their utilization of high-energy ingredients in the ration. O'Donovan and Conway (1968) applied 0, 184, and 368 kg nitrogen per hectare in six dressings per year to ryegrass-white clover pasture and observed that high nitrate levels in the forage did not reduce vitamin A levels in the livers of grazing sheep. They concluded that from a practical standpoint, decreasing returns in ruminant forage yields from nitrogen fertilizer may occur before hazardous levels of nitrate are reached.

3. Effect of Diet Composition on Nitrate Toxicity

Approximately 30 years ago, it was observed in South Africa that ruminants tolerated higher levels of nitrate in their diets when molasses was fed. High-energy nutrients are thought to produce a more intense reducing environment in the rumen, thereby reducing the hazardous nitrite to usable ammonia. Since nitrate reductase contains molybdenum and nitrite reductase contains copper, Buchman *et al.* (1968a) grew millet with high nitrate, with and without molybdenum and copper in the fertilizer. The nitrate in the millet containing 0.6 or 4.3 ppm molybdenum on a dry basis was reduced at the same rate in the rumen of steers. Nitrate in millet with 0.6 or 27 ppm molybdenum was reduced at an equivalent rate in the rumen of sheep (Buchman *et al.*, 1968b). Copper levels in millet that varied from 4 to 99 ppm did not affect the rate of nitrate reduction or ammonia formation in either sheep or cattle. It was concluded that nitrate and nitrite reductase requirements for molybdenum and copper must have high priority, since maximum enzyme activity occurred at the relatively low levels of the elements in the control millets.

In sorghums, there is a greater probability of cyanide than of nitrate poisoning of ruminants, depending on the variety and height of the forage. Toxic levels of cyanide and nitrate coexist in certain forages, and toxicity may result in which methemoglobin arising from nitrite would be a factor in the initial detoxification of the cyanide (Gillingham et al., 1969). Cyanide poisoning should be excluded prior to treating animals for nitrate toxicity. When Sudan grass containing 3.2% potassium nitrate was fed to calves with either urea or soybean meal supplements, the methemoglobin levels indicated that those consuming the urea adapted to nitrate to a greater extent than cattle receiving the soybean meal (Clark et al., 1970).

Nitrates reduced gains in lambs fed soybean meal but increased gains in lamb fed urea (Hatfield and Smith, 1963). When the CP in an 8.04% CP diet fed lambs was increased to 9.54% with either soybean meal, urea, or nitrate, it was found that the three nitrogen sources were equivalent as CP (Hoar et al., 1968a). In younger animals, vitamin A has a protective effect against nitrate toxicity. Hoar et al. (1968b) observed that lambs fed diets containing 2.5% sodium nitrate had reduced weight gains when they were being depleted of vitamin A. The nitrate did not have the same effect during a repletion period when sheep were larger and gaining at a slower rate.

## E. Urea or Ammonia Toxicity

It is generally agreed that urea toxicity is equivalent to ammonia poisoning. Ammonia in the rumen is produced from various sources of nitrogen, and its formation depends largely on the concentration and nature of the CP sources in diets. Soluble proteins, nucleic acids, and urea are generally rapidly degraded in the rumen to ammonia. Toxicity problems are not usually associated with ingestion of protein, but rather with ingestion of excessive levels of urea. The utilization of ammonia depends upon the growth rate of ruminal microbes and is usually limited by the availability of readily fermentable carbohydrates.

Burroughs et al. (1975a,b) and Satter and Roffler (1975) proposed potentially valuable systems for the formulation of diets based on the contribution of energy-containing ingredients in the ruminal ammonia pool. In one system (Burroughs et al., 1975a,b), the excess or lack of nitrogen in feedstuffs is expressed in terms of subtraction or addition of urea (grams per kilogram of dry matter). In the other system (Satter and Roffler, 1975), percentages of CP and TDN determine how much urea can be added to diets.

Dietary urea, if consumed in large quantities in a short time, is well known to be toxic to ruminants. Stiles et al. (1970) observed that an extrusion-processed mixture of grain and urea (Starea) was less toxic to cattle than unprocessed mixtures of grain and urea. However, in processing, attention needs to be paid to conditions of moisture, temperature, pressure, and residence time in the extruder if a nontoxic product is to be obtained. Bartley et al. (1976) evaluated the toxicity of several extrusion-cooked mixtures of ground grain (sorghum, corn, barley, wheat) and urea in fistulated adult dairy cattle. On the test day, each animal was given by rumen cannula enough product to provide 0.5 g urea per kilogram of body weight. Ammonia toxicity expressed as muscle tetany appeared in about half of the animals within 53 minutes. Rumen pH and blood ammonia, but not rumen ammonia, were positively correlated with toxicity. Apparently, a high rumen ammonia concentration can exist without producing toxicity if the ration has readily fermentable carbohydrates and the pH of the rumen fluid is below 7.4. Drenching animals having toxic symptoms with acetic acid did not lower the blood ammonia level during the first 120 minutes, and 20% of the treated cattle died. Emptying the rumen of animals having toxic symptoms resulted in a rapid decrease in blood ammonia and recovery within 2 hr.

Biuret (40% nitrogen) and dicyanodiamide (66% nitrogen) are slowly hydrolyzed to ammonia in the rumen and are not toxic (Fonnesbeck et al., 1975). Davidovich et al. (1977) reported that extruded grain with dicyanodiamide and unprocessed grain with biuret produced almost no change in rumen and blood ammonia and rumen pH. Processed grain with biuret increased these factors and indicated that considerable biuret was converted to urea during extrusion cooking. Toxicity was observed when cattle were dosed with liquid supplements containing urea, but toxicity was prevented when 3% phosphoric acid was incorporated in the liquid supplement. Intravenous administration of an electrolyte solution containing calcium, magnesium, potassium, and dextrose could not prevent or reduce the signs of ammonia toxicity.

Crickenberger et al. (1977) determined the toxicity of fermented ammoniated condensed whey (FACW), ammonium lactate, ammonium acetate, and urea in four feed-lot steers fed 50% corn silage–50% concentrate diets. Intraruminal infusion of FACW and ammonium lactate at a level of 400 mg nitrogen per kilogram of body weight resulted in acute toxicity, as evidenced by muscle tremors. Infusion of 300 mg produced no acute toxic effects. When ammonium lactate was infused at levels of 300 and 400 mg per kilogram of body weight, acute toxicity was observed in one and two steers, respectively. Urea at the 200-mg level resulted in acute toxicity, with elevated ammonia levels in the rumen. It was concluded that FACW could be used as supplemental nitrogen for feedlot cattle with little chance of ammonia toxicity.

## F. Toxins in Soybeans

Numerous papers have been published on the deleterious properties of raw soybeans, as well as on the high biological value of the heated beans. Osborne and Mendel (1917) apparently were the first to report on the improved growth-promoting property of cooked compared to raw soybeans. Chicks and rats have been used in most of the toxicity studies with raw soybeans, but all monogastric animals are sensitive to the toxic substance.

Robison (1930) reported that swine responded to raw and heated soybeans similarly to rats and chicks. Pigs fed raw soybean diets gained 0.36 kg/day, while those fed diets containing cooked soybeans gained 0.56 kg/day. The pigs fed the raw soybean–containing diet had less feed efficiency (4.7 versus 3.5 kilograms of feed per kilogram of gain). These responses to cooked soybeans were confirmed by Becker et al. (1953) and Jimenez et al. (1963).

Calves do not readily consume either heated or unheated soybeans or soybean meal–containing diets (Hilton et al., 1932; Shoptaw et al., 1937). Cows showed no change in milk production or milk flavor when fed raw soybeans or soybean hay (Olson, 1925; Bartley and Cannon, 1947). However, heating soybean meal improves its utilization by ruminants. Sherrod and Tillman (1962) extracted lipids from soybean meal with cold hexane and then autoclaved the extracted meal under 15 lb steam pressure per inch at 121° for either 45 or 90 min. The heated soybean meal markedly decreased the ruminal ammonia level and improved the daily gain and feed efficiency in sheep. Differences in the time of autoclaving did not affect the meal.

Tagari et al. (1962) heated soybeans to 70–80°C for 20 min, followed by dehulling, flaking, and defatting with petroleum ether for 3 hr. Then the extracted soybean meal was treated by (1) evaporation of the solvent at room temperature, (2) solvent removal by heating at 80°C for 10 min, and (3) solvent steaming at 120°C for 15 min. The soybean meals were then fed in diets to rams. Apparent digestibility of protein and nitrogen retention were greater for the treated soybean diets, followed by the second and first soybean diets.

Wilgus et al. (1936) reported that both temperature and length of heating were factors in commercial soybean meal production. A high temperature applied for a short time in the expeller process produced meal with the same biological value as that produced by the hydraulic process at a lower temperature for a longer time. However, solvent-extracted soybean meal had a high biological value despite the fact that a low temperature and short period of heating were used.

Ham and Sandstedt (1944) reported a trypsin inhibitor in raw soybeans that appeared to explain the mechanism by which heat treatment improved the nutritional value of the beans. Autoclaving destroyed the activity of this inhibitor. A crystalline globulin protein isolated from raw soybeans (Kunitz, 1945) formed an irreversible stoichiometric compound with trypsin (Kunitz, 1947). Kassell and Laskowski (1956) reported that pepsin slowly digested the undenatured trypsin inhibitor. Using protein utilization efficiency studies with rats on autoclaved soybean meal, Rackis (1965) estimated that the trypsin inhibitor accounted for 30–50% of the growth inhibitory effect of raw soybean meal and for nearly all of the pancreatic hypertrophy.

A potent hemagglutinating agent was isolated from raw soybean meal by

Liener and Pallansch (1952). When this substance was fed to rats in a diet containing autoclaved soybeans, the animals gained at 75% of the rate of control rats, with a depression in appetite (Wada et al., 1958). Although the agent readily agglutinated the red cells of rabbits, it had no effect on the blood cells of sheep and calves.

## G. Cottonseed Meal

The level of gossypol in cottonseed meal is associated with toxicity primarily when the meal is fed to nonruminants. Pigment glands in the cottonseed contain practically all of the gossypol. This substance appears to be synthesized in the roots. Glandless varieties of cotton appear to prevent the transport of gossypol from the roots to the seeds, since their roots have no reduction in gossypol concentration (Smith, 1962).

The commercial method of reducing gossypol in cottonseed meal involves using suitable amounts of heat and pressure to sheer the glands. Gossypol thus released from the pigment glands appears to combine with protein (primarily with the epsilon-$NH_2$ of lysine) during the heating period (Lyman et al., 1959). The "bound" gossypol is less toxic than the free form (Altschul et al., 1958). The toxicity of gossypol may be decreased by adding iron salts to cottonseed meal containing–diets fed poultry (Evans et al., 1958; Schaible et al., 1934). Ruminants, especially after their rumen microflora have been established, appear to be immune to gossypol toxicity. The mechanism by which ruminants detoxify gossypol is not known, but in vitro studies with rumen fluid indicate that the substance combines with protein. Reiser and Fu (1962) mixed gossypol in rumen fluid, and upon incubation with a number of proteolytic enzymes, the gossypol–protein complex was not degraded to free gossypol.

## XXV. ENZYME SUPPLEMENTS AND DIGESTIBILITY

Enzymes are complex proteinaceous substances normally secreted into the digestive tract of animals to catalyze the digestion of proteins, carbohydrates, and lipids without being changed themselves. Microbes in the rumen and cecum also provide enzymes for degradation and utilization of feedstuffs. The enzymes normally available in the alimentary tract of livestock are usually adequate to give maximum productivity for given diets. However, somewhat successful attempts have been made to increase the digestibility of diets fed poultry, swine, and ruminants.

Burroughs et al. (1960) evaluated a commercial enzyme mixture that contained both amylolytic enzymes in 10 feeding trials with steers and heifers. Liveweight gains were increased by 7% by the enzymes incorporated in their diets, with little or no increase in feed consumption. Feed efficiency was increased by 6%. Organic matter, protein, and cellulose digestibilities were not influenced by enzyme supplementation. No consistent improvement was observed in dressing percentage or federal carcass grades due to the supplementary enzymes.

Van Walleghem *et al.* (1964) conducted two digestibility trials with lambs fed diets that contained different levels of protein (6.5 versus 12.0%), principally from citrus pulp and soybean meal, with and without six different commercial enzyme supplements. The supplements contained the following enzymes: (1) dried *B. subtilis* fermentation extract containing amylase and protease activity; (2) fungal-bacterial fermentation extract containing pectinase, cellulase, and protease activity; (3) fungal-bacterial extract containing cellulase and protease activity; (4) dried *B. subtilis* fermentation extract containing amylase, protease, and gumase activity; and (5) a bacterial source of amylase and protease activity plus added sources of cellulase and pectinase activity. The digestibilities of protein and energy in the diets were not affected by any of the enzyme treatments.

## XXVI. GLUCONEOGENESIS IN CATTLE

Gluconeogenesis is of great importance in ruminants because almost all dietary carbohydrates are fermented to VFA in the rumen. Propionate is the only major VFA that contributes to gluconeogenesis. The rate of gluconeogenesis is lowest in nonruminants after feeding, when glucose is being absorbed, and highest during fasting, when no exogenous glucose is provided. Since the dietary carbohydrates are fermented in the rumen, less than 10% of the required glucose is absorbed from the ruminant tract and gluconeogenesis must provide up to 90% of the necessary glucose (Bensadoun *et al.*, 1962; Otchere *et al.*, 1974).

For high-producing dairy cows, meeting glucose needs can be a tremendous metabolic challenge. According to Young (1977), a dairy cow at peak production gave 89 kg of milk daily. The author postulated that if the milk was 4.9% lactose, she secreted 4.43 kg of lactose daily. Bickerstaffe *et al.* (1974) found for cows, and Bergman and Hogue (1967) found for sheep, that about 60% of the glucose entry is utilized for lactose. Thus, the cow mentioned above required about 7.4 kg total glucose per day and needed to synthesize 90% of it, or 6.6 kg. Steers that weigh 160–205 kg fed slightly above maintenance need to synthesize about 600 g of glucose daily (Otchere *et al.*, 1974; Young *et al.*, 1974). It may be concluded that even with nonlactating ruminants, the gluconeogenic metabolic mechanisms need to operate continually at an efficient rate, and any breakdown can result in serious disorders.

The liver in ruminants releases glucose during both the fed and fasted states, with the greatest production during feeding (Ballard et al., 1969). In nonruminants, the liver has a net uptake of glucose during the fed state but a net release of glucose during fasting. Another difference between ruminants and nonruminants is that both gluconeogenic (Leng, 1970a) and lipogenic rates (Pothoven and Beitz, 1975) are increased when feed intake is high in ruminants, but only lipogenesis (Young et al., 1964) is increased in nonruminants when intake is abundant. Gluconeogenesis is greater in the nonruminant during fasting periods than during normal feed intake.

Research on gluconeogenesis has a number of practical applications. Both ketosis in bovines and pregnancy toxemia in sheep appear to be related to lack of glucose or gluconeogenic precursors. Milk fat depression, which occurs when cows are fed high-grain diets, may be related to an overabundance of glucose or propionate for metabolism. It is possible that in some cows, milk production may be limited by the supply of glucose (Armstrong, 1965) or by an inadequate gluconeogenic capacity to support maximum milk production. Certain feed additives such as monensin, which increase propionate production, may, in turn, increase gluconeogenesis (Young, 1977).

## A. Propionate Conversion to Glucose

Bergman et al. (1966) demonstrated the conversion of propionate to glucose by making a continuous infusion of (2-$^{14}$C) propionate into the rumen vein of sheep. Approximately 50% of the absorbed propionate was converted to glucose, and 25% was oxidized to $CO_2$; conversions to other substances were not identified. Leng et al. (1967) infused (2-$^{14}$C) propionate continuously into the rumen of sheep. They found that rumen fluid propionate plateaued between 4 and 9 hr, and glucose in plasma plateaued between 6 and 9 hr after infusions were started. The data indicated that 54% of the glucose synthesized was derived from propionate. It is not clear what intermediates are produced in the conversion of propionate to glucose.

## B. Amino Acids as Glucose Precursors

The contribution of amino acids to gluconeogenesis varies widely, depending on the nutritional and physiological status of the ruminant (Bergman, 1973). Hunter and Millson (1964) estimated that about 12% of milk lactose was derived from amino acids in cows during lactation when they hydrolyzed a $^{14}$C-labeled protein and intravenously administered the resulting mixture of labeled amino acids. Black et al. (1968) made single injections of $^{14}$C-labeled amino acids in order to follow their conversion to glucose in cows and goats. The time intervals between injection of each labeled amino acid and the maximum specific activity in plasma glucose varied widely with different amino acids. It was concluded that amino acids released from protein result in a prolonged availability of glucose precursors. This should ensure a more continuous supply, aiding their more efficient utilization.

Egan and Black (1968) studied glutamate metabolism by giving single intravenous injections of $^{14}$C-labeled glutamate to a lactating cow and followed the course of the isotope in plasma glucose and milk constituents. It was concluded that the amount of carbon from the plasma glutamate pool utilized for gluconeogenesis exceeds that incorporated into the milk protein, and that gluconeogenesis involves a larger portion of the glutamate carbon than does protein synthesis during milk formation. At least 8% of the carbon in glucose was derived from plasma glutamate. Egan *et al.* (1970) observed the metabolic fate of labeled glutamate, valine, and arginine injected into the portal vein of lactating goats. Glutamate was the most glucogenic of the three amino acids, since more of its $^{14}$C appeared in milk lactose than in milk protein.

Wolff and Bergman (1972) infused $^{14}$C-labeled alanine, aspartate, glutamate, glycine, and serine into the vena cava of sheep over a 6-hr period. Alanine contributed 5.5% to the glucose turnover compared to 0.6% for aspartate, 3.4% for glutamate, 0.9% for glycine, and 0.7% for serine. These values were not corrected for crossover.

# 4

# Monensin and Other Antibiotics Fed to Ruminants

## I. INTRODUCTION

Many antibiotics have been fed to ruminants as growth stimulants. Monensin (Rumensin, and other carboxyllic polyester ionophores (lasalocid, salinomycin and narasin) are antibiotics derived from *Streptomyces* organisms, and have been found to alter rumen fermentation and to increase feed efficiency of both feedlot and grazing cattle. They increases the production of propionic acid while reducing that of acetic and butyric acids in the rumen. Fermentation altered in this manner decreases the energy losses associated with conversion of dietary energy sources to volatile fatty acids (VFA). The VFA make up much of the intermediate compounds supplying the ruminant energy for maintenance and production. Amicloral and avoparcin also increase propionate in rumen fermentation. These antibiotics, as well as monensin, depress methane production. Tylosin and chlortetracycline have no effect on VFA levels but decrease liver abscesses. Ruminants gain faster if they are free of such abscesses.

## II. MONENSIN

Raun *et al.* (1976) conducted trials with fattening cattle fed diets containing 0, 2.7, 5.5, 11, 22, 33, 44, and 88 ppm monensin. All levels of monensin except 88 ppm produced daily gains equal to or greater than those of control cattle, with feed consumption decreasing progressively with increasing levels of monensin. Feed consumption was decreased by 3.5% at 11 ppm and by 13.1% at 33 ppm monensin. At 11 and 33 ppm, feed efficiency was increased by 10 and 17%, respectively. Monensin increased the molar proportions of propionate and isovalerate but decreased those of acetate and butyrate in rumen fluid. Glucose, urea, and insulin tended to increase in the blood at all dietary levels of monensin below 44 ppm. Potter *et al.* (1976b) made four studies with feedlot steers and heifers fed diets containing varying levels of monensin and evaluated their carcasses for possible effects of the additive. Monensin had no effect on moisture, fat, and protein of rib eye muscle. The cattle fed monensin consumed less feed than the controls, but the percentage of dietary energy and protein retained in the carcass was increased.

Growing-finishing beef steers starting at an initial weight of 270 kg were fed diets that contained 33 ppm monensin for 232 days (Perry *et al.*, 1976a). The steers with monensin gained at the same rate (0.89 kg/day) as the controls but required 10% less dietary dry matter per kilogram of gain. VFA in the rumen fluid at 56 days showed a 16% decline in acetic acid, a 76% increase in propionic acid, and a 14% decline in butyric acid due to monensin treatment. Boling *et al.* (1977) fed Angus steers either 0, 25, 50, or 100 ppm monensin per head daily in 0.91 kg ground shelled corn as carrier while they were grazing Kentucky bluegrass-clover pasture. Daily weight gains of 0.55, 0.55, 0.73, and 0.68 kg were obtained for the four groups, respectively. In a second trial, Boling *et al.* (1977) fed steers either 0, 100, 200, or 300 mg monensin per steer daily in a 4.54-kg grain mixture plus corn silage *ad libitum* and obtained daily gains of 1.14, 1.26, 1.23, and 1.18 kg, respectively. While feed intake tended to decrease, feed efficiency was improved in all groups of steers fed monensin. Steers fed the 300-mg level of monensin tended to have lower marbling scores, smaller rib eye, area and less fat over the rib.

Feedlot steers were fed diets that contained 14, 30, or 75% of their dry matter from corn silage, with and without 300 mg monensin per head daily, and were evaluated for performance and carcass data (Gill *et al.*, 1976). Feed efficiency was improved by 6.0% at all silage levels with monensin, but gains and carcass parameters were unaffected. Cattle fed the 75% corn silage ration had 14% lower gains and were fed 28 days longer than either of the other groups. Davis and Erhart (1976) conducted a study with steers fed diets containing 0, 11, or 33 ppm monensin (90% dry matter basis), with supplemental urea withdrawn from rations after 0, 42, 84, or 120 days to determine whether cattle should be gradually adapted to monensin and whether monensin has a urea-sparing effect. Feed efficiency was improved by monensin, and feeding urea with monensin improved daily gains. Neither monensin nor urea improved carcass characteristics.

Heifers were fed diets containing rolled dry corn or propionic acid–treated high-moisture corn with and without monensin by Utley *et al.* (1977). The heifers gained 11% faster on the diets without monensin but required 13% more feed per unit of gain than those provided monensin. The addition of monensin to the diets did not affect their digestibility but did result in higher concentrations of propionic acid and lower concentrations of butyric and valeric acids.

Two cattle growth trials were conducted by Hansen and Klopfenstein (1979) to evaluate their performance with monensin; the diets were supplemented with various levels and sources of protein. In the first trial, diets were supplemented with brewer's dried grains (BDG) or urea at crude protein levels of 10.5 or 12.5% on a dry matter basis, and included either 0 or 200 mg of monensin per head daily. The added monensin improved feed efficiency by 16.3% at the 10.5% protein level and by 8.7% at the 12.5% protein level. The lower feed efficiency with urea in the diets may indicate that microbial protein synthesis is inhibited by monensin. Greater propionate production occurred with both protein treatments when monensin was provided. In the second trial, two soybean meal protein levels (11.1 and 13.1%), with and without monensin (0 and 30 g per 908 kg of feed), were fed to steers. The monensin addition resulted in a feed/gain improvement of 8.1% with the 11.1% crude protein ration but only 3.2% with the 13.1% protein diets. Monensin did not appear to increase the protein requirement for growing steers.

Hereford steers were fed various levels of crude protein diets (9, 11, or 15% continuously; 12% for 63 days and then 10.5 or 13% for 42 days; 11% for 42 days and then 9% until slaughter), with or without 200 mg monensin daily, using soybean meal as the source of supplemental nitrogen in all but the 9% crude protein diet (Thompson and Riley, 1980). Steers fed the 9% crude protein diet gained the least; daily gains and feed efficiency were similar for the other four protein treatments. However, feed efficiency was improved by an average of 7.4% in steers fed 11, 12–10.5, and 13–11% crude protein diets. Monensin increased crude protein digestibility (42.3 versus 36.9%) but had no effect on the digestibility of dry matter or the amount of starch in the feces. Muntifering et al. (1980) fed monensin at levels of 0 or 33 ppm to determine its effect on high-grain (90% corn, 10.5% crude protein) diets on ruminal parameters of yearling steers. Monensin tended to increase (63.4 versus 61.3%) crude protein digestibility and to decrease (2.5 versus 6.5 mg/100 ml) the ruminal ammonia concentration. In metabolism trials with 76% sorghum grain diets, monensin improved the digestibility of crude protein, but not the dry matter or gross energy (Muntifering et al., 1980).

Horton and Nicholson (1980) conducted trials to determine the effect of monensin (0 to 33 ppm) on barley (30, 50, and 70% for lambs and 30, 50, 70, and 90% for steers) diets. They observed that organic matter digestibility in lambs and steers was increased by 0.30 and 0.25% for each percentage increase in barley, respectively. Monensin increased organic matter and crude protein digestibilities in lambs but not in steers. The propionic acid concentration in rumen fluid was increased and that of acetic and butyric acids was decreased in steers fed monensin, but there were no effects on VFA in lambs. In a similar trial with steers and lambs fed barley-containing diets with and without monensin, the feed efficiency was improved in lambs and steers by 27 and 4%, respectively (Horton et al., 1981).

A feeding trial with steers fed 0 to 33 g monensin per ton in diets containing either dried citrus pulp or corn as the main energy source and either cage layer poultry manure or soybean meal as a source of supplementary nitrogen was made by Vijchulata et al. (1980b). Neither monensin nor the dietary energy source had any influence on weight gains, but steers fed soybean meal gained faster than those fed cage layer manure. Monensin decreased the feed : gain ratio from 8.4 to 7.2. In a feedlot trial in which steers were fed once daily a supplement consisting of soybean meal, corn meal, minerals, and vitamins to provide either 1.4 or 2.2 kg supplement per head, or an isonitrogenous supplement that provided 643 g crude protein per head daily, along with cottonseed hulls *ad libitum* for 126 days with and without monensin, feed intake was decreased but feed efficiency was improved by the addition of monensin (Vijchulata et al., 1980a).

Martinsson and Lindell (1981) fed bulls alfalfa silage *ad libitum* with concentrates (about 1.2% of live weight), with and without monensin (50 ppm of concentrate). They observed that the bulls consumed 4.1 kg less silage on a dry matter basis per day but gained 58 g more daily, with feed conversion of 76.5 MJ of metabolizable energy compared to 84.4 MJ per kilogram of live weight gain for the controls. There were no differences between the two groups in dressing-out percentages or contents of lean, fat, and bone in the carcass. Goodrich et al. (1984) reviewed the performance data of approximately 16,000 head of feedlot cattle fed monensin.

## A. Monensin and Forages

Potter et al. (1976a) allotted mixed Brahman, Angus, and Charolais crossbred cattle (steers and heifers) to four pastures that contained orchardgrass, alfalfa, brome grass, and ladino clover with 0, 100, 200, or 400 mg of monensin per head per day fed in a limited amount (0.45 kg daily) of concentrate. Monensin intakes of 100 mg per head daily improved live weight gain (17%) and feed efficiency (20%). Dinius et al. (1976) fed fistulated steers diets that contained 90% chopped orchardgrass, 7% sugarcane molasses, 1% urea, minerals and vitamins, with monensin at levels of 0, 11, 22, and 33 ppm. They observed no differences in cotton cellulose digestibility *in vitro* in rumen fluid from the steers, or in *in vivo* digestibilities of dry matter, crude protein, hemicellulose, and cellulose in the forage diet. There was an increase in the molar proportion of propionate in rumen fluid due to the monensin. There were no effects on the number of protozoa, total bacteria, or cellulolytic bacteria in the rumen due to feeding up to 33 ppm monensin.

For two winters, gravid-mature Hereford cows were fed meadow hay free choice plus 0.45 kg barley, with and without monensin (200 mg daily), by Turner et al. (1977). They observed that the monensin-fed cows outgained the controls (0.43 to 0.23 kg) on less hay intake (11.0 versus 11.4 kg) the first winter compared to daily gains of 0.47 and 0.28 kg and hay intakes of 12.7 and 13.1 kg, respectively, during the second winter. The calving interval from the first estrus was shortened by 12 days with monensin the first year, but there was no difference during the second winter.

A study was made with range beef cows to determine the effects of monensin on forage intake and lactation while grazing low-quality dry winter range grass supplemented with soybean meal and 0, 50, or 200 mg monensin per head daily (Lemenager *et al.*, 1978a). Cow weight was not affected by monensin, but forage intake was reduced 13.6 and 19.6% when 50 or 200 mg monensin was fed, respectively, and grazing time was decreased by 14.6% at the 200-mg level of monensin. Calves reared by monensin-fed cows gained weight more rapidly. In a similar trial (Lemenager *et al.*, 1978c), beef cows grazing low-quality winter range grass and provided with a 30% soybean meal–based supplement were compared to those that received either a 15% soybean meal–based supplement or an extruded urea-grain mixture, with and without monensin. Cow weight, ruminal nitrogen, ammonia, and total VFA were not affected by the addition of monensin. However, propionate and potassium were increased in rumen fluid by the ionophore.

Rouquette *et al.* (1980) compared steer calves provided with Bermuda grass pasture only, pasture plus 0.91 kg/head/day of 14% protein supplement, or pasture plus 0.91 kg/head/day of 14% protein supplement plus 200 mg monensin/head/day, and observed average daily gains of 0.45, 0.47, and 0.68 kg/day, respectively. Heifers that grazed wheat pasture only, pasture plus a pelleted supplement, or pasture plus a pelleted supplement that contained sufficient monensin to provide 100 mg/head daily, gained 0.08 kg more per day due to the antibiotic (Horn *et al.*, 1981).

Wilkinson *et al.* (1980) summarized the results of feeding monensin in 12 trials to grazing cattle under European conditions. In each trial, a group of cattle were given 200 mg monensin per head daily in 0.5–1.0 kg of carrier supplement over an average of 119 days. The average daily gain for the monensin-fed cattle was 0.893 kg compared to 0.786 kg for the controls. Spears and Harvey (1984) reported that supplemental ground corn (0.91 kg/steer/day) containing monensin (0, 200, 300 mg/head/day) gave improved gains (0.50, 0.60, 0.57 kg/head/day) over 120 days while grazing mixed grass–clover pasture.

## B. Monensin and Reproduction

Hereford, Brahman, and crossbred heifers grazing wheat-oats-ryegrass pastures, with and without monensin (0 versus 200 mg), in 0.91-kg range cubes fed daily were observed by Moseley *et al.* (1977). More heifers given monensin reached puberty during the 172-day test than did controls (92 versus 58%), and the corresponding conception rates were 55 and 47%. All pregnant heifers produced normal calves, with no incidence of dystocia. McCartor *et al.* (1979) evaluated the effect on age and weight at puberty of diets consisting of 80% alfalfa hay plus 20% concentrates, 80% alfalfa hay plus 20% concentrates plus

200 mg monensin per head daily, and a diet of 50% alfalfa hay and 50% concentrates fed to Brangus heifers. All heifers had equivalent rates of daily gain (0.60 kg) and similar increases in body condition score (3.46, where 1 = very thin and 10 = very fat). Monensin-fed heifers reached puberty 29.5 days earlier and weighed 17.2 kg less than controls.

The effect of 0, 50, 200, and 300 mg of monensin in a 0.45-kg barley supplement with meadow hay fed to gravid spring-calving Hereford cows 3–10 years of age was evaluated by Turner et al. (1980). Treatments were terminated 30 days after calving. Daily weight losses averaged 0.12, 0.05, 0.10, and 0.17 kg for the 0-, 50-, 200-, and 300-mg monensin levels, respectively, with corresponding hay consumption of 92, 88, and 90% of control cows. Monensin had no effect on the weaning weight of calves or the time required for the first estrus of cows after calving. Walker et al. (1980a) evaluated the effect of 250 mg monensin fed per head daily to Hereford and crossbred cows synchronized with Syncro-Mate-B and inseminated once after removal of the implant. All cows were fed 2.27 kg per head daily of a corn-based supplement plus corn silage *ad libitum*. Monensin had no effect on conception rates. Monensin fed beef cows at levels of 50, 200, or 300 mg daily during late gestation and early lactation resulted in feed savings of 3.2, 10.5, and 13.5%, respectively, when intakes were regulated to give similar changes in weights of cows and condition scores during a 168-day feeding period. However, monensin had no effect on calf birth and weaning weights or on first service conception rates of cows (Walker et al., 1980b).

Clanton et al. (1981) studied the effect of 0, 50, 200, and 300 mg of monensin fed to beef cows daily that were limited to 95, 90, and 90%, respectively, of the sorghum-sudan hay fed the control cows during 194 days prior to calving and 82 days after the start of calving. Monensin had no effect on cow weights before or after calving, or on calf weight gains, onset of estrus after calving, or pregnancy rates. Nonlactating cycling Brahman crossbred and Angus cows were evaluated by Smith et al. (1980) for effects of monensin, body condition, and supplemental energy level on pregnancy rate. The cows were separated into moderate or poor body condition at the start of the study and were fed either 9 or 18 Mcal of supplemental metabolizable energy (ME) per head daily, with or without 125 mg monensin, during 21 days prior to and throughout a 45-day breeding season. The pregnancy rate and the percentage of conceptions per estrus were increased when cows were fed the higher-energy supplement, but body condition of cows or addition of monensin had no effect on reproduction indexes.

## III. LASALOCID

Lasalocid and monensin at 6 µg/ml reduced the *in vitro* fermentation rate of glucose, fructose, galactose, sucrose, lactate, mannose, ground corn, ground

## III. Lasalocid

sorghum, and ground wheat (Dennis et al., 1981b). Both ionophores decreased total lactate production. Fuller and Johnson (1981), using a continuous-flow fermentor, observed that in high-grain and roughage substrates with 32.5, 65, or 130 ppm lasalocid or 33 or 44 ppm monensin added, methane production was reduced from 4.1 to 2.7 kcal/day in the high-grain rations and from 3.0 to 2.2 kcal/day in the roughage fermentations. Both ionophores generally increased propionate production with high-grain but not with roughage substrate. Neither ionophore inhibited total bacterial growth. Energy digestibility was unchanged, but nitrogen digestibility was markedly depressed by the two ionophores, resulting in decreases of 29 and 48% in fermentor fluid ammonia by lasalocid and monensin, respectively. The general effects of the two ionophores were similar, but higher levels of lasalocid were required for equivalent biological effects. Ricke et al. (1984) compared the influence of lasalocid and monensin on nutrient digestion, metabolism, and rate of passage in sheep.

In one growth and two finishing trials, pure lasalocid, mycelia-cake lasalocid, and monensin at intake levels of 100 mg daily of each additive were fed to steers by Berger et al. (1981a,b). The treatments resulted in no differences in feed intake, daily gain, feed efficiency, or dry matter digestibility. Both forms of lasalocid and monensin decreased the concentration of coccidia oocysts in steers fed on the ground but not that of steers fed on slatted floor pens. Steers fed mycelia-cake lasalocid had higher dressing percentages than those fed pure lasalocid or monensin. In another trial, lasalocid at 30 and 45 g/ton and monensin at 30 g/ton increased feed efficiency by 7.5, 11.0, and 8.2%, respectively.

Bartley et al. (1979), using in vitro studies, tested the effect on rumen fermentation of monensin, lasalocid, and Polyether A (an experimental antibiotic) with and without niacin or amicloral. Monensin (11 ppm) and lasalocid (66 ppm) decreased protein synthesis by microbes, but all three antibiotics at 176 ppm severely inhibited the synthesis of protein. Niacin increased but amicloral decreased microbial protein synthesis. None of the three antibiotics affected total VFA production, but monensin and lasalocid increased propionate and decreased acetate. Monensin increased lactate. Lasalocid and monensin decreased methane production.

Lasalocid, monensin, or thiopeptin were administered intraruminally to cattle and evaluated for their effectiveness in preventing induced lactic acidosis (Nagaraja et al., 1982). The investigators induced acidosis by intraruminal administration of 12.5 g glucose per kilogram of body weight. In three trials, lasalocid resulted in higher rumen pH and lower lactate concentrations than those found in control cattle or cattle given thiopeptin or monensin. Higher pH and lower lactate levels were observed with monensin only at doses of 0.65 and 1.3 mg per kilogram of body weight. Thiopeptin was similarly effective only at a dose of 1.3 mg per kilogram of body weight.

Gutierrez et al. (1982) fed three levels of lasalocid (0, 33, and 49 ppm) in whole plant grain sorghum silage to stocker calves. Lasalocid at 33 ppm depressed feed intake of steers but increased feed efficiency proportionately. Lasalocid at 49 ppm did not improve feed conversion but increased total VFA production in the rumen. Bergen and Bates (1984) reviewed the general effects of ionophones on rumen function and animal production.

## IV. AMICLORAL

Amicloral increased propionate and almost completely inhibited methane production, with an increase in hydrogen accumulation during *in vitro* fermentation (Chalupa *et al.*, 1980). Horton (1980a) fed steers monensin and amicloral at 33 and 1500 mg per kilogram of diet, respectively. Both agents, fed alone or in combination, increased organic matter and crude fiber digestibilities. Crude protein digestibility was enhanced by monensin. Amicloral and monensin, alone or in combination, increased molar proportions of propionic acid by 22%, and monensin depressed butyrate levels by 37%. Steers fed monensin consumed 6% less feed, gained 9% more weight, and had 14% more feed efficiency than controls. Amicloral lowered feed intake by 11% and improved feed efficiency by 12% but had no effect on daily gain. Carcass parameters were not affected by any treatment.

## V. AVOPARCIN

Avoparcin (an antibiotic produced by *Streptomyces candidus*) was fed to steers at levels of 0, 16.5, 33, and 66 ppm together with monensin (33 ppm) in diets that contained 50% pelleted alfalfa and 44.4% steam-rolled barley. Avoparcin at all levels improved feed efficiency and produced daily gains greater than those of controls and monensin-treated steers. Propionate in the rumen was increased by avoparcin. Carcass parameters were not affected by either avoparcin or monensin. Heifers fed similar levels of avoparcin and monensin in feedlot barley (77%)–based diets had responses similar to those of the steers above (Dyer *et al.*, 1980).

## VI. TYLOSIN

Angus Hereford crossbred steers were fed diets that contained monensin (0, 100, or 300 mg per head daily), with and without 75 mg tylosin per head daily (Pendlum *et al.*, 1978). Over a 140-day feeding period, daily gains averaged 1.10, 1.12, and 1.10 kg for steers fed 0, 100, or 300 mg monensin per head per day, respectively, compared to daily gains of 1.17 and 1.04 kg for steers fed 0 or

75 mg tylosin per head daily, respectively. Feed : gain ratios were higher with steers fed 75 mg tylosin compared to those fed no tylosin, and lower carcass weight and smaller rib eye area were obtained with steers fed tylosin. Heinemann *et al.* (1978) fed Hereford, Angus, and crossbred steers a high-energy diet composed of corn, dehydrated sugarbeet pulp with molasses, cubed alfalfa hay, vitamins and minerals with tylosin (0 or 11 ppm), and monensin (0, 5.5, 11, or 33 ppm). Daily gains were depressed by 33 ppm monensin but were unaffected by tylosin. Average feed dry matter conversion was increased by 5.5% by all levels of monensin but not by tylosin. There was no interaction of the two additives. The incidence and severity of liver abscesses were markedly reduced by tylosin.

Barley-alfalfa meal diets that contained tylosin (11 ppm), monensin (33 ppm), or a combination of the two were fed to steers (Horton and Nicholson, 1980). Both tylosin and monensin increased acid detergent fiber digestibility. Monensin increased propionate and reduced acetate and butyrate, but tylosin had no effect on these ruminal acids. Free amino acids in the rumen were lower in steers fed tylosin, both alone and with monensin. The treatments had no significant effects on feedlot performance or carcass measurements. Tylosin decreased the incidence of liver abscesses, and steers without such abscesses gained 5% more rapidly. Brown *et al.* (1973) evaluated two forms of tylosin activity (50, 75, and 100 mg per head of feedlot cattle per day of tylosin phosphate or tylosin urea adduct) in high-concentrate diets. Both forms of tylosin improved daily gains (5.9%) and feed efficiency (3.1%) and reduced the incidence of liver abscesses (81.8%).

## VII. CHLORTETRACYCLINE

The effect of 70 mg chlortetracycline and 75 mg tylosin per head of feedlot cattle in medium- and high-concentrate diets was evaluated by Brown *et al.* (1975). Liver condemnation was 56.2, 44.2, and 18.6% for control, chlortetracycline-, and tylosin-treated cattle, respectively. Average daily gains for the corresponding treatment groups were 1.10, 1.14, and 1.17 kg. Corresponding feed : gain ratios were 8.21, 8.14, and 7.87, respectively. Severely abscessed livers reduced the total gain by 12.7% compared to that of cattle with no abscessed livers.

# 5

# Nutritional Energetics

## I. INTRODUCTION

Nutritional energetics necessarily follow basic rules of energy. The First Law of Thermodynamics or Law of Conservation of Energy was first formulated in the 1840s by Julius Mayer and since that time has been extended from its original applications to all forms in which energy is manifested. In metric terms, work is expressed in ergs or in joules. A calorie in nutrition is equivalent to 4.18 joules. The following compilation shows how the joule should be used as the unit of energy per se (Kleiber, 1975):

|  | Heat (kcal) | Chemical energy (kcal) | Work (m-kg) | Electrical energy (W-sec) |
|---|---|---|---|---|
| Joules/unit | 4.18 | 4.18 | 9.81 | 1.0019 |

Some nutritionists prefer joules to calories, since the joule is the fundamental unit for expressing energy and since some journals allow only joules; others prefer calories and some allow scientists to use either unit for expressing energy related to animal nutrition.

While modern physics holds that a distinction between matter and energy is untenable, it is nevertheless convenient and useful in nutritional energetics to consider energy as a property of matter. The heat of combustion value, i.e., gross energy, represents the most common characteristic to which organic substances can be reduced. Since organic nutrients can be equated in terms of energy, the energy value becomes a common basis for expressing the nutritive value of various feedstuffs and for describing the response of the animal. Since all significant sources of energy in nutrition can be converted to heat, it is convenient to describe nutritional energetics in terms of heat energy.

## II. NUTRITIONAL AND ENERGY TERMS

The National Research Council (NRC, 1981) published the second revised edition of "Nutritional Energetics of Domestic Animals and Glossary of Energy Terms." My objective is to provide a brief review of terminology that should be helpful in understanding dietary energy and its relationship to nitrogen utilization in ruminants. The author believes that the term *calorie* serves the purposes of the present work better than the joule, especially since so many of the publications on livestock nutrition involving energetics are expressed on the kilocalorie (kcal) or megacalorie (Mcal) basis. Moreover, it requires only a simple calculation if one would rather think in terms of the fundamental unit, i.e., the joule. Investigators in the field of nutrition generally standardize their bomb calorimeters using a thermochemical standard (purified benzoic acid) whose heat of combustion has been determined in electrical units and computed in terms of joules per gram mole.

Some of the energy terms and definitions that nutritionists need to utilize readily are as follows (NRC, 1981):

1. The joule (J) is equivalent to $10^7$ ergs, where 1 erg is the amount of energy expended in accelerating 1 g by 1 centimeter per second. The international joule (j) is defined as the energy liberated by 1 international ampere flowing through a resistance of 1 international ohm in 1 sec. It is considered to be the fundamental unit of energy.

2. The calorie (cal) is defined as 4.184 J and is the amount of energy required to raise the temperature of 1 g of water from 14.5 to 15.5°C. Both the calorie and the joule are so small that, in practice, nutritionists work with multiple units.

3. The kilocalorie (kcal) and the kilojoule (kJ) are 1000 times greater than the calorie and the joule, respectively. *Kilocalorie* has replaced the term *Calorie,* with a capital "C."

4. The megacalorie (Mcal) and megajoule (MJ) are 1000 times greater than the kcal and kJ, respectively. In the older literature, the term *Therm* was used for 1 Mcal.

5. The British thermal unit (Btu) is the amount of heat required to raise the temperature of 1 lb of water 1°F; 1 BTU = 0.2520 kcal and 1 kcal = 3.968 BTU.

6. Gross energy ($E$) is the amount of energy released when an organic substance is completely oxidized to carbon dioxide and water.

7. Metabolic body size ($W^{0.75}$) is the body weight in kilograms of an animal raised to the three-fourths power. It is useful in comparing the metabolic rates of mature animals of different body sizes. Generally, the exponent 0.75 is used, but other values may be more appropriate in some situations.

## A. Free Energy Change

The energy from biological reactions is called *free energy* ($\Delta F$). Reactions in the body that release energy are called *exergonic*, and those reactions that cannot proceed unless the needed free energy is released simultaneously by exergonic reactions are called *endergonic*. Exergonic and endergonic reactions occurring concurrently are called *coupled reactions*.

For example, in order for glucose to be used for the synthesis of glycogen, energy is required:

$$\text{Glucose} + (+\Delta F) \rightarrow \text{glycogen (endergonic reaction)}$$

In the body, energy ($+\Delta F$) required to drive this reaction is supplied by conversion of adenosine triphosphate (ATP) to adenosine diphosphate (ADP), a hydrolytic reaction that breaks down the high-energy bond of ATP to provide the necessary energy as follows:

$$\text{ATP} + H_2O \rightarrow \text{ADP} + H_3PO_4 + (-\Delta F) \text{ (exergonic reaction)}$$

During catabolism or breakdown of organic metabolites, free energy is released and captured in the terminal high-energy bond as ATP. During oxidative metabolism of dietary organic matter, the following general reaction occurs:

$$\text{Diet} + O_2 + \text{ADP} + P \rightleftharpoons CO_2 + H_2O + \text{ATP}$$

The ATP is involved in providing energy for synthesis of peptide bonds, osmosis, nerve conduction, muscular contraction, and various synthetic reactions. The ATP is then utilized to push endergonic reactions, many of which are involved in the synthesis of body substances.

## B. Second Law of Thermodynamics

This law is involved with the driving force of reactions; chemical affinity; loss of energy in reactions; size of molecules and their electronic, vibrational, rotational, and translational energy; and limitations of conversion of heat to other forms of energy. These characteristics involve *entropy* ($S$), which is the heat (H) content that is not available for useful work. The product of entropy (S) times absolute T equals energy wasted as the result of random motions of molecules. Therefore, oxidation of complex molecules as glucose (T $\Delta S$ = +15 kcal/mole) results in more change in entropy (T $\Delta S$) than does the oxidation of a smaller molecule like acetic acid, where T $\Delta S$ = −2.3 kcal/mole. In terms of entropy, the Second Law of Thermodynamics states that any reaction will change in the direction that results in an increase in entropy, if required conditions exist. When entropy reaches a maximum, equilibrium has been obtained and no further change will occur unless additional energy is provided from outside.

Reactions, proceeding as they must in the direction of equilibrium and at a given temperature, are of principal interest. Changes in free energy ($\Delta F$) provide the energy available for useful work if the system contains an appropriate means for it to proceed to equilibrium as $\Delta F = \Delta H - T \Delta S$ at a given temperature. This means that systems not at equilibrium proceed spontaneously only toward a negative change in free energy, i.e., $-\Delta F$. Once equilibrium is attained, no further change in free energy may spontaneously take place, and a system at equilibrium can be changed only if free energy can be made available to it. Free energy utilized for this purpose is work performance. An example of this situation is the oxidation of 1 mole of solid glucose by gaseous oxygen at 1 atmosphere of pressure, in which water and $CO_2$ are produced and 688 kcal of free energy must be absorbed. This change in free energy is not influenced by intermediate steps in its oxidation, since the same amount of $\Delta F$ from direct combustion of glucose to $CO_2$ and water is required, as occurs with glucose in some 30 distinct enzymatic steps in the body.

Despite the usefulness of free energy in isolated systems involving a specific reaction, it cannot be employed as a parameter of overall energy transformations in the total animal for practical and technical reasons. For this reason, energy exchanges in the whole animal must be measured and expressed in terms of changes in heat. Dietary energy cannot be expressed in terms of potentially useful energy, i.e., $\Delta F$ from complete oxidation. It needs to be measured and expressed in terms of total heat ($\Delta H$) produced from complete oxidation. So, in studies of the whole animal, $\Delta H$ is more useful than $\Delta F$. The general viewpoints in the above discussion were expressed by Wilkie (1960).

## C. Energy Partition on a Biological Basis

Animals have a flow of energy through many intermediate steps in the utilization of dietary nutrients. Most of the intermediate and output forms are presented in Fig. 5.1 (NRC, 1981).

In order to clarify the terms shown in Fig. 5.1, the following definitions are presented (NRC, 1981):

1. Intake of food energy (IE) is the gross energy in the food consumed and is equal to the weight of feed intake times the gross energy of a unit weight of food.

2. Fecal energy (FE) is the gross energy in the feces and is calculated as the weight of the feces times the gross energy from undigested food ($F_iE$) and energy from substances of metabolic origin ($F_mE$).

3. Apparently, digested energy (DE) is energy in food consumed less energy in feces: $DE = IE - FE$.

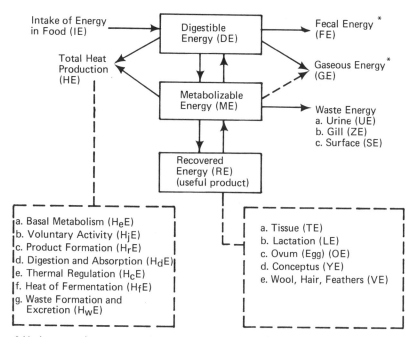

Fig. 5.1. The idealized flow of energy through an animal (From NRC, 1981).

4. True digested energy (TDE) is the intake of food energy minus the fecal energy of food origin ($F_iE = FE - F_eE - F_mE$) minus heat of fermentation and digestive gaseous losses, i.e., $TDE = IE - FE + F_eD + F_mE - H_fE - GE$.

5. Gaseous products of digestion (GE) include combustible gases produced in the digestive tract incident to digestion of food by microbes. Methane is the principal gas but hydrogen, carbon dioxide, and hydrogen sulfide are generally produced in small amounts and can reach significant levels under certain dietary conditions.

6. Urinary energy (UE) is the total gross energy in urine and includes energy from nonutilized absorbed compounds from food ($U_iE$), end products of metabolic processes ($U_mE$), and end products of endogenous origin ($U_eE$).

7. Metabolizable energy (ME) is the energy from feed less energy lost in feces, urine, and gases.

8. Nitrogen-corrected metabolizable energy ($M_nE$) is ME adjusted for total nitrogen retained or lost from body tissues and may be calculated as $M_nE = ME - (k \times TN)$. The correction for mammals is generally $k = 7.45$ kcal per gram of nitrogen retained in body tissue (TD).

9. True mtabolizable energy (TME) is the intake of true digestible energy minus urine energy of food origin, i.e., TME = TDE − UE + $U_eE$.

10. Total heat production (HE) is the energy lost in a form other than as a combustible compound and may be measured by either direct or indirect calorimetry. The commonly accepted equation for indirect calculation of heat production from respiratory exchange is HE (kcal) = 3.866 (liters of $O_2$) + 1.200 (liters of $CO_2$) − 1.431 (grams of UN) − 0.0518 (liters of $CH_4$). Heat production may also be measured by difference from a comparative slaughter trial.

11. Basal metabolism ($H_eE$) is the energy needed to sustain the animal's life processes in the fasting and resting states and is referred to as the *basal metabolic rate (BMR)*. The absence of methane production in ruminants is often considered the time at which they are in the postabsorptive state. In general, fasting metabolism is considered to be taking place when the respiratory quotient becomes equivalent to the catabolism of fat or near 0.7, and may be achieved in 48–144 hr after the last meal.

12. Heat of activity ($H_jE$) is heat production resulting from muscular activity required for such actions as getting up, standing, grazing, drinking, and lying down.

13. Heat of digestion and absorption ($H_dE$) is heat produced by the action of digestive enzymes in the food and by the digestive tract in moving digesta as well as moving nutrients through the wall of the alimentary tract.

14. Heat of fermentation ($H_fE$) is heat produced as a result of microbial action in the digestive tract. In ruminants, $H_fE$ is a major component often included in the heat of digestion ($H_dE$).

15. Heat of product formation ($H_rE$) is the heat produced in association with metabolic processes during product formation from absorbed metabolites. In its simplest form, it is heat produced by a biosynthetic pathway.

16. Heat of thermal regulation ($H_cE$) is the additional heat required to maintain body temperature when the environmental temperature drops below the zone of thermal neutrality or the additional heat produced by an animal's efforts to maintain body temperature when environmental temperature is above this zone.

17. Heat of waste formation and excretion ($H_wE$) is additional heat production associated with synthesis and excretion of waste products. An example is the synthesis of urea from ammonia, which is costly energy largely due to the requirement for carbon and results in a measurable increase in total heat production.

18. Heat increment ($H_iE$) is the increase in heat production following feed consumption in a thermoneutral environment. $H_iE$ is usually considered a nonuseful energy, loss but during cold stress it helps to maintain body temperature. It includes heat of fermentation ($H_fE$), energy expenditure of the digestive process ($H_dE$), and heat produced as a result of nutrient metabolism ($H_rE + H_wE$).

19. Recovered energy (RE), or *energy balance,* is that portion of feed energy

retained as part of the body or voided as a useful product. In animals raised for meat, RE = TE (tissue energy); in lactating animals, RE is the sum of tissue energy, lactation energy, and energy in products of conception, i.e., RE = TE + LE + YE.

## III. FERMENTATION PATHWAYS

Typical ratios of volatile fatty acids (VFA) and fermentation pathways for their formation are briefly described by the following reactions (Hungate, 1966):

1. Acetate formation

   1 hexose (contains 672 kcal/mole)
   $\downarrow$ enzymes
   $2CH_2OHCHOHCHO$ (triose)
   $\downarrow$ enzymes
   $3CH_3COCOOH \xrightarrow{+2H_2O} 2CH_3COOH + 2CO_2 + 2H_2$
   acetic acid

But if the $H_2$ is used to reduce $CO_2$ to $CH_4$, i.e., $CO_2 + 4H_2 \rightarrow CH_4 + 2H_2O$, then 1 hexose $\xrightarrow{enzymes}$ $2CH_3COOH + CO_2 + CH_4$ (contains 420 kcal per 2 moles of acetate) and $672 - 420 = 252$ kcal *lost* in fermentation of 1 hexose mole to 2 moles of acetate.

2. Propionic acid formation

   1 hexose (672 kcal/mole)
   $\downarrow$ $2H_2$, circuitous pathway
   $2CH_3CH_2COOH + 2H_2O$ (734 kcal per 2 moles of propionate)

and $734 - 672$ kcal $= +62$ kcal *gained* with fermentation of 1 hexose mole to 2 moles of propionate.

3. Butyric acid formation

   1 hexose (672 kcal/mole)
   $\downarrow$ $2H_2O$
   $2CH_3COOH + CO_2 + 4H_2$
   $\downarrow$ condensation of acetate
   $CH_3COCH_2COOH + H_2O$
   $\downarrow$ $2H_2$
   $CH_3CH_2CH_2COOH + H_2O$ (524 kcal per mole butyrate), butyric acid

and $672 - 524 = 148$ kcal *lost* in fermentation of 1 hexose mole to 1 mole of butyrate.

These reactions show only why investigators seek diets that yield maximum yields of propionic acid in rumen fermentation, since propionate gains 62 kcal per mole of hexose fermented compared to losses of −252 and −148 kcal per mole of acetate and butyrate formed, respectively.

If the amounts of acetic, propionic, and butyric acids are known, the theoretical amounts of $CO_2$ and $CH_4$ expected in fermentation in the rumen can be calculated. Production of acetate will lead to the greatest relative yield of $CH_4$ since $4H_2$ could arise from fermentation of 1 mole of hexose, compared to $2H_2$ for butyrate and $1H_2$ in conversion to propionate. Relative high production of propionate is associated with low $CH_4$ production, while high acetate and butyrate production is accompanied by increased methane production.

Overall, in fermentation of hexose, the following reactions occur:

58 hexose → 62 $CH_3COOH$ + 22 propionic + 16 butyric + 33.5$CH_4$ + 60.5$CO_2$
38,976 kcal   13,000 kcal      8070 kcal     8340 kcal    7030 kcal + 27 $H_2O$

In summary, 38,976 kcal − 36,400 kcal = 2,536 kcal dissipated as heat, or 6.5% of the initial hexose energy.

Overall products of fermentation include $CO_2$, $CH_4$, $H_2$, formate, acetate, propionate, butyrate, lactate, succinate, and ethanol, but of these, only $CO_2$, $CH_4$, acetate, propionate, and butyrate are final products (Hungate, 1966). The $H_2$, ethanol, formate, lactate, and succinate do not accumulate in the rumen to an appreciable extent, except that occasionally large quantities of lactate are found. Figure 5.2 illustrates schematically the synthetic pathways of most of these fermentation intermediates to end products (Hungate, 1966).

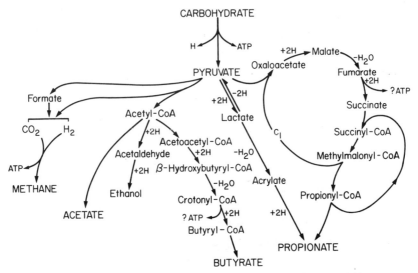

**Fig. 5.2.** Carbohydrate conversions in the rumen (From Hungate, 1966).

# IV. RATIONALE OF ENERGY SYSTEMS FOR RUMINANTS

According to Moe and Tyrrell (1973), differences among the net energy (NE) systems are largely those of interpretation rather than of scientific validity. NE systems have been proposed by Lofgreen and Garrett (1968), Blaxter (1962), Schiemann et al. (1971), and Moe et al. (1972), and some confusion is due to different assumptions used in their development and differences in terminology.

## A. Assumptions Involved in Net Energy (NE) Systems

Generally, the systems recognize that gross efficiency is influenced by level of production and body size, and that these two factors have the most influence on gross energetic efficiency. If production is zero, then gross efficiency is zero as well. Therefore, total requirements of animals have generally been divided into requirements for maintenance and production. It is also generally recognized that dietary energy is not used with equal efficiency for all physiological functions. Moe and Tyrrell (1973) summarized the approximate range for efficiency of utilization of metabolizable energy (ME) as follows: cold stress, 100%; maintenance, 70–80%; lactation, 60–70%; growth, 40–60%; and pregnancy, 10–25%. There is no variation in efficiency in cold stress, since all of the ME is used to produce heat. ME is a better measure of feed value in this situation than NE. Digestible energy (DE) is GE in diet minus energy lost in feces. Digestible energy minus energy lost in urine and methane results in ME. Net energy for production ($NE_p$) is ME minus heat production. $NE_p$ is unbiased in describing animal response, since it accounts for energy lost in feces, urine, gases, and heat. For this reason, the net energy concept has been regarded as theoretically the most precise system for the formulation of livestock rations. While the NE system appears ideal, practical problems in its application require some compromises.

According to Moe and Tyrrell (1973), one of the first compromises concerns the requirement of energy for maintenance. For instance, measurement can be made rather precisely for maintenance of an animal in terms of ME or DE, but not for NE. This is because the energy used for maintenance and the energy that is wasted are both ultimately lost as heat. Since the useful portion cannot be separated out, it is necessary to obtain the NE requirement for maintenance indirectly, and it is in the manner that the maintenance requirement is determined or expressed that most energy systems differ.

The relationship between energy output and ME intake, as ME is increased from zero to *ad libitum* intake, is shown in Fig. 5.3 (Moe and Tyrrell, 1973). Body tissue is mobilized to meet the energy needs for maintenance at zero ME intake (fasting). The fasting metabolism or fasting heat production is often used

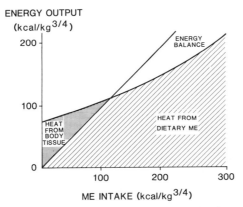

**Fig. 5.3.** Relationship between ME intake and heat output, showing a greater change in heat production per unit increase in ME intake above maintenance. The point where the lines cross is maintenence (From Moe and Tyrrell, 1973).

as an expression of the NE requirement of maintenance ($NE_m$). As feed is increased, heat production also increases until, at maintenance, total heat production equals intake of ME. The change in heat production between fasting and maintenance is a measure of the relative value of body tissue and dietary energy in meeting maintenance energy needs. Increasing ME intake above maintenance results in a positive energy balance and an increase in heat production; the increase at this stage is termed *heat increment* or *calorigenic effect*. The increase in energy balance per unit increase in ME intake is the partial efficiency of production.

The relationship between NE terminology and change in energy balance per unit increase in dry matter intake is shown in Fig. 5.4 (Moe and Tyrrell, 1973). Slopes of the solid lines represent the change in energy balance per unit change in dry matter intake and are equivalent to NE values; the change in energy balance between fasting and maintenance is $NE_m$. The change in energy balance above maintenance is termed $NE_p$. A net energy value calculated as the difference between fasting metabolism and energy balance at some intake greater than for maintenance is indicated by the term $NE_{m+p}$. This term indicates net energy for both maintenance and production. The dashed lines in Fig. 5.4 indicate $NE_{m+p}$ values at successively lower levels of feed intake, and the resulting values will gradually approach the $NE_m$ value. Since $NE_{m+p}$ is influenced by the level of intake, it is not acceptable as an expression of the nutritive value of a feedstuff. For this reason, Lofgreen and Garrett (1968) proposed the California system for cattle utilizing $NE_m$ and $NE_g$ as separate terms rather than the combined term $NE_{m+p}$. The system proposed by Rattray and Garrett (1971) for sheep is also of this type. These two systems are the only ones in common use, with a separate term for maintenance. The British system proposed by Blaxter (1962) and

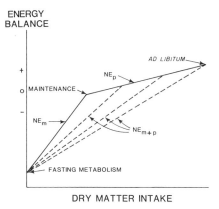

**Fig. 5.4.** Relationship between energy balance and net energy terminology. As dry matter intake is reduced from *ad libitum* to maintenance, $NE_{m+p}$ approaches $NE_m$. It is therefore of limited use as an expression of the energy value of feedstuffs (From Moe and Tyrrell, 1973).

adopted by the Agricultural Research Council (ARC) (1965) is also of this type, but the terminology is different. The British system is based on the ME content of feeds, expressed as efficiency of utilization for maintenance and production.

The assumption that the NE requirement for maintenance is equal to the fasting metabolism is a common assumption of the California and British systems. In an alternative description of maintenance, the amount of energy required is expressed in terms of production units. This is done in the Rostock system described by Nehring *et al. (1971) and Schiemann et al.* (1971), in which the amount of energy required for maintenance as well as production is described in terms of $NE_{fat}$. Within each species, a single term is used to describe the requirements for maintenance, growth, fattening, and milk production.

The maintenance requirement is also expressed in production units in the Beltsville system for dairy cattle (Moe *et al.*, 1972); here the energy requirement for both maintenance and milk production is expressed in terms of $NE_{milk}$ or $NE_{lact.}$, and the energy value of feedstuffs is expressed in this single unit. The California and British systems assume that the appropriate expression of the NE requirement for maintenance is equal to the fasting metabolism and that the variation in efficiency of energy used for maintenance is less than that for production. On the other hand, the Beltsville and Rostock systems describe maintenance requirements in terms of energy value for production. The difference is in application rather than in the fundamental concept.

## B. Limitation of Feeding Systems Utilizing NE

Some of the limitations of the NE system were summarized by Moe and Tyrrell (1973). Due to variations in the chemical and physical properties of feedstuffs, no single NE value is constant for a particular feed. Even corn grain has variations in NE, since different varieties are grown under different climatic and soil conditions and processed in different ways.

VFA are absorbed directly from the rumen, whereas glucose, amino acids, and other nutrients are absorbed from the lower intestinal tract. The amount of each substance absorbed determines the actual energy value of a feed or combination of feedstuffs. It is important to be able to predict how much of a feedstuff will be fermented in the rumen and whether it will be absorbed as VFA or glucose. The various methods of processing feedstuffs may affect the extent and type of fermentation. Corn grain may be fed whole, ground, cooked, popped, rolled, flaked, fermented, extruded, or pelleted, for example. Feedstuffs may also be fed with chlorotetracycline, oxytetracycline, zeranol, sulfamethazine, or methionine hydroxy analog, which may influence energetic efficiency. Animal productivity may be influenced by feed intake level, digestibility, and efficiency of use of absorbed nutrients due to the nutritional history and genetic background of animals.

The composition of the product formed is not described in NE systems, and this is a limitation in their use for $NE_g$ or $NE_{milk}$. There may be considerable variation in the ratio of milk fat to protein due to depression of the milk fat percentage, and if the potential efficiency of energy use for fat and protein synthesis is different, it limits the use of the NE concept. Similarly, $NE_g$ does not distinguish between efficiency of protein and fat synthesis, and there are differences (Orskov and McDonald, 1970; Thorbek, 1970; Peterson, 1970) because the efficiency of fat deposition is greater than that of protein deposition. Tyrrell *et al.* (1971) observed that as feed intake increased from maintenance to *ad libitum,* there was relatively little change in the total amount of protein deposited in tissues and the increased energy retained at *ad libitum* feed intake was due to increased fat deposition. Net energy per kilogram of feed is decreased at high feeding levels, and values on feedstuffs at near maintenance intakes would be overestimated.

## V. THE CALIFORNIA NET ENERGY SYSTEM (CNES)

Lofgreen *et al.* (1963) developed a NE system that makes possible the separation of ME into net energy for maintenance ($NE_m$) and net energy for gain ($NE_g$). Gross energy of diets is partitioned into ME and utilizes comparative slaughter feedlot trials for determining energy retention. In these studies,

$$ME = GE - (FE + UE + GPD) \qquad (1)$$

where ME = metabolizable energy, GE = gross energy of the diet, FE = fecal

## V. The California Net Energy System (CNES)

energy excreted, UE = urinary energy excreted, and GPD = gaseous products of digestion. Lofgreen (1965) based the system on the following relationships:

$$ME = HI + NE \qquad (2)$$

and

$$NE = M + P \qquad (3)$$

where NE = net energy, M = energy expended for maintenance, P = net increase in the energy of products (new tissue) and HI = heat increment. Maintenance energy (M) includes basal heat produced (B) and heat from activity (A):

$$M = B + A \qquad (4)$$

Total heat (H) produced by an animal is the sum of the heats generated from standing, walking, etc., (A), basal heat (B), and the calorigenic effect of the feed intake (HI): H = B + A + HI. The various energy factors noted above may be combined as: ME − HI = M + P [Eqs. (2) and (3)]; B + A + HI = ME − P [Eqs. (2)–(4)]; and H = ME − P [Eqs. (2)–(5)]. In the equation H = ME − P, the ME and P are actually determined and H is calculated by difference.

The CNES involves determination of ME in feedstuffs by a conventional metabolism trial (Lindahl, 1959) and utilizing the value in a comparative slaughter feedlot trial with cattle. The energy gained in the empty body from the feed intake is determined by specific gravity measurements on carcasses of representative initial steers (or heifers) at the beginning of the feedlot period and corresponding cattle after approximately 120 days on diets. Calories accumulated in the fat and protein of the empty bodies may be calculated by an equation established by Garrett and Hinman (1969) based on the specific gravity of the chilled carcasses and chemical analyses of tissues.

### A. Metabolism Phase

In order to partition the $NE_m$ and $NE_g$ of a feedstuff, it is necessary to determine the ME of the feed in a conventional metabolism trial in which energy in the feces, urine, and methane ($CH_4$) is subtracted from the feed intake. Energy lost as $CH_4$ by cattle may be calculated from digested carbohydrates in the diet that need to be determined in the metabolism trial. The equation recommended by Bratzler and Forbes (1940) may be used for calculation of energy lost in methane ($CH_4$) as:

$$E, g\ CH_4 = 4.012 \times (\text{no. of 100s of grams of carbohydrates digested}) + 17.68$$

There are 13.2 kcal per gram of $CH_4$. Metabolizable energy, Mcal/kg = Mcal per kilogram of diet intake − (Mcal in feces, urine, and $CH_4$ per kilogram of feed intake.

## B. Feedlot Phase

The total NE of a feed for maintenance and energy retention is:

$$NE_{m+g} = NE_m + NE_g \tag{5}$$

When the diet is fed at two levels and energy retention is measured, the NE of the feed increment is considered equal to the energy retention at the higher feed intake ($P_2$) minus the energy retention at the lower feed intake ($P_1$). If both levels are above maintenance, the NE of the feed increment is the NE of that feed for production:

$$NE_g = P_2 - P_1 \tag{6}$$

In the comparative slaughter phase, different animals are fed the two levels of intake, and those fed more liberally will become larger and will expend more energy for maintenance. In such a situation, $M_1$ ($NE_m$) and $M_2$ ($NE_m$) are not equal and the excess energy expended for maintenance by animals fed at the higher level must be added to the difference in energy retention. Therefore, Eq. (6) would need to be changed thus:

$$NE_g = (M_2, NE_m - M_1, NE_m) + (P_2, NE_g - P_1, NE_g) \tag{7}$$

However, Lofgreen and Otagaki (1960) pointed out that when values are expressed on the basis of $W^{0.75}$ (metabolic body weight), then $M_1$ and $M_2$ are equal and Eq. (6) is valid, and one does not need to use Eq. (7).

In the comparative slaughter procedure, all steers are fed the basal diet for approximately 3 weeks. Then approximately 10 representative animals are selected and slaughtered; empty body weight is determined by removal of the contents of the alimentary tract, and protein and fat (total calories) are determined by specific gravity of the chilled carcass.

$$\frac{\text{Specific gravity}}{\text{of carcass}} = \frac{\text{weight of carcass in air (kg)}}{\text{weight of carcass in air - weight of carcass in water}}$$

These values are assigned as the initial composition of animals that remain in the feedlot, which are then placed on the various diets to be evaluated and continued in the feedlot for approximately 120 days. The finished cattle are then slaughtered and empty body weight and specific gravity are determined on the chilled carcasses, as was done with the initial group.

One may save the labor of manually removing the contents of the alimentary tract to obtain the empty body weight in the comparative slaughter phase by utilizing the weight of the animal's carcass in the appropriate equation in Table

## V. The California Net Energy System (CNES)

### TABLE 5.1.
### Relation between Carcass and Empty Body Weight of Beef Steers[a]

| Component | Carcass (X) | Empty body (Y) | Estimating equation[b] | Correlation coefficient |
|---|---|---|---|---|
| Empty body wt., kg | 216.5 | 325 | $Y = 1.362X + 30.3$ | 0.99 |
| Water, % | 52.3 | 54.6 | $Y = 0.9702X + 3.92$ | 0.99 |
| Fat, % | 27.9 | 25.1 | $Y = 0.9246X + 0.647$ | 0.99 |
| Nitrogen, % | 2.54 | 2.69 | $Y = 0.7772X + 0.713$ | 0.96 |
| Ash, % | 4.23 | 3.76 | $Y = 0.6895X + 0.844$ | 0.96 |
| Energy, kcal/g | 3.48 | 3.27 | $Y = 0.9400X + 0.003$ | 0.99 |

[a] From Garrett and Hinman (1969).
[b] The first equation is useful in determining the fill of cattle at slaughter.

5.1 (Garrett and Hinman, 1969), i.e., $Y$, empty body weight, kg = 1.362 × carcass weight, kg + 30.3. Estimating the equations in Table 5.2 also shows the relationship of water, fat, nitrogen, ash, and energy between the carcass and empty body weights of cattle.

Essential relationships for calculating fat, protein, and calories in the empty bodies of initial and finished cattle of the comparative slaughter trial are presented in Table 5.2. The estimating equations relating specific gravity of the carcasses to ether extract, nitrogen, calories, and water are based on actual chemical analyses and specific gravity determinations of cattle by Garrett and Hinman (1969). Using the equations in Table 5.1, very thin cattle with 12% fat or less

### TABLE 5.2.
### Relationship between Components of the Beef Carcass (Y) and the Empty Body (Y') with Carcass Density (X)[a]

| Item | Correlation coefficient | Replicates to measure differences | | Estimating equation |
|---|---|---|---|---|
| | | 5% | 10% | |
| Carcass (Y) | | | | |
| Ether extract, % | −0.95 | 17 | 6 | $Y = 587.86 - 530.45X$ |
| Water, % | 0.91 | 7 | 3 | $Y = 375.2X - 343.8$ |
| Nitrogen, % | 0.93 | 6 | 3 | $Y = 20.0X - 18.57$ |
| Energy, kcal/g | −0.94 | 9 | 4 | $Y = 49.54 - 43.63X$ |
| Empty body (Y') | | | | |
| Ether extract, % | −0.96 | 13 | 5 | $Y' = 551.38 - 498.5X$ |
| Water, % | 0.93 | 5 | 3 | $Y' = 378.74X - 345.18$ |
| Nitrogen, % | 0.92 | 5 | 3 | $Y' = 15.97X - 14.17$ |
| Energy, kcal/g | −0.95 | 8 | 4 | $Y' = 47.58 - 41.97X$ |

[a] From Garrett and Hinman (1969).

have slightly more nonfat organic matter (protein) than those with more than 12% fat in carcass. The caloric value of fat determined in the studies of Garrett and Hinman (1969) was 9.385 ± 0.082 kcal/g. Carcass fat and offal fat had identical means. Protein had a value of 5.539 kcal/g and a nitrogen content in protein of 16.33%. The estimated errors of the prediction equations indicate that six to eight or more animals should be used per treatment.

During the approximately 120 days in the feedlot, it is necessary to keep account of the amount of feed consumed while fat and protein (calories) are being deposited by the cattle. In order to determine the amount of energy from feed consumed that was utilized for body maintenance plus activity and heat increment of steers, it is necessary to calculate heat production (HP), expressed as kilocalories of ME intake per day of average metabolic body weight ($W^{0.75}$) for each level of feeding as: Log HP, kcal/$W^{0.75}$ = 1.8851 (which is log HP of fasting animals, i.e., 77) + 0.00166 ME average daily intake in kilocalories per $W^{0.75}$.

In order to determine the amount of energy utilized by cattle for body maintenance plus activity plus heat increment, one needs to plot the HP (kcal/$W^{0.75}$) on the Y axis of semilog paper and the average ME (kcal/$W^{0.75}$) daily intake on the X axis for each level of feeding and draw a line through these points and the 77-kcal point on the Y axis (see Fig. 5.5). Then the point is determined on the line where ME intake (kcal/$W^{0.75}$) is equal to HP (kcal/$W^{0.75}$). This equilibrium point gives the kilocalories of ME intake per day/$W^{0.75}$ body weight for maintenance of the animal plus its activity. The equilibrium point value makes possible the determination of the amount of the diet required per day for maintenance and activity. By dividing this equilibrium point value, i.e., 131 kcal, by the kilocalories of ME per gram diet (which was determined in metabolism trials to be 2.04

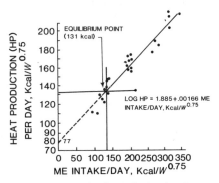

**Fig. 5.5.** Determination of (a) heat production at fasting (need to extrapolate the line to the X axis) and (b) the equilibrium point, i.e., the point on the line where kilocalories of ME intake/day/$W^{0.75}$ is equal to daily heat production (kcal/$W^{0.75}$). This value is needed to determine $NE_m$ + activity energy (From Lofgreen and Garrett, 1968).

## V. The California Net Energy System (CNES)

**TABLE 5.3.**

**Calculation of Net Energy for Gain (NE$_g$) of a Diet Utilizing Data From CNES**[a]

| Level of feeding | Feed intake (g/$W^{0.75}$/day) | Energy gain (kcal/$W^{0.75}$/day) |
|---|---|---|
| *Ad libitum* | 146 | 41 |
| Equilibrium point | 64 | 0 |
| Difference | 82 | 41 |

[a]From Lofgreen and Garrett (1968).

kcal/g), one determines the number of grams of diet required per day per $W^{0.75}$ for maintenance plus activity, i.e., 64.2 g.

Then the NE$_m$ for the diet can be calculated as:

$$NE_m, \text{kcal/g or Mcal/kg} = \frac{77 \text{ kcal}}{\text{no. g ration required/day/}W^{0.75} \text{ for maintenance + activity, i.e., 64.2 g}}$$

In order to calculate the NE$_g$ of the diet, it is necessary to utilize the calories measured in fat and protein gains of cattle fed above maintenance levels. Data in Table 5.3 illustrate the calculation of NE$_g$ of ration from data. Therefore, NE$_g$ = 41 ÷ 82 = 0.5 kcal/g or 0.5 Mcal/kg diet.

Figure 5.6 illustrates the relationship between empty body weight gain and energy retained per day. From these data, it was found that for any size animal and energy stored in the empty body weight gain, the NE$_g$ requirement could be expressed for steers by the following equation:

$$NE_g, \text{kcal}/W^{0.75} = (52.7X \text{ or empty body weight gain in kg} + 6.84X^2)$$

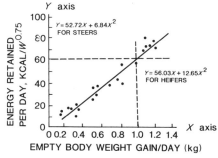

**Fig. 5.6.** Relationship of weight gain and energy gain of steers (From Lofgreen and Garrett, 1968).

For heifers, the equation is:

$$NE_g, \text{kcal}/W^{0.75} = (56.03X + 12.65X^2)$$

According to the National Research Council (NRC, 1976a), expected daily body weight gain from the amount of daily energy retained can be calculated as follows:

$$\text{Steers gain, kg} = 73.099 (0.002779 + 0.02736\ NE_g/W^{0.75})^{1/2} - 3.8538$$

$$\text{Heifers gain, kg} = 39.526 (0.003139 + 0.0506\ NE_g/W^{0.75})^{1/2} - 2.2146$$

## C. Tables of Energy Requirements

Equations such as those above (NRC, 1976a) can be used to calculate the energy requirements of any body weight cattle and for any weight of gain per day. However, tables have been prepared for $NE_m$ and $NE_g$ requirements for various weights of cattle and gains expected per day based on the equations, and they are quite convenient to use (Table 5.4). For greater variations in values, one should see the NRC (1976a), where the information is expanded to include nine different body weights of beef cattle as well as dry matter consumption requirements per kilogram of gain, percentage of roughage in characteristic rations for desired gains, percentage of protein, percentage of digestible protein, etc.

Table 5.5 illustrates how to formulate a diet that has a number of ingredients, their percentages in the diet, $NE_m$ and $NE_g$ values per kilogram of ingredient,

**TABLE 5.4.**

**$NE_m$ and $NE_g$ Requirements of Various Weights of Cattle and Rates of Gain**[a]

| Daily gain (kg) | Body weight (kg) | | | |
|---|---|---|---|---|
| | 150 | 300 | 450 | 500 |
| | | $NE_m$ required (Mcal/day) | | |
| 0 | 3.30 | 5.55 | 7.52 | 8.14 |
| | | $NE_g$ required (Mcal/day) | | |
| 0.1 | 0.23 | 0.39 | 0.52 | 0.56 |
| 0.2 | 0.46 | 0.78 | 1.06 | 1.14 |
| 1.0 | 2.55 | 4.29 | 5.82 | 6.29 |
| 1.5 | 4.05 | 6.81 | 9.23 | 9.98 |

[a]From Lofgreen and Garrett (1968).

**TABLE 5.5.**

**Formulation of Cattle Diet with Various Ingredients Utilizing Their NE$_m$ and NE$_g$ Contributions**[a]

| Feed ingredients | % ingredient | Mcal/kg ingredient | | Mcal/kg diet | |
|---|---|---|---|---|---|
| | | NE$_m$ | NE$_g$ | NE$_m$ | NE$_g$ |
| Alfalfa, 24% fiber | 14 | 1.31 | 0.69 | 0.16 | 0.07 |
| Oat hay | 6 | 1.31 | 0.70 | 0.06 | 0.02 |
| Cottonseed hulls | 5 | 1.03 | 0.19 | 0.05 | 0.01 |
| Barley grain | 28 | 2.13 | 1.40 | 0.54 | 0.36 |
| Wheat, mill run | 15 | 2.15 | 1.42 | 0.26 | 0.15 |
| Beet pulp, molasses dried | 10 | 2.03 | 1.34 | 0.18 | 0.12 |
| Cottonseed meal | 10 | 1.70 | 1.11 | 0.14 | 0.09 |
| Molasses, cane | 8 | 1.78 | 1.18 | 0.09 | 0.06 |
| Fat | 2 | 4.57 | 2.62 | 0.09 | 0.06 |
| Additives | 2 | | | | |
| | 100 | | | 1.62 | 0.97 |

[a]From Lofgreen and Garrett (1968).

and the NE$_m$ and NE$_g$ contributions of each ingredient per kilogram of diet. The diet contains 1.62 Mcal of NE$_m$/kg and 0.97 Mcal of NE$_g$/kg. The following is a check on the amount of the diet required per day for maintenance and for a gain of 1 kg per day of a 350-kg steer, as the NRC (1976a) gives requirements of 6.24 Mcal/day for maintenance and 4.82 Mcal/kg gain/day of NE$_g$ for this weight steer:

6.24 Mcal needed for NE$_m$ ÷ 1.62 = 3.85 kg for maintenance per day
4.82 Mcal needed for NE$_g$ ÷ 0.97 = 4.96 kg for gain per day
                                       8.81 kg total diet requirement per day, dry basis

The NRC (1976a) indicates that a 350-kg calf should gain 1 kg/day when fed 6.24 Mcal/day for NE$_m$ and 4.82 Mcal/day for NE$_g$; approximately 8.8 kg feed can be expected to be consumed per day.

Shirley (1980) proposed a procedure to generate a single net energy value for a cattle ration that simplifies computer least-cost ration formulations.

## D. Associative Effects on NE of Diets

Kriss (1943) concluded, largely from earlier studies by Forbes and others, that NE values of individual feeds are fundamentally variable in that their value depends on the combination with other feedstuffs with which they are fed. Vance *et al.* (1972a) and Byers *et al.* (1976) attributed variability in NE values of corn and corn silage diets to associative effects. However, Garrett (1979) conducted a study with a wide range of roughage : concentrate ratios in diets fed steers in a comparative slaughter feedlot trial, and concluded that associative effects may often be due to technical and animal problems of the experiments. He found that ME (kilocalories/gram) was linearly related to the percentage of roughage in the diets ($r = -.96$) and to NE ($r = -.99$ for $NE_m$ and $r = -.97$ for $NE_g$), with no significant curvilinear components in any of the regressions. Checking his data by the Blaxter (1962) ME efficiency system, he found no curvilinear relationship between the proportion of roughage in the diets and $K_m$ and $K_f$ ($r = .96$ and $-.93$, respectively). The diets fed by Garrett (1979) contained 11.3% crude protein and were fortified with vitamin A, calcium, and phosphorus, and the cattle were implanted with Synovex-S. Some of the associative effects reported in the literature certainly could be due to other ingredients helping to meet dietary requirements or diluting their concentrations.

A criticism of the NE system is that no allowance is made for the associative effect of different ingredients in the rations. Blaxter *et al.* (1956) and Lofgreen and Otagaki (1960) demonstrated that the nutritive value of a food varies with the diet to which it is added. The reduction of fiber digestibility when starch or any soluble carbohydrate is added to a diet is well established (Swift and French, 1954). Lack of essential nutrients and physiological factors such as lactation, growth, fattening, physical exercise, sex, and species variations influence NE values of feeds (Blaxter *et al.*, 1956). Kromann (1967), using simultaneous equations, demonstrated that the $NE_{m+p}$ of a basal ration increased as molasses was increased in the diet, and that $NE_{m+p}$ in alfalfa hay decreased as milo was added to the diet.

Vance *et al.* (1972b) fed steers five different corn grain-corn silage diets that varied in the percentage of concentrate dry matter from 36 to 97% in a comparative slaughter trial. The $NE_m$ values of diets increased linearly, indicating that the $NE_m$ of each feed stayed constant while the $NE_g$ value increased curvilinearly, with high levels of concentrate resulting in greater $NE_g$ values. The $NE_g$ of corn grain decreased while that of corn silage increased as the level of corn grain in the diet was decreased. The data indicate that the $NE_m$ of a feed tends to remain constant with varying proportions in the diet but that the $NE_g$ may vary depending on the proportion of the feed in the total diet. Woody *et al.* (1983) suggested from data obtained with Charolais crossbred and Hereford cattle that discounting whole plant corn silage and corn grain $NE_m$ values by 4.3%, and $NE_g$ by 8.9%, would adjust for associative effects when total diet corn grain content is between 50 and 80%.

Knox and Handley (1973) concluded that the use of the CNES in feedlots is an improvement over other systems used but pointed out that where environments are cold and stressful, the accuracy of the system can be increased by applying appropriate correction factors to $NE_m$ requirements. Maxson *et al.* (1973) applied the CNES for comparing the $NE_m$ and $NE_g$ values of steer feedlot diets based on corn, non-bird-resistant (NBR), or bird-resistant (BR) sorghum grains. The $NE_m$ values for the grains were 2.40, 1.98, and 1.78 Mcal/kg and the $NE_g$ values were 1.49, 1.15, and 0.73 Mcal/kg dry weight when the corn, NBR,and BR sorghum diets, respectively, were fed *ad libitum*. The predicted gains for steers fed *ad libitum* using the CNES agreed with those actually obtained within 0.08 kg/day. Brommelsiek *et al.* (1979) used the CNES to evaluate the effect of methionine hydroxy analog (MHA) on energy utilization of BR sorghum grain diets fed steers. The levels of MHA fed were 0, 7, or 30 g per steer daily. Predicted gains per day were 0.03–0.06 kg greater than actual gains utilizing ME determined with *ad libitum*-fed steers and 0.05–0.07 kg/day greater than actual gains utilizing ME values determined at 60% *ad libitum* feed intake.

### E. Frame Size Nutrient Adjustment

The animal maintenance requirement equation $NE_m = 0.077 \times W^{0.75}$ is a function of actual body weight, but the gain requirement is a function of the proportion of fat and protein in the tissue gain (Fox *et al.*, 1977). The proportion of fat and protein in body weight gain is influenced by frame size and the use of growth stimulants. For example, the NE required per kilogram of gain is lower at a particular weight for steers that will be graded low choice at 1320 lb than those that reach low choice at 880 lb. McCarthy *et al.* (1985) observed that large frame steers had greater daily protein gain than did small frame steers, and small frame steers had a greater percentage of fat gain and were more energy efficient.

The estimate for $NE_g$ per pound of gain for an 880-lb small-framed steer gaining 2.2 lb/day is 2.86 Mcal compared to 2.13 Mcal for an 880-lb large-framed steer (Fox *et al.*, 1977). An assumption was made that steers with the same empty body chemical composition, i.e., the same percentage of fat and protein, have the same NE requirement per pound of gain. Table 5.6 gives estimates of the weights at which cattle of different frame sizes have equivalent percentages of protein and fat. When a steer varies from average frame, the weight at which an average-framed steer is equivalent to him in body composition is used in gain projection by the following equation:

$$\text{Gain, projection for steers} = \frac{\{0.002779 + 0.020736\,[(NEFG)/(\text{breed})(WE^{0.75})]\}^{1/2} - 0.05272}{0.01368}$$

In this equation, Breed is an adjustment factor for Holsteins relative to British and British × Exotic breed crosses, and WE is an adjustment (equivalent) weight measured in kilograms for frame size and use of stimulants. For example, a 900-

**TABLE 5.6.**
**Estimated Weights for Animals of Alternative Frame Sizes at Which the Proportions of Body Fat and Protein Are Similar**[a,b]

| | Shrunk body weight (lb) | | | | | | | | | |
|---|---|---|---|---|---|---|---|---|---|---|
| **Steers** | | | | | | | | | | |
| Small frame | 320 | 400 | 480 | 560 | 640 | 720 | 800 | 880 | 960 | 1040 | 1120 |
| Average frame | 400 | 500 | 600 | 700 | 800 | 900 | 1000 | 1100 | 1200 | 1300 | 1400 |
| Large frame | 480 | 600 | 720 | 840 | 960 | 1080 | 1200 | 1320 | 1440 | 1560 | 1680 |
| **Heifers** | | | | | | | | | | |
| Small frame | 260 | 330 | 390 | 460 | 530 | 590 | 660 | 720 | 790 | 860 | 925 |
| Average frame | 320 | 400 | 480 | 560 | 640 | 720 | 800 | 880 | 960 | 1040 | 1120 |
| Large frame | 390 | 470 | 560 | 660 | 750 | 840 | 940 | 1030 | 1130 | 1220 | 1315 |
| | Chemical composition (% of empty body weight) | | | | | | | | | |
| Fat | 12.6 | 14.9 | 17.2 | 19.5 | 21.8 | 24.2 | 26.5 | 28.8 | 31.1 | 33.3 | 35.6 |
| Protein | 19.9 | 19.5 | 19.1 | 18.6 | 18.1 | 17.6 | 17.1 | 16.5 | 16.0 | 15.4 | 14.9 |

[a] From Fox et al. (1977).
[b] Small-, average-, and large-frame steers reach a fatness of low choice, yield grade 2.5 to 3.0, in the weight range 800–880, 1000–1100, and 1200–1320 lb, respectively. The weights for heifers are 660–720, 880, and 940–1000 lb, respectively.

lb steer would be used in the gain projection equation above for a 700-lb small-framed steer and for a 1080-lb large-framed steer. Garrett (1976) suggested that requirements for breeds maturing at heavier weights and some heavy, short yearlings may be adjusted by multiplying the equation $NE_g = (0.05272G_{kg}/day + 0.01265G^2)$ ($W^{0.75}$) by 0.92 or some other factor.

Variation in frame size in heifers is treated comparably with that in steers, except that a different equation is used for heifers, as follows (Fox *et al.*, 1977):

$$\text{Gain, projection for heifers} = \frac{\{0.003139 + 0.0506\,[NEFG/(breed)WE^{0.75}]\}^{1/2} - 0.05603}{0.0253}$$

Klosterman and Parker (1976) observed heifers to have a feed efficiency similar to that of steer herd mates when both were fed to the same body composition. However, Harpster *et al.* (1976) found that heifers rejected as herd replacements had 10% less feed efficiency than steers when fed to similar body composition. They pointed out that heifers are culled for reasons other than sex, i.e., light weaning weight, sickness, injury, and poor nutrition causing stunting, which are factors in the feedlot performance of heifers. In preliminary work, an adjustment factor of 0.84 was used to give an equivalent weight of bulls relative to steers, since a bull is equivalent in body composition to a steer weighing 84% as much.

## F. Breed

Adjustments for differences in energetic efficiency were not believed necessary for different British and British × exotic breed crosses, but adjustments were made for Holsteins (Fox *et al.*, 1977). Holsteins have greater $NE_m$ and $NE_g$ requirements than the others when compared at equivalent body composition. The $NE_m$ and $NE_g$ requirements for Holsteins were set at multiplier factors of 1.12 and 1.12, respectively, and those for Holstein × British crosses at 1.06 and 1.06, respectively.

## G. Environment

Using a scale of 1 to 7 to adjust the $NE_m$ for environment, typical values are shown in Table 5.7 for periods covering 2–3 months. This was done to correct the CNES $NE_m$ values that were obtained in a thermal neutral environment (Fox *et al.*, 1977).

## H. Growth Stimulants

Growth stimulants generally improve gain and/or feed efficiency, as shown in Table 5.8 for the same length of time on diets (Fox *et al.*, 1977). The growth

### TABLE 5.7.

**Estimated Impact on $NE_m$ Requirements**[a]

| Scale value | Lot condition | Multiplier for $NE_m$ |
|---|---|---|
| 1 | Outside lot with frequent deep mud | 1.30 |
| 5 | Outside lot, well mounded, bedded during adverse weather | 1.10 |
| 7 | No mud, shade, good ventilation, no chill stress | 1.00 |

[a] From Smith (1971).

stimulants increase weight, since the percentages of fat and protein in the empty body are similar to those of untreated cattle. Animals receiving growth stimulants need to be fed to heavier weights in order to obtain the same carcass grade as those not receiving the stimulants. Diethylstilbestrol (DES)-treated cattle were estimated to deposit 21% more protein and 11% more fat per day than untreated animals. The CNES energy requirements for gain are based on the use of a growth stimulant, and the NE requirement for gain for a steer not receiving DES or Synovex-S are similar to those of a treated animal that is 18% heavier. In comparison, energy requirements for untreated steers are similar to those of steers treated with Ralgro that are 12% heavier. With heifers, a 13% adjustment for untreated animals is used when DES and MGA are utilized and a 8.5% factor is used with Ralgro. Absolute feed intake is slightly increased when a growth stimulant is used, but when data are adjusted for the fact that treated cattle are fed to heavier weights and compared to other animals fed the same length of time, the dry matter intake/$kg^{0.75}$ shows little or no increase.

### TABLE 5.8.

**Effect of Growth Stimulants on Performance and Carcass Quality of Cattle Fed for Same Length of Time**[a]

| Sex and diet | Improvement | | Change in DM intake per day (%) | Change in quality grade of carcass |
| | Daily gain (%) | Feed/ gain (%) | | |
|---|---|---|---|---|
| Steers | | | | |
| DES in finishing diets | 13.0 | 9.5 | +2.2 | Lower |
| DES implants | 14.8 | 10.3 | +3.0 | Lower |
| Ralgro | 10.4 | 6.8 | +3.0 | None |
| Heifers | | | | |
| DES in finishing diets | 11.7 | 11.3 | −0.9 | None |
| Ralgro | 6.6 | 5.9 | None | None |
| MGA | 11.2 | 7.6 | +1.0 | None |

[a] From Fox et al. (1977).

**TABLE 5.9.**

**Estimated Impact of Feed Additive on Nutrient Values**[a]

| Additive | Multiplier | |
|---|---|---|
| | $NE_m$ | $NE_g$ |
| Without Rumensin | | |
|   Without growth stimulants | | |
|     Without antibiotics | 1.000 | 1.000 |
|     With antibiotics | 1.040 | 1.040 |
|   With growth stimulants | | |
|     Without antibiotics | 1.000 | 1.000 |
|     With antibiotics | 1.030 | 1.030 |
| With Rumensin | | |
|   Without growth stimulants | 1.110 | 1.110 |
|   With growth stimulants | 1.100 | 1.100 |

[a]From Fox et al. (1977).

Schanbacher (1984) reviewed the utilization of various hormonal products (Ralgro, Synovex-S, Compudose, and others) as implants for steers, the feeding of melengesterol and the implanting of heifers with synovex-II, Ralgro, and Compudose, and the implanting of zeranol in bulls.

## I. Digestive Stimulants

"Multipliers" are given in Table 5.9 to estimate the influence of antibiotics on utilization of feed. The values were estimated by working backward to find the $NE_m$ and $NE_g$ values that would have had to increase in order to generate the gains and feed efficiencies found in feeding trials; it was assumed that the impact was similar for $NE_m$ and $NE_g$.

## J. Previous Nutritional Treatment

The nutritional treatment of cattle prior to being placed in the feedlot has an impact on the utilization of dietary energy, as shown in Table 5.10. A multiplier scale ranging from 1 to 9 based on body composition and previous rate of gain at the time the animal was placed on feed was devised. The multiplier for $NE_m$ differs from that for $NE_g$.

## K. Daily Dry Matter Intake

The dry matter intake must be consistent with the energetic components for accuracy in projections of gain and feed conversion. Intake is influenced by age,

**TABLE 5.10.**

**Estimated Impact of Previous Nutritional Treatment on Nutrient Values**[a,b]

| Value | Body Condition | Previous rate of gain (lb/day) | Multiplier | |
|---|---|---|---|---|
| | | | $NE_m$ | $NE_g$ |
| 1 | Very fleshy | 2.3 | 0.955 | 0.90 |
| 5 | Average | 1.5 | 1.000 | 1.00 |
| 9 | Very thin | 0.7 | 1.045 | 1.10 |

[a]From Fox et al. (1977).
[b]Based on average-frame calf.

weight, frame size, breed, energy density of diet, physical form, and degree of fermentation. It is assumed to be limited by higher fiber content, but as diets become less fibrous, fiber no longer restricts intake. As energy density increases further, a point is reached (probably 60–70% corn in corn-corn silage diets, dry basis) at which chemostatic and thermostatic controls take over. Then daily energy intake and gain are about constant, and intake falls as the energy density is increased. Beyond about 90% concentrate, daily gain is depressed and daily dry matter intake falls faster.

The dry matter intake equation is:

Dry matter intake = $(0.100)$ (breed) (age) $(W^{0.75})$ for equivalent weights of 800 lb

$(0.095)$ (breed) (age) $(W^{0.75})$ for equivalent weights of 950 lb

$(0.090)$ (breed) (age) $(W^{0.75})$ for equivalent weights of 1050 lb

where $W$ is empty body weight in kilograms and for energy densities where intake is limited by neither the fiber content nor the energy density of the diet. The breed and age multipliers are given in Table 5.11.

Daily dry matter intake is not adjusted upward for cattle placed on feed in "thin" condition. They may eat more when first placed in the feedlot, but over the finishing period, intake does not differ significantly from that of cattle whose growth was not retarded.

## VI. NET PROTEIN ($NP_m$ AND $NP_g$) REQUIREMENTS

After the diet has been evaluated for $NE_m$ and $NE_g$ by the CNES system and adjustments have been made for various factors discussed above, the $NP_m$ and $NP_g$ can be determined for cattle of given weights and expected daily gains by

## VI. Net Protein (NP$_m$ and NP$_g$) Requirements

**TABLE 5.11.**

**Estimated Impact of Various Factors on Daily Dry Matter Intake**[a,b]

| Item | Multiplier |
|---|---|
| Age | |
|   Started on feed as calf | 1.00 |
|   Started on feed as yearling | 1.10 |
| Breed | |
|   British | 1.00 |
|   Exotic | 1.00 |
|   British × Exotic | 1.00 |
|   Holstein | 1.17 |
|   Holstein × Exotic | 1.09 |
| Feed additives | |
|   Without Rumensin | |
|     Without antibiotics | 1.00 |
|     With antibiotics | 1.00 |
|   With growth stimulant | |
|     Without antibiotics | 1.00 |
|     With antibiotics | 1.00 |
|   With Rumensin | |
|     Without growth stimulant | 0.91 |
|     With growth stimulant | 0.91 |

[a]From Fox et al. (1977).
[b]Feedlot diets. Antibiotics have shown an impact on intake in some backgrounding diets for calves.

using equations (Fox et al., 1977). The NP$_m$ was calculated utilizing the equation by Smuts (1935) as follows: NP$_m$ = 70.4 $W^{0.734}$ × 0.0125, where $W$ is empty body weight. Then the empty body protein at specific weights of average-frame steers was calculated as = $0.255W - 0.00013W^2 - 2.418$ (Reid, 1974), which estimated the NP$_g$. Since it is time-consuming to make calculations using formulas, Fox and Black (1977) tabulated NP$_m$ and NP$_g$ requirements for steers and heifers at various equivalent body weights to yield body weight gains ranging from 1 to 3.5 lb/day, as shown in Table 5.12. In formulating diets, the NP$_m$ requirement for the actual weight of cattle is added to the NP$_g$ requirement for the particular equivalent weight. For instance, the equivalent weights of small-, average-, and large-frame steers in column 1 of Table 5.12 are 264, 330, and 396 lb, respectively. The NP$_{m+g}$ requirement of these three different weight and frame size steers is the same. That is, if a 2.5 lb/day gain is expected from their energy intake, then their NP$_m$ + NP$_g$ requirement = 0.08 + 0.49 = 0.57 lb/day.

Limited data on NP are available for cattle in feedstuffs, but some values are presented in Table 5.13 (Fox and Black, 1977). An interesting discussion on the

## TABLE 5.12.

### Net Protein Requirements for Maintenance and Gain of Growing and Finishing Cattle[a]

| | Net protein requirements for maintenance (lb/day) Body weight (lb) | | | | | | |
|---|---|---|---|---|---|---|---|
| | 330 | 440 | 550 | 660 | 770 | 880 | 990 |
| | | | | NP (lb/day) | | | |
| | 0.08 | 0.09 | 0.11 | 0.13 | 0.14 | 0.16 | 0.17 |
| | Net protein requirements for weight gain (lb/day) Empty body weight (lb) | | | | | | |
| Steers | | | | | | | |
| Small | 264 | 352 | 440 | 528 | 616 | 704 | 792 |
| Average | 330 | 440 | 550 | 660 | 770 | 880 | 990 |
| Large | 396 | 528 | 660 | 792 | 924 | 1056 | 1238 |
| Heifers | | | | | | | |
| Small | 215 | 286 | 363 | 429 | 506 | 572 | 644 |
| Average | 264 | 352 | 440 | 528 | 615 | 704 | 792 |
| Large | 313 | 418 | 528 | 627 | 726 | 836 | 941 |
| | | | | $NP_g$ (lb/day) | | | |
| Average daily gain (lb) | | | | | | | |
| 1.0 | 0.20 | 0.18 | 0.17 | 0.16 | 0.14 | 0.13 | 0.12 |
| 1.1 | 0.22 | 0.20 | 0.19 | 0.17 | 0.16 | 0.14 | 0.13 |
| 1.2 | 0.24 | 0.22 | 0.20 | 0.19 | 0.17 | 0.16 | 0.14 |
| 1.3 | 0.25 | 0.24 | 0.22 | 0.20 | 0.19 | 0.17 | 0.15 |
| 1.4 | 0.27 | 0.26 | 0.24 | 0.22 | 0.20 | 0.18 | 0.17 |
| 1.5 | 0.29 | 0.27 | 0.26 | 0.24 | 0.22 | 0.20 | 0.18 |
| 1.6 | 0.31 | 0.29 | 0.27 | 0.25 | 0.23 | 0.21 | 0.19 |
| 1.7 | 0.33 | 0.31 | 0.29 | 0.27 | 0.24 | 0.22 | 0.20 |
| 1.8 | 0.35 | 0.33 | 0.31 | 0.28 | 0.26 | 0.24 | 0.21 |
| 1.9 | 0.37 | 0.35 | 0.32 | 0.30 | 0.27 | 0.25 | 0.22 |
| 2.0 | 0.39 | 0.37 | 0.34 | 0.31 | 0.29 | 0.26 | 0.24 |
| 2.1 | 0.41 | 0.38 | 0.26 | 0.33 | 0.30 | 0.28 | 0.25 |
| 2.2 | 0.43 | 0.40 | 0.37 | 0.35 | 0.32 | 0.29 | 0.26 |
| 2.3 | 0.45 | 0.42 | 0.39 | 0.36 | 0.33 | 0.30 | 0.27 |
| 2.4 | 0.47 | 0.44 | 0.41 | 0.38 | 0.35 | 0.31 | 0.28 |
| 2.5 | 0.49 | 0.46 | 0.42 | 0.39 | 0.36 | 0.33 | 0.29 |
| 2.6 | 0.51 | 0.48 | 0.44 | 0.41 | 0.37 | 0.34 | 0.31 |
| 2.7 | 0.53 | 0.49 | 0.46 | 0.42 | 0.39 | 0.35 | 0.32 |
| 2.8 | 0.55 | 0.51 | 0.48 | 0.44 | 0.40 | 0.37 | 0.33 |
| 2.9 | 0.57 | 0.53 | 0.49 | 0.46 | 0.42 | 0.38 | 0.34 |
| 3.0 | 0.59 | 0.55 | 0.51 | 0.47 | 0.43 | 0.39 | 0.35 |
| 3.1 | 0.61 | 0.57 | 0.53 | 0.49 | 0.45 | 0.41 | 0.37 |
| 3.2 | 0.63 | 0.59 | 0.54 | 0.50 | 0.46 | 0.42 | 0.38 |
| 3.3 | 0.65 | 0.60 | 0.56 | 0.52 | 0.48 | 0.43 | 0.39 |
| 3.4 | 0.67 | 0.62 | 0.58 | 0.53 | 0.49 | 0.45 | 0.40 |
| 3.5 | 0.69 | 0.64 | 0.59 | 0.55 | 0.50 | 0.46 | 0.41 |

[a] From Fox et al. (1977).

## TABLE 5.13.

### Conversion Factors for Various Feeds[a]

| Ingredient | % CP[b] | % NP[c] | NP/CP[d] |
|---|---|---|---|
| Dry corn | 10.0 | 3.5 | 0.35 |
| High-moisture corn | 10.0 | 3.0 | 0.30 |
| Corn silage, untreated | 8.0 | 1.6 | 0.20 |
| Corn silage, NPN treated | 12.0 | | |
| For cattle under 600 lb | | | |
| Equivalent weight | | 3.0 | 0.25 |
| For cattle over 600 lb | | | |
| Equivalent weight | | 4.2 | 0.35 |
| Soybean meal | 45.0 | 15.7 | 0.35 |
| Urea | | | |
| For cattle under 600 lb | | | |
| Equivalent weight | 281 | 70.2 | 0.25 |
| For cattle over 600 lb | | | |
| Equivalent weight | | 98.4 | 0.35 |

[a]From Fox et al. (1977).
[b]Crude protein percent of dry matter.
[c]Net protein, percent of dry matter
[d]These factors can likely be applied to the CP of similar feeds until NP values are determined for the feed in question.

utilization of NPN (mainly as urea) for the generation of NP is given by Bergen et al. (1979).

## VII. THE BLAXTER (ARC) ME SYSTEM

Blaxter (1962) proposed an ME system applied to cattle diets that was also outlined by the ARC (1965). The ME in the system is used with different partial efficiencies for maintenance, growth, gain, and milk and wool production. In Fig. 5.7, effects of diet composition and physiological function on efficiency of utilization of ME are shown (Smith, 1971) for finishing steers. Maintenance needs of ME are based on fasting metabolism and on the percentage of the gross energy (GE) of the diet that is ME. This percentage is designated as $Q_m$, the efficiency of utilization for maintenance as $k_m$, and the efficiency of utilization for finishing of cattle as $k_f$. As schematically illustrated in Fig. 5.7, the line for maintenance is based on the equation $k_m = 54.6 + 0.30Q_m$. The corresponding line for finishing is based on the equation $k_f = 0.81Q_m + 3.0$.

The expected gain of a 300-kg steer that is consuming 1.8 kg alfalfa hay and 5.4 kg barley daily can be calculated utilizing the data presented in Table 5.14.

5. Nutritional Energetics

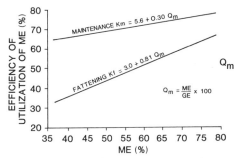

**Fig. 5.7.** Effect of metabolizable energy intake on the efficiency of utilization of ME for maintenance and fattening (From Smith, 1971).

The level of intake is corrected by subtracting $(9.6 - 0.11Q_m)$ from the observed ME value.

## A. Maintenance Requirement

1. The fasting requirement of a 300-kg steer is based on the daily fasting energy expenditure of $0.077$ Mcal/$W^{0.75}$, and since the metabolic body weight ($W^{0.75}$) of a 300-kg steer is 72.07 kg, the maintenance requirement is determined by the equation $0.077 \times 72.07 = 5.55$ Mcal/day.

2. The efficiency of utilization of ME required to meet the fasting energy requirement of 5.5 Mcal/day may be calculated as:

$$k_m = 54.6 + 0.30 Q_m$$
$$= 54.6 + 0.30 \times 61\% = 72.9\%$$

3. Therefore, the ME in the diet required for maintenance is calculated as $5.55 \div 72.9\% = 7.61$ Mcal/day.

**TABLE 5.14.**

**Determination of Percentage of Gross Energy (GE) That Is Metabolizable Energy (ME) or the $Q_m$ of a Diet**[a]

| Feed | DM (kg) | GE (Mcal/kg) | ME (Mcal/kg) | GE (Mcal) | ME (Mcal) | $Q_m$ (%) |
|---|---|---|---|---|---|---|
| Alfalfa hay | 1.8 | 4.28 | 3.06 | 7.70 | 3.71 | |
| Barley | 5.4 | 4.59 | 3.00 | 24.79 | 16.20 | |
| | 8.2 | | | 32.49 | 19.91 | 61% |

[a]From Smith (1971).

## B. Body Weight Gain Requirement

1. Mcal $ME_g$ = 19.91 Mcal of ME intake/day − 7.61 Mcal ME/day = 12.30
2. Efficiency of finishing:

$$k_f = 0.81\ Q_m + 3.0$$
$$= (0.81 \times 61\%) + 3.0 = 49.7 + 3.0$$
$$= 52.7\%$$

3. Expected gain:

$$\frac{12.30\ \text{Mcal} \times 52.7\%}{6.00\ \text{Mcal required/kg gain}} = 1.08\ \text{kg/day gain}$$

Additional observations with cattle and sheep after the ARC (1965) edition resulted in a new equation for $k_f$ (ARC, 1980). The new equation is: $k_f = 0.78\ Q_m + 0.006$.

Harkins *et al.* (1974) described a simplified NE system for ruminants based on Blaxter's (1962) ME system, which enables a noniterative approach to the formulation of diets with appropriate tables for growing cattle. Minson (1981) presented an energy system based on Blaxter's ME system in which he retained NE as the method of defining energy requirements and combined the ME values for different feeds of conversion of ME to NE for maintenance, growth, and lactation. Minson (1981) concluded that Blaxter's (1962) system is easy to use for predicting performance but difficult to use in formulating diets, and that his proposal is equally suitable for predicting performance and formulating diets.

## VIII. NET ENERGY VALUE OF FEEDS FOR LACTATION

Flatt *et al.* (1969) proposed adoption of an NE system for lactating dairy animals. Moe *et al.* (1972) described observations, principles, and assumptions involved in an NE system for lactating dairy cows. Nearly all of their experiments involved the relationship between the nature of the diet and the efficiency of energy utilization for milk production. They calculated total heat production from respiratory exchange according to the formula adopted by the EEAP (Brouwer, 1965); heat production (kcal) = 3.866 $O_2$ + 1.200 $CO_2$ − 0.518 $CH_4$ − 1.431 N where $O_2$, $CO_2$, and $CH_4$ are liters of oxygen consumed, carbon dioxide and methane produced, and N per gram of urinary nitrogen. Energy contents of feeds, feces, urine, and milk were determined by adiabatic bomb calorimetry.

## A. Net Energy Values

Moe *et al.* (1972) obtained the following equation for total NE intake: $NE_{milk}$, Mcal = energy balance, Mcal + $0.085 \times kg^{0.75}$, where energy balance is adjusted to zero tissue loss. The NE value, Mcal $NE_{milk}$ per kilogram of dry matter, was obtained by dividing total $NE_{milk}$ intake by total intake of dry matter. From their pooled data, they found the DE related to $NE_{milk}$ to be: $NE_{milk}$, Mcal/kg DM = 0.840 DE, Mcal/kg DM − 0.773, $r^2$ = .870, and that ME was: $NE_{milk}$ (Mcal/kg DM) = 0.885 ME (Mcal/kg DM) − 0.445, $r^2$ = .895. The corresponding relationship of $NE_{milk}$ to TDN was: $NE_{milk}$, Mcal/kg DM = 0.0352 %TDN − 0.62. Their data were incorporated in the 1971 revision of the "Nutrient Requirements of Dairy Cattle" (NRC, 1971a). In their $NE_{milk}$ system, the maintenance requirement was given as 85 $kcal/kg^{0.75}$, and the requirement for milk production at 0.74 Mcal $NE_{milk}$ per kilogram of 4% FCM and the $NE_{milk}$ values of individual feedstuffs were calculated from the above TDN equation rather than from DE or ME, since both of these were largely derived from TDN data. The resulting $NE_{milk}$ values for feedstuffs were reduced by 7% due to the uncertainty of applying energy balance results under confined and closely controlled conditions to commercial dairy cattle. Further adjustment was made for some diets of corn silage, coarse-textured grains, or forages of high cell wall content to compensate for changes in nutritive value associated with high feed intakes.

## B. Excess Nitrogen Intake Adjustment

This adjustment was based on total heat production partitioned into that attributable to intake of digestible nitrogen in excess of that required for maintenance and production. This resulted in adding 7.2 kcal to the measured energy balance for each gram of digestible nitrogen in excess of requirements when energy balance was related to ME intake. When based on DE intake, the correction is 13.3 kcal per gram of excess dietary nitrogen due to increased urinary excretion of energy.

## C. Energy Adjustment for Pregnancy

This adjustment was made by computing the energy balance of the cow alone by subtracting the amount of energy retained in the fetal tissues. The fetal energy was estimated by differentiating the equation of Jakobsen *et al.* (1957) as follows: Energy retained in fetal tissues = $7.2e^{0.0174t}$ kcal/day where $t$ is the number of days pregnant. The maternal energy balance was then adjusted upward, assuming that energy used to promote growth of fetal tissues would have contributed $44e^{0.0174t}$ kcal to the total energy balance of cows weighing 600 kg, with a linear adjustment for those of other body weights.

## IX. LACTATION AND REPRODUCTION STUDIES

Neville (1974) conducted a study with nonlactating and lactating Hereford cows that were group fed in one trial, but individually fed in two other trials, to maintain body weight during the test period. The TDN requirement based on body weight expressed as maintenance of nonlactating cows = 0.0074 TDN/$W^{1.00}$/24 hr; maintenance of lactating cows was expressed as = 0.0104 TDN/$W^{1.00}$/24 hr, or maintenance of lactating cows = 0.174 Mcal ME/$W^{0.75}$/24 hr, and for 4% fat-corrected milk = 1.122 Mcal ME per kilogram of 4% fat-corrected milk. The results showed that the TDN requirement of lactating cows is approximately 27% greater than for those that are not lactating. The energy in the milk, produced by the cow in proportion to the ME consumed in excess of her nonlactating requirement was 34%. Dairy heifers reared on a subnormal amount of feed produced less milk as cows than their twins that were grown with normal feed intake (Swanson and Hinton, 1964). Anestrus and delay in puberty were associated with a low plane of nutrition in beef cows by Wiltbank et al. (1965). Buchanan-Smith et al. (1964) observed a "triggering effect" on the incidence of estrus when an all-concentrate diet was consumed. Overfat beef heifers (Arnett, 1963) and dairy heifers (Swanson, 1960) had decreases in milk yield.

Bond and Wiltbank (1970) conducted studies with beef females to observe the effects of feeding diets of different protein and energy levels on the body weight of the dam and the milk production and growth of calves, and to evaluate continuous versus compensatory growth and an all-concentrate versus a conventional diet on the estrous cycle and conception rate. First calves born to heifers fed a low-energy diet weighed less at birth, but the level of protein fed the cows had no effect on the birth weight of calves. However, heifers on low-protein and low-energy diets weighed less and gave less milk during the first lactation than those fed higher levels of nutrients. No effect on the estrous cycles and conception rates of heifers fed for a compensatory growth rate was observed with either all-concentrate or conventional diets.

Warnick et al. (1965) observed that heifers raised on clover-grass pastures had a 96% calving rate compared to 81% for those raised on grass pastures. The average percentage of protein in the grass pasture was 7.1% from February through September versus 2.52% during the winter months, while corresponding protein levels for the clover-grass pastures were 12.3 and 9.14%, respectively. Heifers on the grass pastures were divided into two groups; one group was supplemented during the winter with 2.5 lb of protein supplement and the other group with 1.0 lb of protein supplement per day. Essentially no difference was observed in the calving performance of heifers on grass due to protein intake during the winter. However, Bedrak *et al.* (1969) found that heifers that consumed 0.13 kg crude protein daily had delayed puberty, long anestrous periods, and failure to show estrus and to ovulate.

Chapman *et al.* (1965) observed that beef cows fed an average of 5 lb of blackstrap molasses per day had higher conception rates and produced heavier, higher-grading calves than cows receiving no supplement on pasture. Seasonal feeding of molasses for 133 days during the winter months resulted in about the same performance by cows as continuous molasses supplementation on pasture.

The plane of energy offered Angus heifers markedly affected the pregnancy rates, and the level of nutrition postweaning had more influence on reproductive performance than did the preweaning gain (Hill and Godley, 1974). Pendlum *et al.* (1977) studied the relationship of growth and conception rates of Hereford heifer calves to the level of dietary energy with diets that contained none, 1.36 kg, or 2.72 kg ground shelled corn per head daily. The control group, which was fed only corn silage supplemented with protein, vitamins, and minerals, was adequate in energy for conception rates equivalent to those of the groups supplemented with energy, but the supplementary corn provided energy for fatter carcasses when the heifers were slaughtered at 18 months of age.

Ferrell *et al.* (1976) studied the utilization of dietary energy during the pregnancy of Hereford heifers. The fasting heat production of pregnant heifers estimated from carbon dioxide production was 500, 698, 1779, and 4613 kcal/day greater at 120, 160, 200, and 240 days of gestation, respectively, than that of corresponding nonpregnant heifers. Pregnant and nonpregnant heifers were fed either a high-feed (215 kcal ME/$W^{0.75}$/day) or a low-feed (150 kcal ME/$W^{0.75}$/day) intake in a comparative slaughter trial in which pregnant and nonpregnant heifers at both energy intake levels were slaughtered at times corresponding to 134, 189, 237, and 264 days of gestation for pregnant heifers. The nonpregnant heifers had a fasting heat production of 68 kcal/$W^{0.75}$/day and utilized ME with an efficiency of 73% for maintenance and 39% for gain. Efficiencies of ME utilization for energy retention in the gravid uterus, conceptus, and fetus were 14, 12.5 and 12.2%, respectively. Daily ME requirements for pregnancy were calculated to be 237, 1021, 3264, and 8336 kcal on days 100, 160, 220, and 280 of gestation, respectively (Ferrell *et al.*, 1976).

The level of nutrition affects the age at which heifers reach puberty. Short and Bellows (1971) fed British crossbred heifers during winter on low, medium, and high levels of nutrition, which resulted in an age of puberty of 433, 411, and 338 days, respectively. The body weights at which the heifers reached puberty when fed the low, medium, and high levels of nutrition were 238, 248, and 259 kg, respectively. Lemenager *et al.* (1978b) supplemented heifers during winter with

0, 1.22, and 2.54 kg of ground ear corn daily and observed conception rates of 69.2, 73.9, and 83.5% for the dietary groups, respectively.

High energy intake by cows following calving results in greater conception rates, as was demonstrated when they were fed two levels of TDN the last 140 days of gestation (Wiltbank, 1973). After the cows calved, each TDN level group was subdivided into two subgroups: 4 and 2 kg versus 9.3 and 3.6 kg TDN/day. Conception rates of the various groups were as follows: moderate-moderate, 95%; low-moderate, 95%; moderate-low, 77%; and low-low, 20%. Only 10% of the cows fed the low-low level of TDN got pregnant during the first 20 days of breeding compared to 43% of those fed the moderate-moderate amount of TDN.

Estimates of efficiency of ME utilization for pregnancy vary considerably but are generally low, in the range of 10–20% (Graham, 1964; Moe and Tyrrell, 1971; Rattray *et al.*, 1974). It has been well documented that there is a large increase in heat production during gestation; the source of this heat increment is not clear (Brody, 1945; Graham, 1964; Moe and Tyrrell, 1971; Rattray *et al.*, 1974).

Meacham *et al.* (1964) fed beef bulls diets that contained 8.09, 5.10, and 1.35% crude protein for 84, 112, and 170 days, respectively, and observed that the gross weights of sex glands and organs were reduced by protein deficiency in proportion to decreases in body weight. Histological examination showed a decrease in the size and development of seminiferous tubules and ducts, as well as reduced thickness of epithelium in the testes, epididymides, seminal vesicles, and the Cowper's and prostate glands.

## X. CONDITIONS AFFECTING DE AND ME OF DIETS

Particle size of forage may markedly influence the effect of increasing feed intake on DE (Blaxter and Graham, 1956). Digestible energy in forage with a particle size of 1 cm or larger is affected only slightly, but DE decreases with increasing inputs of finely ground forage. The effect of level of intake on dietary DE is most evident with high-producing dairy cows, which may consume several times their maintenance requirements. Cows that ingested seven different forage-concentrate diets at levels of up to five times the maintenance requirement had DE values of diets depressed by 2.1–6.2%, with an average of 4%, per each maintenance equivalent intake increment (Moe *et al.*, 1965; Reid *et al.*, 1966). Lactating cows fed mixed diets that contained 50% ground corn or ground barley had reductions in certain composition fractions per multiple of ME intake, as follows: hemicellulose, 8.2%; cellulose, 8.1%; and soluble cell contents, 3.2% (Tyrrell and Moe, 1972).

Wheeler et al. (1975) found that the digestibility of starch ranged from 84.7 to 88.1% at feed intake levels of 2.5 to 3.2 times maintenance, but values of 96.2 to 96.8 were obtained at maintenance feed intake. The addition of 2.17% limestone to a complete diet fed milking cows at 3.5% of their live weight increased DE from 65.7 to 69.0%, starch from 85.9 to 94.6%, cell wall constituents from 51.3 to 55.1%, and crude protein from 65.7 to 67.8%.

Rapid body weight gain and high milk production require large energy intakes per unit of time; these inputs are generally obtained by feeding high-concentrate diets. With such diets, as intake increases, the rate of passage through the alimentary tract increases; this decreases cell wall digestion, since fermentation in the rumen is slow. Diets high in concentrates result in the production of VFA, which lowers the pH of the rumen and intestines to 5.0–6.0. At a pH of less than 6.0, cellulolytic bacterial action is inhibited (Slyter et al., 1970). Armstrong and Beever (1969) summarized the pH optima for several carbohydrates of the small intestine of ruminants as follows: pancreatic α-amylase, 6.9; intestinal maltase, 6.8–7.0; and oligo-1:6 glucosidase, 6.2–6.4. These pH levels indicate that ruminants consuming high-concentrate diets would have sufficient acidity to impair enzyme activity for digestion of cellulose, starch, and other nutrients.

The metabolizable energy value of feedstuffs is affected by the amount of methane generated with diets. Blaxter and Clapperton (1965) observed ruminants to lose 5–12% of dietary energy as methane. Inhibiting methane formation would be helpful only if other processes, such as digestibility of energy, synthesis of microbial protein, and accumulation of hydrogen, were not also inhibited (Czerkawski, 1971, 1972, 1974). Methane production is suppressed by a number of compounds such as chloroform, methylene chloride, carbon tetrachloride, long chain fatty acids, bromochloromethane, and others (Bauchop, 1967; Clapperton, 1977). These substances generally increase the proportion of propionic acid and reduce acetic acid in fermentation, thereby not only decreasing methane energy loss but increasing the efficiency of ME utilization in diets for growing-finishing cattle (Reid et al., 1980). The benefits of methane suppressants would be expected to be different for lactating cattle than for growing and finishing ruminants. However, methane depressants are not always effective and have not yet found practical application in increasing DE and ME utilization.

The requirement of ME for maintenance and the utilization of ME for certain production functions by ruminants are influenced by environmental conditions such as temperature, humidity and wind, extent of physical activity, sex, nature of the absorbed products of digestion, and a variety of dietary characteristics. Insufficient amounts of protein, certain vitamins, and mineral elements in diets will reduce the efficiency of ME utilization, both for maintenance and for production. Grazing ruminants require 10–90% more maintenance ME than those confined to stalls (Reid et al., 1958), and the energy cost of locomotion increases with increasing gradients of locomotion (Clapperton, 1964; Ribeiro et al., 1977).

The maintenance requirement for ME has been found to increase with ambient temperature changes above and below the zone of thermal comfort (Joyce and Blaxter, 1964) and with increases in dietary protein above the requirement (Garrett, 1970; Tyrrell et al., 1970). In dairy cows, the ME expense of metabolizing amino acids consumed in excess of the requirement for protein was 7.2 kcal per gram of digestible nitrogen (Tyrrell et al., 1970). The proportion of GE that is ME influences the net efficiency with which ME is utilized. Van Es (1976) proposed that as the ME:GE ratio increases by 1%, the ME requirement for maintenance, milk production, and body weight gain in cattle should be decreased by about 0.4%.

## XI. EFFECT OF VFA ON ME

High heat increment of ruminants is associated with the synthesis and utilization of VFA. Infusing acetic, propionic, and butyric acids into the rumen of sheep fed a maintenance diet of hay demonstrated that the net efficiencies with which the ME of the VFAs infused singly was utilized for body energy gain were: acetic acid, 32.9%; propionic acid, 56.3%; and $n$-butyric acid, 61.9% (Armstrong and Blaxter, 1957; Armstrong et al., 1958). When glucose was infused into the rumen, abomasum, or jugular vein, the net efficiencies were found to be 54.5, 71.5, and 72.8%, respectively, indicating that rumen fermentation reduced the net efficiency of glucose utilization (Armstrong and Blaxter, 1961). These observations seemed to indicate clearly that increased proportions of propionic acid relative to acetic acid gave increased efficiency of ME utilization for gains in body weight.

However, Bull et al. (1970) failed to confirm that the ruminal ratio of propionic acid to acetic acid per se influenced the net utilization of ME for energy gain by sheep. Their diets were different from those of previous studies in that the hay fed was of very high quality (74% DE), feeds were finely ground and pelleted, and the acetic acid was administered as triacetin, which provided the glucose precursor glycerol. Orskov et al. (1978) infused molar ratios of 0.64:1 to 17:1 propionic and acetic acids into the rumen of sheep and observed the net utilization of ME for body weight gain to range from only 60 to 65%. They concluded that since the VFA proportions from practical diets are within the range that they worked with, the differences in utilization of ME could not be explained by differences in efficiency with which VFA are utilized.

## XII. EFFECT OF SEX ON UTILIZATION OF ENERGY

Rams and ewes were found to have equivalent ME requirements for maintenance, but net efficiency for ME utilization for body energy gain was higher for ewes (67%) than for rams (57%) (Bull *et al.*, 1970). Ewes weighing 20–50 kg contained 30–35% more fat and 3–17% less protein than rams and, during the 175-day feeding period, gained 30% more fat and 31% less protein than the rams. Similar observations were made with heifers, bulls, and steers (Reid *et al.*, 1980). Webster (1977) associated the greater leanness of bulls than steers with a significant increase in the daily loss of ME as heat. Steers given DES after 27 weeks of feeding were observed to have greater carcass gains, which consisted of 33% more protein, 76% more ash, and 4% more energy, but 1% less fat than pair-fed controls (Rumsey *et al.*, 1977).

Preston and Willis (1970) cited 30 reports demonstrating that bulls have a 10–15% faster growth rate than steers and 8 sources indicating that carcasses of bulls contain 5–13% less fat and 5–10% more lean than steer carcasses.

## XIII. RELATIVE EFFICIENCY OF FAT AND PROTEIN SYNTHESIS

The observations reported imply that, in the usual growth of animals, ME is utilized more efficiently in the net synthesis of fat than in the net synthesis of protein (Pullar and Webster, 1972; Reid *et al.*, 1980). However, this may depend on the level of feeding above maintenance and on ingredients in the diet of feedlot cattle. Utilizing the data presented by Shirley (1975a), it can be calculated that *ad libitum* feeding of corn-containing diets to feedlot steers resulted in the deposition of approximately two times more calories as fat per kilogram of feed than occurs with limited feeding of diets (61–67% of *ad libitum*), and also that DES decreased the efficiency of energy deposition as fat relative to energy deposition as protein. Milligan (1971) observed the efficiency of peptide bond synthesis to be only 3% and, due to turnover, the same protein was resynthesized a number of times. The cost of resynthesis, synthesis of nonessential amino acids, and transport of nutrients contribute to the low energetic efficiency of net protein synthesis for the animal. Geay (1984) reviewed studies on dietary energy and protein utilization by European grown cattle. He discussed differences between the calorimetric and slaughter techniques, but concluded that both techniques can correctly describe the effect of breed, sex, weight, or daily gain on energy retained and its distribution between fat and protein deposition.

## XIV. DE COST OF PROTEIN PRODUCTION

Since the DE values of the usual diets vary among different species of animals, DE serves as a biological equating parameter of dietary energy for the various kinds of livestock. The outputs in grams of protein in various food products of animal origin per megacalorie of DE ingested for several species are presented in Table 5.15.

## TABLE 5.15.

### Overall Efficiency with Which Animals Produce Food Protein[a]

| Food product | Level of output and(or) intensity of production | Protein production efficiency (g protein/Mcal DE) |
|---|---|---|
| Eggs | 200 eggs/year | 10.1 |
| | 236 eggs/year[b] | 12.6 |
| | 250 eggs/year | 13.7 |
| Broiler | 1.6 kg/12 weeks; 3.0 kg feed/1 kg gain[b] | 11.9 |
| | 1.6 kg/10 weeks; 2.5 kg feed/1 kg gain | 13.7 |
| | 1.6 kg/8 weeks; 2.1 kg feed/1 kg gain | 15.9 |
| Pork | 91 kg/month; 6 kg feed/1 kg gain | 5.0 |
| | 91 kg/6 months; 4 kg feed/1 kg gain[b] | 6.1 |
| | 91 kg/month; 2.5 kg feed/1 kg gain | 8.7 |
| | Biological limit (?); 2 kg feed/gain; no losses | 12.1 |
| Milk | 3,600 kg/year; no concentrates | 10.5 |
| | 4,944 kg/year; 22% of energy as concentrates[b] | 12.4 |
| | 5,400 kg/year; 25% of energy as concentrates | 12.8 |
| | 9,072 kg/year; 50% as energy as concentrates | 16.3 |
| | 13,608 kg/year; 65% of energy as concentrates | 20.5 |
| Beef | 500 kg/15 months; 8 kg feed/1 kg gain[b] | 2.3 |
| | 500 kg/12 months; 5 kg feed/1 kg gain | 3.2 |
| | Highly intensive system; no losses | 4.1 |

[a]From Reid and White (1978a).
[b]Data represent approximately average U.S. management conditions.

Efficiency of production in animals increases with increasing inputs of DE per unit of time. The efficiency of protein production is shown for each of several rates of egg, broiler, pork, milk, and beef production in Table 5.15. Since the DE requirement for maintenance is constant per unit of $W^{0.75}$/day, the increasing inputs of DE give increases in proportion to the total energy for production. Efficiency is greatest for protein production as milk, broiler meat, and eggs, intermediate for pork, and lowest for beef. Protein production efficiency increases with intensity of DE input and with decreases in losses from infertility and mortality. Efficiency of protein production is highest in uncastrated males, intermediate in castrated males, and lowest in females.

## XV. PREFORMED PROTEIN SOURCES IN FINISHING DIETS

A number of investigators have suggested that dietary proteins differ in their degree of degradation in the rumen (McDonald, 1954; Ely *et al.*, 1967; Hume, 1970), and this influences the quantity and quality of amino acids reaching the abomasum (Little *et al.*, 1968; Potter *et al.*, 1969; Amos *et al.*, 1971). For this reason, animal performance may be affected by the extent to which dietary protein escapes degradation in the rumen. Burris *et al.* (1974b) conducted a study with yearling steers fed corn-silage with protein supplements of varying solubility (corn, soybean, fish, and linseed protein). Soybean meal, fish meal, and linseed meal provided 23.5, 21.2, and 24.7% of the total daily dietary crude protein intake. Glycine, valine, leucine, lysine, arginine, and total amino acids were highest in the plasma on day 69 for steers fed fish meal, and total amino acids were higher in steers fed fish meal than in those fed soybean meal. Daily weight gains were lowest for steers fed corn as the only protein source. Carcass observations were not altered by the source of supplemental protein.

Loerch and Berger (1981) fed three slowly degraded protein sources—blood meal, meat and bone meal, and dehydrated alfalfa—to steers and lambs and compared them to soybean meal as supplemental protein sources. No differences in feed intake or feed efficiency were observed between steers or between lambs consuming diets containing different protein sources. Horton and Nicholson (1981) fed yearling steers diets composed of barley grain and wheat straw (35.7–40.5%) that contained no protein supplement, 2.15% urea or 19% soybean meal, and a diet containing 34% barley and 60% alfalfa meal. Digestibilities of organic matter, crude protein, and acid detergent fiber were increased by urea and soybean meal but not by alfalfa meal. Urea supplementation had no significant effect on daily gains or feed efficiency compared to increased gains of 8 and 24% for those supplemented with soybean meal and alfalfa meal, respectively. A study was conducted by Utley and McCormick (1980) in which whole shelled corn-based steer finishing diets supplemented with a pelleted 50% cruce protein concentrate containing urea and cottonseed meal, whole raw cottonseed, or high-quality dehydrated Bermuda grass pellets were evaluated. Nitrogen balance data were similar for steers in all three treatments. Dry matter digestibility and calculated TDN were less for diets that contained dehydrated Bermuda grass pellets, but in a 120-day feeding trial with yearling steers, those fed the Bermuda grass–containing diet gained faster. Average daily gains with heifers in a similar trial were not different due to the diet.

## XVI. EFFECT OF WITHDRAWAL OF PROTEIN IN DIETS

The NRC (1970) recommended 10.9% crude protein for finishing steers weighing 350 kg and 10.5% for steers weighing 400 kg (on a dry matter basis). Preston and Cahill (1972, 1973, 1974) decreased the crude protein levels of diets fed steers weighing 345 kg to approximately 8.4% and observed their performance to be equivalent to that of steers consuming supplemental protein for the entire feeding period. However, Preston *et al.* (1975) later observed that when supplemental protein was withdrawn at 84 days, steer feedlot performance was depressed due to a lower level of protein in corn than in previous years. Heifers fed no supplemental protein for the last 91 days of a 161-day finishing period performed as well as those fed supplemental crude protein throughout the trial.

Thomas et al. (1976b) conducted two feedlot trials with Hereford calves to evaluate the effect of feeding supplemental protein for different lengths of time during the finishing period. When supplemental protein was withdrawn from steers at 70 days (368 kg body weight) in the first trial, the daily gain was 0.17 kg compared to 1.03 kg for the next 84 days, with feed conversions ranging from 28.7 to 6.67 kg per kilogram of weight gain. In the second trial, steers withdrawn from supplementary protein at 56 days (397 kg body weight) gained less and required 20% more feed per kilogram of gain in the first 28-day weight period after protein withdrawal. However, feeding no supplemental protein during the final 88 days of this trial had no effect on the feedlot performance of steers. The investigators concluded that the primary effect of withdrawal of supplemental protein in diets fed growing-finishing steer calves was to depress feed intake (Thomas et al., 1976b).

## XVII. INSOLUBILITY OF PROTEIN IN RUMINANT FEEDS

The degree of degradability of protein during rumen fermentation has long been recognized as a factor affecting protein and nonprotein nitrogen (NPN) utilization by ruminants (Pearson and Smith, 1943). Burroughs et al. (1975a) assigned rumen protein degradation values for 100 feedstuffs for utilization in calculations of metabolizable protein and urea fermentation potentials of diets for cattle. The system for making decisions on NPN use in ruminant diets based on ammonia accumulation in the rumen proposed by Satter and Roffler (1975) is dependent on the solubility and degradation of proteins. The importance of separating protein into degradable and undegradable fractions was recognized by Chalupa (1975). Undegradability values for selecting protein sources or treating proteins with heat or formaldehyde require a measurement of insolubility.

Different assays of protein insolubility have been utilized, but no single method has been generally accepted. Little et al. (1963) used cold water and 0.02 N sodium hydroxide (NaOH). The salt solution of Burroughs et al. (1950) was utilized by Wohlt et al. (1973) and Crawford et al. (1978). Other solvents have been utilized, such as phosphate buffer (MacRae, 1976) and autoclaved rumen fluid (Little et al., 1963; Wohlt et al., 1973; Crawford et al., 1978). In vitro ammonia accumulation was used to determine protein degradation by Little et al. (1963). Investigators with silage have utilized cold water (Bergen et al., 1974) and boiling water (McDonald et al., 1960; Goering and van Soest, 1970).

Waldo and Goering (1979) determined protein insolubility by four methods (autoclaved rumen fluid, hot water, Burroughs' solution, and sodium chloride) on 15 feeds (oats, wheat middlings, malt sprouts, corn gluten feed, wheat bran, soybean meal, cottonseed meal, barley, dehydrated alfalfa meal, hominy feed, corn, corn distiller's dried grains, brewer's dried grains, dried beet pulp, and corn gluten meal). The mean insolubilities for feeds across all methods ranged from 49% of total nitrogen for corn gluten feed to 94% for dried beet pulp. Mean insolubilities for the four methods across all feeds ranged from 67% for autoclaved rumen fluid to 80% for Burroughs' solution. It was concluded that (1) significant interactions and variable within-feed correlations imply that the four insolubility methods used must vary in their usefulness for predicting the undegradability of protein in the rumen and that (2) autoclaved rumen fluid may be superior to the other three methods tested, but that its common use would require standardization of pH and time of solubilization.

Mahadevan et al. (1980) incubated various soluble and insoluble proteins at 37°C with partially purified protease in potassium phosphate buffer, pH 7.6, for 2–18 hr and determined the amino acids liberated. The results showed that (1) although soluble, the serum albumin and ribonuclease A were resistant to hydrolysis; (2) soluble and insoluble proteins of soybean meal were hydrolyzed at almost identical rates; (3) soluble proteins of soybean meal, rapeseed meal, and casein were hydrolyzed at different rates; and (4) treatment of resistant proteins (serum albumin, ribonuclease A and insoluble fish meal, and rapeseed meal protein) with mercaptoethanol in 8 $M$ urea or oxidation with performic acid rendered these proteins susceptible to hydrolysis. It was concluded that the solubility or insolubility of a protein is not in itself an indication of the protein's resistance or susceptibility to hydrolysis by bacterial protease in the rumen, and that structural characteristics of the proteins as disulfide bonds are likely determining factors.

## XVIII. STARCH UTILIZATION BY RUMINANTS

Forage-based programs of feeding ruminants are dependent largely on microbial digestion of cellulose and hemicellulose as sources of energy. About 90% of digestion of forage occurs in the rumen and the remainder primarily in the lower intestine. The major portion of the protein and energy needs for forage-fed ruminants is supplied by the microbial cells of the rumen and the VFA byproducts of bacterial and protozoal activity. Efforts to increase the efficiency of meat, milk, and wool production of ruminants have led to feeding more concentrates that contain high levels of starch. The starch is subjected initially to fermentation in the rumen, which results in microbial growth and VFA production. Unfermented starch from the rumen is then subjected to degradation to glucose by enzymes in the intestines.

## A. Consumption of Starch in Grains and Silage

According to the review by Waldo (1973), total consumption of grain starch by ruminants in 1970, expressed as millions of metric tons, was as follows: cattle on feed, 19; other beef cattle, 5; milk cows, 10; other dairy cattle, 0.75; and sheep, 0.25. In 1970, corn silage production was estimated to be 85 million metric tons, which provided an additional 9.2 million metric tons of corn starch. This was fed as follows (millions of metric tons): cattle on feed, 1.9; other beef cattle, 2.2; milk cows, 4.6; other dairy cattle, 0.5; and sheep, 0.03. Similar estimations with sorghum silage indicated that there were 0.6 million metric tons of sorghum starch available yearly. Grains vary considerably in their percentage of starch on a dry basis, as follows: corn, dent, yellow, all analyses, 71.9 (range, 63.7–78.4); sorghum, all analyses, 63.8 (range, 54.2–71.1); oats, all analyses, 44.7 (range 34.4–70.0); and barley, all analyses, 64.6 (range, 52.2–71.7) (NRC, 1958).

## B. Starch Digestion in the Entire Alimentary Tract

Waldo (1973) did recalculations of published data on 51 starch digestion trials on corn, barley, and sorghum to get them in comparable form. The grains had starch digestibilities of 99 ± 1.2% in the total ruminant alimentary tract and indicated that very little starch escaped digestion with cattle and sheep. Total tract digestibility of starch was not affected in sheep fed corn having different particle sizes (Orskov et al., 1969; Adeeb et al., 1971). However, feeding of larger particle sizes of corn to cattle decreased starch digestibility (Adeeb et al., 1971). The DE of a 60% corn diet fed to dairy cattle increased from 2.56 with whole grain to 2.94 Mcal per kilogram of dry matter when corn was ground through a 0.64-cm screen (Moe and Tyrrell, 1972).

## C. Starch Digestion in Various Segments of the Alimentary Tract

Most investigations with concentrate diets high in starch content have been directed to digestibility of starch in the entire tract. However, there has been considerable interest in determining how much starch is utilized in the various segments of the alimentary tract due to significant amounts of starch containing feedstuffs bypassing the rumen. Larsen et al. (1956) studied the digestion of starch, glucose, maltose, and corn in ruminal and fecal fistulated dairy calves when the carbohydrates were introduced into the omaso-abomasal cavity. Digestibility data and blood-reducing sugar levels indicated that glucose was readily absorbed and maltose was readily hydrolyzed and absorbed, but there was little digestion of starch and carbohydrates of corn.

Karr et al. (1966) fed diets to abomasal cannulated steers that contained 20, 40, 60, and 80% corn (diets contained ground corn, alfalfa hay, soybean meal, animal fat, and minerals, with about 12% crude protein) in a digestibility trial. The effect of varying levels of starch intake on starch digestibility in various segments of the alimentary tract is shown in Table 5.16.

Recovery of starch from the abomasal contents indicated that considerable starch escaped ruminal fermentation, and the amount was influenced by the level in the diet. The data indicate that efficiency of starch digestion in the rumen decreased as the daily starch intake exceeded 2000 g. The amount of starch that reached the posterior ileum was relatively small when up to 60% corn was in the diet but was larger for the diet that contained 80% corn. This difference was a

**TABLE 5.16.**

**Effect of Varying Levels of Dietary Starch on Daily Starch Recovery and Apparent Digestion in Various Areas of the Digestive Tract**[a]

| Item | Corn (%)[b] | | | |
|---|---|---|---|---|
| | 20 | 40 | 60 | 80 |
| Daily consumption, g[c] | | | | |
| Total diet | 5448 | 5448 | 5061 | 4256 |
| Starch | 1002 | 1948 | 2438 | 2684 |
| Starch recovered, g | | | | |
| Abomasum | 357[1] | 542[1] | 778[1,2] | 982[2] |
| Posterior ileum | 26 | 79 | 169 | 358[4] |
| Feces | 12[1] | 17[1] | 40[2] | 62[3] |
| Starch disappearance, g | | | | |
| Rumen | 640[4] | 1423 | 1592 | 1655 |
| Small intestine | 331 | 463 | 609 | 624 |
| Large intestine | 14[1] | 61[1,2] | 129[2] | 296[3] |
| Apparent starch digestion, % | | | | |
| Rumen | 64.2 | 72.6 | 67.2 | 63.0 |
| Small intestine | 92.7[1] | 85.4[1] | 78.3[1,2] | 64.3[3] |
| Large intestine | 53.8[4] | 78.5 | 76.3 | 82.7 |
| Mouth to: | | | | |
| Posterior ileum | 97.2 | 96.0 | 93.4 | 85.3[3] |
| Rectum | 98.8 | 99.0 | 98.4 | 97.7 |

[a]Karr et al. (1966).
[b]Means on the same line bearing different superscript numbers are different ($P < .05$).
[c]Averages for both intestinal and abomasal fistulated steers.

reflection of the amount of starch digested in the small intestine. Apparently, there is a limit to the amount of digestion of starch in the small intestine, since an average of 358 g passed this segment daily on the 80% corn diet. More starch was digested in the large intestine of steers fed the 60 and 80% corn diets, compensating partially for the less efficient digestion in the small intestine. However, more starch was excreted in the feces with the 60 and 80% corn diets.

Waldo (1973) summarized the trials of seven groups of investigators on the digestibility values of corn starch in sheep and cattle fed grain that was ground, cracked, or flaked. Thirty observations on corn starch digestibility at the abomasum or duodenum averaged 78 ± 12.5%. Waldo et al. (1971a) found in one trial that different levels of ground corn in pelleted diets with alfalfa and soybean meal gave 99% starch digestibility in the entire tract of cattle but only 58.7, 59.4, 50.0, and 73.5% in the rumen for the 20, 40, 50, and 80% corn diets, respectively. In another trial with a different lot of corn, corresponding digestibilities for the various levels of corn in the diets were 99% for the entire tract and 89.6, 84.7, 76.2, and 84.2% in the rumen. Ground corn and flaked corn gave digestibilities of starch in the rumen of 82.8 and 95.1%, respectively (Orskov et al., 1969; Beever et al., 1970).

Galyean et al. (1976) utilized steers fitted with rumen and abomasal cannulae to determine the site and extent of starch digestion when 78% corn grain–containing diets were fed. The corn had been either dry rolled, steam flaked, high-moisture harvested-ground prior to ensiling, or high-moisture harvested-whole corn treated with propionic acid prior to storage. Digestibilities of starch in the rumen were 89.3, 82.9, 77.8, and 62.8% for the high-moisture harvested-whole grain treated with propionic acid prior to storage, respectively. Total tract digestibilities of starch in the corresponding corn diets were 99.1, 99.1, 96.3, and 95.8%, respectively. There were no significant differences in intestinal tract digestibility of starch. Total dry matter and organic matter digestibilities in the rumen, intestine, and total alimentary tract followed the patterns of starch.

Galyean et al. (1981b) reported that dry rolled corn grain, when sieved to give particle sizes of 6000, 3000, 1500, and 750 μm and incubated in rumen fluid, showed little differences in starch and dry matter digestibilities at the 6000- and 3000-μm particle sizes but increased with the 1500- and 750-μm particle sizes. When corn was steam flaked and then ground, much higher digestibilities were observed for starch and dry matter within each particle size than with dry rolled corn. Starch disappearance was much lower in dry ground than in high-moisture corn within each particle size. Galyean et al. (1979) fed crossbred steers fitted with rumen cannulae 84% corn-containing diets at 1.00, 1.33, 1.67, and 2.00 times maintenance intakes and determined the site and extent of starch and dry matter digestion. Total alimentary tract starch digestion was greater for steers fed at maintenance (99.6%) than for those fed 1.67 maintenance (93.8%) and 2.00 times maintenance (90.4%) intakes of diet. Less starch was digested in the rumen as intake increased from maintenance (94.5%) to 2.00 times maintenance (89.6%).

Waldo (1973) summarized 23 observations made by five groups of investigators on barley starch digestion at the abomasum or duodenum of sheep and cattle fed whole, ground, rolled, or pelleted barley diets. Mean digestion of starch was 94 ± 2.4%, indicating that little variation occurred with different lots of barley, type of ruminant fed, processing method, or variation in the level of barley starch in the diet. Mean sorghum starch digestibility was 76 ± 22.4% at the abomasum in eight observations of sheep and cattle fed diets that contained sorghum grain steamed at 3.5 kg per square centimeter of pressure, steamed at atmospheric pressure, ground, reconstituted, steam flaked, or micronized (Waldo, 1973). Sorghum grains were reported to vary in ruminal starch digestion from 80 to 75, 68, and 48% for floury, waxy, normal, and corneous endosperm types, respectively (Samford et al., 1971).

Holmes et al. (1970) compared starch digestion in abomasal fistulated cattle fed diets containing 80% milo that had been steamed at atmospheric pressure or at 3.5 kg per square centimeter of pressure before rolling. About 90% ruminal starch digestion was estimated for the steamed grain and 95% for pressure-steamed grain. Corresponding total digestibility values of 97.3 and 97.6% were observed for starch in the two milo treatments. These data indicate that moist heat treatment of sorghum grain makes it equivalent to barley in starch digestibility.

## D. Effect of Starch on Amylase, Maltase, and pH

The postruminal digestion of starch is of considerable importance when concentrate diets are fed, since the amount of starch that escapes ruminal fermentation increases as starch consumption increases. Trends toward reduced digestibility of starch in the small intestine occur with increased starch consumption (Karr et al., 1966; Galyean et al., 1979). This depression of starch digestion in the small intestine may involve inadequate pancreatic amylase production (Karr et al., 1966), a suboptimal pH for amylase activity (Wheeler and Noller, 1977), inadequate maltase activity (Mayes and Orskov, 1974), inadequate glucose absorption (Orskov et al., 1971b), structural effects of the grain and/or starch kernel (Harbers, 1975; Galyean et al., 1979), and increased rate of passage on full feeding (Sutton, 1971).

J. R. Russell et al. (1981) reported that the pH optimum of pancreatic $\alpha$-amylase activity was 6.9 and that of intestinal maltase activity 5.8. Intestinal maltase activity was highest in the jejunum of steers and decreased toward the ileum. Increasing starch levels in corn diets fed steers at one, two, or three times maintenance levels did not increase maltase activity in various sections of the small intestine. Digesta pH was higher at locations farther down the small intestine in steers fed an alfalfa-hay diet than in steers fed concentrate diets. It was concluded that pH dropped along the intestine due to fermentation of the concentrate digesta.

## XIX. DIETARY FAT FOR RUMINANTS

Fat is frequently added at levels of 3–4% in finishing diets for cattle to control dust, decrease the segregation of micronutrients, increase caloric density, or protect protein from ruminal degradation. Levels of fat above 5% generally result in decreased animal performance (NRC, 1976a). The problem with higher levels of dietary fat is sometimes presumed to be largely due to interference with rumen microbes functioning in an aqueous medium. Addition of fat in ruminant diets has produced conflicting effects ranging from improved ether extract digestibility (Erwin *et al.*, 1956; Bradley *et al.*, 1966) to no effect on nutrient digestibility (Esplin *et al.*, 1963; Hatch *et al.*, 1972) to depressed digestibility of nutrients other than fat (Erwin *et al.*, 1956; Bradley *et al.*, 1966; Dyer *et al.*, 1957; Kowalczyk *et al.*, 1977).

Additions of fat to diets of finishing cattle gave increased gain (Erwin *et al.*, 1956; Bohman *et al.*, 1957). However, a number of investigators found that fat additions to diets resulted in no improvement in gains (Bradley *et al.*, 1966; Dyer *et al.*, 1957; Roberts and McKirdy, 1964), and others observed a depression in weight gain when fat was added to diets (Bradley *et al.*, 1966; Hatch *et al.*, 1972).

Inclusion of low levels of fats in diets fed ruminants depresses cellulose digestibility (Brooks *et al.*, 1954; Davidson and Woods, 1960). Davidson and Woods (1963) showed that depression of cellulose digestion in diets fed lambs could be partially alleviated by increasing the calcium level in the diet. Polan *et al.* (1970) reported that 2% fat ensiled with corn plant material decreased dry matter intake by dairy cows and decreased calcium and magnesium digestibility.

Johnson and McClure (1973) determined the effects of 0, 4, 8, and 12% partially hydrolyzed animal and vegetable fat additions to corn silage, with and without 1% limestone, on the intake and digestibilities of diets in sheep and intake trials with steers. Fat additions did not alter fermentation of silage. Digestibilities of dry matter, organic matter, and cellulose in the 4% fat silage was lower than those of the control silage, but digestibilities of the 8 and 12% fat silages were not different from the control values. The highest digestibilities were with the 12% fat–1% limestone silage. The amount of fecal soaps increased with increasing fat levels in the silage but was not affected by the limestone. Voluntary feed intake by steers was decreased by fat additions, and the lowest intake occurred with 12% fat silage; however, 1% limestone improved intake.

Dinius *et al.* (1974) continuously infused 280 g of safflower oil daily via abomasal cannula, offered 280 g of oil daily mixed in the diet, or gave no oil in diets fed growing steers. Perianal fat was 8.8, 1.9, and 1.8% linoleic acid and plasma cholesterol was 268, 138, and 103 mg/100 ml in steers on the three treatments, respectively. Cooked steaks of steers on the various treatments had equivalent aroma, flavor, and desirabilities of fat and lean on taste panel tests.

Since bile salts are known to emulsify fat and aid in its absorption, Perry *et al.* (1976b) evaluated the emulsifying action of lecithin in added fat diets fed beef cattle and lambs. The feeding of 192 g feed-grade fat per head daily in high-moisture corn-corn silage diets did not improve the gains or feed efficiency of beef cattle. Feeding lambs 3% feed-grade fat in a dry diet resulted in depressed crude fiber digestibility. Prior treatment of the fat with lecithin had no effect on crude fiber digestibility but depressed protein digestibility. These observations with lambs were confirmed by Perry and Stewart (1979). They also reported that cattle fed diets that contained 3% added fat with two levels of shelled corn (11% corn stored in a bin and 26% ensiled high-moisture corn) and two levels of corn silage (64% September harvest and 50% October harvest) had depressed protein and ash digestibilities but increased fiber digestibility.

Buchanan-Smith *et al.* (1974) fed finishing steers diets containing alfalfa-brome grass at a constant level with varying dietary treatments of 0 or 5% animal fat. Soybean meal or urea, and zero or added MHA, or L-lysine and ground shelled corn were utilized to maintain a constant gross energy : protein ratio of approximately 0.31 Mcal GE per percent of crude protein. Animal fat in soybean meal or urea-containing diets had no effect on weight gains or feed efficiency. A nitrogen source × animal fat interaction on feed intake resulted in fat causing the intake of soybean meal diets to be increased by 10% but depressing the intake of urea diets by 5%. An improvement in daily gains by feeding soybean meal compared to urea was greater for diets in which the fat and amino acid supplements were provided separately, rather than in combination or omitted. Diets supplemented with fat resulted in greater carcass backfat and liver fat.

## A. Encapsulation of Fat

Microbes in the rumen are able to hydrolyze glyceridic lipids and hydrogenate unsaturated fatty acids when limited levels of fat are present in diets. High levels of fat are believed to delay the action of microbes in the aqueous medium of the rumen. In order to bypass the rumen microbes, high-fat supplements for beef cattle have been encapsulated with a formaldehyde–protein complex (Cook and Scott, 1970; Scott *et al.*, 1971). Scott *et al.* (1970) observed that a formaldehyde-treated, spray-dried homogenate of equal parts of lineseed oil and casein was not degraded by rumen fermentation, as measured by an increase in polyunsaturated fatty acids in the milk fat of cows. A formaldehyde-treated, spray-dried casein-safflower oil emulsion was effective in increasing the linoleic acid in depot fat of steers (Faichney et al., 1972).

Cuitun et al. (1975) fed finishing steers diets that contained either (1) 6% formaldehyde-treated casein, (2) 6% formaldehyde-treated casein plus 6% safflower oil, or (3) 12% formaldehyde-treated, spray-dried casein-safflower oil homogenate. The diet containing 6% formaldehyde-treated casein plus 6% safflower oil had the lowest digestibilities of dry matter, lipids, and gross energy. Safflower oil digestibility calculated by difference was 41% for the diet containing 6% formaldehyde-treated casein plus 6% safflower oil compared to 69% digestibility for the diet containing 12% formaldehyde-treated, spray-dried casein-safflower oil homogenate. The diet without added safflower oil had the most DE, dry matter, feed intakes, and daily weight gains of the three dietary treatments. Steers fed the protected safflower oil diet during finishing had an average lipid level of about 930 mg per 100 ml serum, which was 2.1 times greater than that of the control steers and 1.7 times greater than that of the steers supplemented with unprotected safflower oil (Dryden et al., 1975).

Garrett et al. (1976) conducted a comparative slaughter trial with beef steers to determine the energy utilization of diets containing polyunsaturated oils (PO) or tallow (PT) encapsulated with a formaldehyde–protein complex. The supplements replaced a portion of a high-energy barley-based feedlot diet (15–30% for PO and 15–25% for PT). The fat-supplemented steers had 30–32% carcass fat when slaughtered compared to 26.5% for controls and graded one-third grade higher than control steers. The efficiency of ME utilization for energy deposition was: basal, 44%; 15% PO, 47%; 15% PT, 48%; 30% PO, 53%; 25% PT, 59%. Cholesterol content of the meat and fat of steers was not influenced by the different levels and types of dietary fat. Dinius et al. (1978) fed steers either ground alfalfa or ground orchardgrass hay with 6% added animal fat or with a formaldehyde–protein–lipid complex added at a level to provide 6% dietary animal fat. Steers fed alfalfa gained more weight than those fed orchardgrass (1.12 versus 0.56 kg/day), but gains of steers were equivalent on the different fat treatments. Rhodes et al. (1978) fed prepubertal heifers coastal Bermuda grass hay plus concentrate containing zero or 50% of a microencapsulated tallow for 168 days. Protected lipid-fed heifers tended to be more efficient in converting energy to weight gain (7.7 versus 7.9 Mcal $NE_g$ per kilogram of gain) than controls, which required 13.7% more total feed per kilogram of gain. Fewer of the heifers that were fed lipids reached puberty than control heifers (34 versus 80%) while on the test.

McCartor and Smith (1978) grazed steers on winter pasture of wheat, oats, and ryegrass for 145 days prior to a 56-day comparative slaughter feedlot trial in which they were fed diets with and without protected tallow (0.8%). The control and tallow-containing diets had calculated $NE_g$ values of 6.03 and 5.15 Mcal/kg, respectively. Projected gains per steer were 0.98 and 0.87 kg/day, but actual gains were 0.92 and 1.10 kg/day, respectively. Steers fed protected tallow-containing diets had greater feed efficiency, a higher marbling score, and higher USDA grades than control steers.

Haaland et al. (1981) fed feedlot steers flaked corn and corn silage diets with either 0, 5, or 10% protected tallow adjusted to equivalent amounts of ME. Steer performance was superior with the 5% level of protected tallow compared to the 10% tallow level in regard to gain and feed efficiency. Percentages of carcasses that graded choice or higher did not differ among treatments. Specific gravity measurements of steers fed 0, 5, and 10% fat diets indicated 29.1, 30.7, and 30.4% empty body fat, respectively.

## B. Effect of Fat on Dietary Protein

Ruminal ammonia levels of sheep were reduced by the addition of fat to peanut meal and coconut meal (Jaysainghe, 1961). A coating of linseed meal with corn oil increased nitrogen retention, while a coating of coconut oil decreased it in lambs (Glen et al., 1977). Digestibilities of dry matter and cellulose were lower in diets containing linseed meal treated with coconut oil than in those containing equivalent amounts of corn oil or lard (0.3 kg oil per kilogram of linseed meal). Treatment of proteins with VFA decreased ammonia levels in the rumen of cattle and sheep (Atwal et al., 1974).

Stanton et al. (1981) evaluated the effects of adding tallow to cottonseed and soybean meals with heat and pressure or vacuum on ruminal ammonia concentrations of growing lambs. *In vitro* trials showed that pressure and steam heat treatment of the fat and meals reduced ammonia-nitrogen accumulation, but vacuum with dry heat had no similar effect. Dry matter digestibility and nitrogen retained by lambs tended to be reduced by diets containing soybean meal or 1% urea plus 15% tallow heated to 121°C at 1.36 atmospheres of pressure for 5 min.

## C. Fat in Diets Fed Lactating Cows

In a review, Emery (1978) concluded that protein in milk increased 0.015 percentage unit per megacalorie of increased daily energy intake. The increased protein concentration in milk obtained by feeding of concentrate diets was attributed to an increased supply of glucose to the mammary gland. Milk protein levels decreased when oil or fat was added to the diet. Numerous reports show that feeding high-fat diets to lactating cows decreases the percentage of milk protein (Storry et al., 1974; Banks et al., 1976; Bines and Hart, 1978; Anderson et al., 1979; Macleod et al., 1977; Sharma et al., 1978). Yield of milk is also depressed by high-fat levels in diets of lactating cows (Banks et al., 1976; Bines and Hart, 1978; Macleod et al., 1977; Storry et al., 1974).

Encapsulation of lipids in a formaldehyde-treated protein allows feeding of considerable amounts of fat in ruminant diets without direct adverse effects on rumen fermentation (Scott *et al.*, 1970; Scott and Hills, 1975). Due to their high energy value, it is believed that the encapsulated products can be utilized to increase energy consumption in ruminants when the energy density of conventional diets limits feed intake (Garrett *et al.*, 1976).

N. E. Smith *et al.* (1978) fed alfalfa hay, barley, corn, and soybean meal-based diets with 0, 15, or 30% fat supplement (40% tallow and 60% formaldehyde-treated soybean meal) to Holstein cows during the first 15 weeks of lactation. The calculated energy densities of the diets were 1.85, 1.85, and 2.15 Mcal $NE_{lactation}$ per kilogram of dry matter. Treatments had no effect on milk yield. Both protected tallow-containing diets resulted in increased yields of fat and fat-corrected milk and energy efficiency and decreased yields of solids-not-fat. Palmquist and Moser (1981) found that feeding fat in a protected supplement did not change the amount of milk or milk protein produced but did depress the quantity and concentration of milk fat. Total lipid concentration was increased but glucose and insulin were reduced in blood plasma by protected fat.

## D. Lipotropic Factors for Beef Cattle

Lipotropic factors that facilitate mobilization of fat from the liver into blood and adipose tissue are generally synthesized by microbes of the digestive tract in adequate amounts (Porter, 1961). However, high-concentrate diets have been tested in regard to the possibility that lipotropic factors may be needed at levels above the usual requirements. Choline supplementation of steer finishing diets gave improved performance (Dyer, 1969). Lipotropic function might be involved in the beneficial effects of methionine supplementation of ruminants (McLaren *et al.*, 1965; McCarty *et al.*, 1968; Mowat and Deelstra, 1972).

Smith *et al.* (1974) administered lipotropic factors (choline, m-inositol, folacin, and vitamin $B_{12}$) or an "anti-lipotropic factor" (niacin) intraperitoneally or intraruminally to steers at intervals during the finishing period. The cattle were fed a high-concentrate, barley-based diet. The lipotropic factors and niacin tended to improve body weight gains, and the lipotropic factors improved carass yield grade.

# 6

# Amino Acids

## I. INTRODUCTION

The general consensus has been that ruminants are not dependent on direct dietary sources of amino acids, since microbes of the rumen degrade both organic and inorganic nitrogen compounds and resynthesize them into microbial protein. Nevertheless, research indicates that ruminants do respond to amino acid supplementation. While microbial synthesis is beneficial from the standpoint of utilizing low-quality proteins and nonprotein nitrogen, the same microbes may reduce the biological value of high-quality proteins through degradation of essential amino acids. Research on protein nutrition of ruminants has been directed to upgrading the quality of amino acids that reach the abomasum. When available dietary sources of nitrogen are of low amino acid quality, research has been directed to encouraging rumen microbes to utilize it for the synthesis of fairly high-quality microbial protein. When the dietary source is of very high-quality protein, techniques have been directed to bypassing the rumen. Efforts to obtain maximum production of meat, milk, or wool demand more protein than can be provided by microbes. If this level of nitrogen is included in the ruminant diet, the excess may be degraded to ammonia and lost, if conditions are not favorable for bypassing the rumen in going to areas of absorption in the lower digestive tract.

All protein is not degraded in the rumen, as indicated by the fact that 40–60% zein bypasses the rumen of sheep (McDonald, 1954), showing that the amino acid composition of that portion would not be altered by rumen microbes. The amount of degradation depends on the solubility of protein in the rumen (Little *et al.*, 1963; Glimp *et al.*, 1967; Hudson *et al.*, 1970) or on the disulfide linkages of the constituent amino acids (Mahadevan *et al.*, 1979). Little *et al.* (1963) found unheated soybean meal, linseed meal, and casein to be rapidly converted to ammonia by *in vitro* rumen microbes in comparison to heated soybean meal, corn gluten, and zein. Sherrod and Tillman (1962) observed that heat treatments of soybean and cottonseed meals decreased protein digestibility but increased nitrogen retention in sheep. Peter et al. (1970) reported that soybean meal treated with either formaldehyde, tannic acid, glyoxal, or glutaraldehyde increased the rate at which protein bypassed the rumen and produced greater feed efficiency when fed to steers.

Burris et al. (1976) infused lysine continuously into the abomasum of Angus steers fed a high-urea diet at levels of 0, 12, 24, and 36 g/day. Maximum nitrogen retention (165% of infused nitrogen) occurred with 24 g lysine infusion per day, indicating that the effect was apparently due to lysine per se. Lyman et al. (1956) analyzed by microbiological assays 115 different feedstuffs for 10 essential amino acids. These feedstuffs included animal by-products, fermentation feeds, fish by-products, grains, grain by-products from milling and processing, oilseed meals, peas, beans, and algae.

## II. BEEF CATTLE FED SUPPLEMENTARY AMINO ACIDS

Hale et al. (1959) fed 4 and 10 g of supplemental L-lysine per day with 8.6 or 11.5% protein in practical diets to steers and observed that the 10-g level of lysine increased the daily gain by 15%, compared to no improvement with the 4-g level of lysine in the high-protein diet. The 10-g level of lysine improved feed efficiency at both protein levels. Cattle fed diets containing 10 g added lysine per head daily with different levels of urea increased the daily gain from 5 to 13% and the feed efficiency from 1 to 4%; however, adding lysine to the supplement containing only natural protein had no beneficial effect (Gossett et al., 1962; Perry et al., 1960).

Gossett et al. (1962) found that 10 g of methionine or 10 g of lysine and a combination of 10 g of each amino acid per steer daily improved weight gains slightly with methionine and more with lysine, but no improvement was observed when amino acids were fed in combination. When 5 or 10 g of methionine hydroxy analog (MHA) was fed daily for 161 days to steers receiving a diet of corn silage, rolled shelled corn, and a high-urea supplement, no benefit was obtained at the 5-g level and a significant depression in rate of gain occurred at the 10-g level. Brommelsiek et al. (1979) fed MHA in bird-resistant sorghum grain diets to steers in metabolism and comparative slaughter feedlot trials at levels of 0, 7, or 30 g per steer per day and found that total digestible nutrients (TDN), metabolizable energy (ME), nitrogen balance, and net energy of maintenance ($NE_m$) values were slightly improved by supplemental MHA, but net energy of gain ($NE_g$) values and liveweight gains were unaffected by the analog; however, more protein was deposited in the carcasses with the MHA supplement.

Cross et al. (1974) fed corn silage diets to finishing steers supplemented with soybean meal, urea, soybean meal–low-nitrogen, and urea–low-nitrogen levels that met 100 or 85% of the National Research Council (NRC) requirements for crude protein. Faster weight gains were made by the steers receiving soybean meal at the higher protein level. Lysine was of higher concentration in blood plasma of steers fed higher levels of soybean meal in their diets compared to those supplemented with urea. Lysine, isoleucine, leucine, histidine, and valine occurred at greater concentrations in plasma during the finishing phase (day 123) than during the growing phase (day 23), while the nonessential amino acids glycine and glutamic acid were lower during the finishing phase.

Richardson and Hatfield (1978) attempted to determine the first, second, and third limiting amino acids in abomasally cannulated growing steers fed a semipurified diet essentially protein free but presumed to be adequate in total nitrogen. Urinary nitrogen excretion in steers infused with L-methionine was lower than in steers infused with lysine, threonine, or tryptophan. Infusion of methionine, lysine, and threonine increased nitrogen retention over that obtained when methionine and lysine were infused. Methionine, lysine, and threonine were the first three limiting amino acids in growing steers, as indicated by nitrogen balance and three amino acids in plasma when microbial protein was essentially the sole source of protein.

## III. AMINO ACIDS FOR LACTATING COWS

Chandler (1970), using certain assumptions, calculated the net protein available to the lactating cow at various levels of production. Net protein was converted to net available amino acids based on the average composition of microbial protein, and these values were compared to the amino acids in milk protein at various production levels. Amino acids limiting milk production were methionine, valine, isoleucine, tryptophan, and lysine. Conclusions were that all amino acids should be adequate for the production of 10 kg of milk per day; that methionine and valine would become limiting at the 15-kg level of production and isoleucine at the 20-kg level; and that tryptophan and lysine would become limiting only at higher levels of production.

Griel *et al.* (1968) fed milk cows 0, 40, or 80 g of MHA per day and observed a significant increase in milk production. The 40-g level stimulated production with all breeds, while the 80-g level was stimulatory only to Holsteins and Jerseys and may have depressed milk production in Guernseys and Brown Swiss. Polan *et al.* (1970) fed cows 0, 0.2, 0.4, or 0.8% MHA in diets containing 0.4% urea and observed peak milk production at about 25 g MHA intake per day, increased fat content in milk with increased MHA intake, and depressed feed intake with the highest level of MHA.

Broderick *et al.* (1981) determined the total and individual amino acids in ruminal fluid from cows fed a low-protein (8.3%), high-concentrate diet (79%

TDN) supplemented with 0, 1, 2, 3, 4.5, and 6.5% crude protein as urea. Total amino acids plateaued beyond 2% crude protein as urea. Individual amino acids followed the trends of the total amino acids.

## IV. AMINO ACIDS FOR SHEEP

Cuthbertson and Chalmers (1959) compared the effects of casein administered through rumen and duodenal cannulae and found greater nitrogen retention in sheep that received the casein duodenally. Egan and Moir (1965) confirmed that postruminal administration of casein improved its utilization compared to oral or ruminal administration. Infusion of the abomasum of sheep fed a 10.6% protein roughage diet with cystine, methionine, or casein nearly doubled wool growth (Reis and Schinckel, 1963). Schelling and Hatfield (1967) reported that infusion of 10 essential amino acids into the abomasum of lambs at levels supplied by casein resulted in an increased positive nitrogen balance compared to controls. Schelling and Hatfield (1968) found that when methionine-supplemented casein was infused postruminally, voluntary feed intake increased by about 15%.

## V. DEGRADATION OF AMINO ACIDS IN THE RUMEN

Scheifinger *et al.* (1976) studied amino acid degradation of five major genera of rumen bacteria under *in vitro* conditions. They concluded that not all amino acids are degraded by all strains of rumen bacteria and that degradation occurs at different rates. Chalupa (1976) reported degradation of physiological quantities of amino acids by microbes of the rumen under *in vitro* and *in vivo* conditions. Rate constants for essential amino acids obtained *in vitro* indicated that arginine and threonine were rapidly degraded at 0.5–0.9 m$M$/hr, and valine and methionine were least rapidly degraded at 0.1–0.14 m$M$/hr. Essential amino acid degradation *in vivo* occurred at rates that were approximately 1.5 times greater than those of *in vitro* constants; this indicated that similar degradative pathways occurred in both systems. Combining a mixture of nonessential amino acids (aspartic acid, serine, glutamic acid, alanine, tyrosine, and ornithine) with an essential amino acid mixture had no influence on the degradation rates of the essential amino acids. It was concluded that with the possible exception of methionine, supplements of free amino acids cannot survive degradation in the rumen.

Merricks and Salsbury (1976) investigated several methionine analogs for their ability to inhibit methanethiol production in a protozoal extract from rumen fluid *in vitro*. At least two enzymes in the rumen appear to be capable of producing alkyl mercaptans from sulfur amino acids. Using diaminopimelic acid as a microbial marker, Santos *et al.* (1984) found that protein degradation in the rumen of lactating dairy cows was higher for a soybean meal (70%) diet than for corn gluten meal (45%), wet brewers grains (52%), and distillers dried grains with solubles (46%) diets. The bypassed protein was available in the intestines even though it was more resistant to microbial breakdown in the rumen.

## VI. UPGRADING OF PLANT PROTEIN IN THE RUMEN

Anderson and Anderson (1976) worked with the emission spectra of the elements in amino acids and demonstrated that essential amino acids have higher energy levels than nonessential amino acids. They compared the essential amino acid content of protein synthesized by rumen microbes and grass leaf protein, as shown in Table 6.1. These data indicate that forage protein is "upgraded" in the rumen. This upgrading involves increasing the concentration of chemical energy, and amounts to about 0.500 kcal/g in being changed from grass protein to microbial protein.

## VII. METHIONINE HYDROXY ANALOG

The value of supplementation of ruminant diets with amino acids is restricted due to the deaminating activity of rumen microbes. MHA and methionine were

**TABLE 6.1.**

**Essential Amino Acid Composition of Rumen Microbial Synthesized Protein and Grass Leaf Protein**[a]

| Amino acid | Rumen synthesis | Grass leaves |
|---|---|---|
| Methionine, % | 1.7 | 2.5 |
| Leucine, % | 8.1 | 10.0 |
| Tryptophan, % | 2.2 | 2.2 |
| Histidine, % | 2.1 | 2.0 |
| Phenylalanine, % | 6.0 | 6.0 |
| Lysine, % | 10.3 | 5.6 |
| Isoleucine, % | 9.5 | 5.0 |
| Valine, % | 10.2 | 5.0 |
| Threonine, % | 7.0 | 5.4 |
| Arginine, % | 8.1 | 7.0 |
| Total % essential amino acids | 65.2 | 50.7 |
| Actual spectral energy density (SED) of EAA (kcal/g) | 16.708 | 16.638 |
| Calculated SED of protein (kcal/g) | 15.535 | 15.012 |

[a]From Anderson and Anderson (1977).

reported to increase cellulose fermentation and accelerate rumen bacterial rate of growth *in vitro* (Salsbury *et al.*, 1971; Gil *et al.*, 1973a). By *in vitro* fermentation trials with rumen bacteria utilizing urea as a source of nitrogen and cellulose or glucose as substrates, Gil *et al.* (1973a) found that MHA or DL-methionine accelerated bacterial nitrogen incorporation and the concurrent substrate digestion rate. During shorter periods of fermentation, MHA and methionine had greater stimulatory action than inorganic sulfate. At 18 hr of fermentation, MHA supported more starch digestion than methionine or sulfate.

Gil *et al.* (1973b) fermented glucose with rumen bacteria, utilizing urea as a sole source of nitrogen, or without MHA, and found that the maximum effect of MHA on growth occurred at 0.2 mg per milliliter of medium. The free acid and calcium salt of MHA and DL-methionine produced equivalent stimulation of rumen bacteria. The MHA had no effect on the composition of bacterial protein. In a medium buffered at pH 6.9, lactate was the principal product of glucose fermentation. Brommelsiek *et al.* (1979) observed that MHA fed at a level of 7 g/steer/day with bird-resistant sorghum grain diets in a comparative slaughter trial increased the tissue protein gain. While the TDN, ME, and nitrogen balance were slightly improved by MHA, the $NE_g$ and liveweight gains of the steers were unaffected by the analog.

Varner (1974) and Thomas and Langford (1978) reported that beef cows fed MHA before and after calving increased the weaning weights of their calves. Clanton and England (1980) evaluated beef cows supplemented with MHA between calving in March and April and going on pasture in mid-May. When MHA (8–10 g/head/day) was provided with a 13.3% protein supplement, 33% more cows cycled during the first 21 days of the breeding season, and 40% more conceived during the first 30 days of the breeding season than those that received lesser amounts of MHA.

Burroughs and Trenkle (1969) reported that MHA gave a 13% increase in gains (2.44 versus 2.16 lb/day) and a 10% improvement in feed efficiency when diets were fed to heifer calves over 151 days. The diets contained a high-urea supplement with 3 g MHA per heifer daily. Greater gains and feed efficiency were obtained with high-urea–containing diets than with soybean meal diets when MHA was included. Gossett *et al.* (1962) fed 5 or 10 g of MHA per steer daily for 161 days in a diet consisting mainly of corn silage, rolled shelled corn, and a high-urea supplement. They found no benefit at the 5-g level and a significant depression of rate of gain at the 10-g level.

Stokes *et al.* (1981) studied the effect of MHA and methionine with two levels of dietary protein on milk production of Holstein cows. The MHA and methionine supplementation did not affect milk yield or composition of milk at 80 or 100% protein requirement levels. In some trials, MHA increased milk yield and fat production (Bishop, 1971; Bishop and Murphy, 1972; Chandler and Jahn, 1973; Griel *et al.*, 1968). In other investigations, no such results occurred

(Bouchard and Conrad, 1973; Burgos and Olson, 1970; Fuquay *et al.*, 1974; Hutjens and Schultz, 1971). It may be concluded that stimulation of milk and fat or tissue protein yield occurs only if dietary methionine limits rumen microbial metabolism or a deficiency of absorbed methionine limits tissue metabolism.

## VIII. LABILE PROTEIN AND PROTEIN TURNOVER

The concept of a labile protein reserve in the animal's body is based on the establishment of a new equilibrium in the rate of urinary nitrogen excretion after an abrupt change in dietary protein intake. The evidence indicates that most of this nitrogen is derived from or deposited in proteins of the skeletal muscle. Munro (1964) cited the early work of Voit in 1866 as providing basic data on the concept of protein reserves. Voit observed that dogs that had consumed 2500 g of meat excreted 170 g of urea during a subsequent 8-day fast, while dogs that had consumed 1500 g of meat excreted only 105 g of urea during the fasting period. The amount of nitrogen that apparently had been "stored" was proportional to the protein content of the diet consumed before fasting.

Munro (1964) concluded from a review that the amount of labile protein deposited in the body does not exceed 5% of the total body protein and that, upon fasting, this labile protein disappears within a few days. However, Allison *et al.* (1964) pointed out that there was a larger protein reserve that can be contributed to the maintenance of essential tissue structure and that the protein reserve includes a very labile part.

Paquay *et al.* (1972) conducted experiments with mature, dry, nonpregnant cows to determine their capacity to lose and replete nitrogen and to study the significance of the labile and total protein reserves. It was concluded that such cows can store and lose large amounts of body proteins (more than 15 kg) when nitrogen and energy feed intakes are greatly varied. Repletion and depletion of protein reserves include a very labile part, and this part is lost more rapidly during fasting than during feeding of low-protein diets. The labile part can temporarily meet the animal's energy requirements. Nitrogen balance and gains in body weight were well correlated.

Biddle and Evans (1973) employed a nitrogen depletion-repletion feeding technique to evaluate nitrogen utilization with Jersey steers. Three consecutive 28-day periods were used to standardize, deplete, and replete body nitrogen. Dietary protein was supplied by corn gluten or urea and averaged 15, 5, and 14% per period. Body weights and dry matter intakes were not affected by altering the level of dietary nitrogen offered, but digestibility of dry matter was reduced during the depletion period. Total urinary and fecal nitrogen losses were reduced during the depletion period, and nitrogen retained (percentage of intake) reached a low of −16% after 1 week of restricted intake. However, nitrogen retention

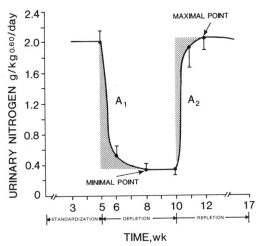

**Fig. 6.1.** Patterns of urinary nitrogen excretion in growing steers during depletion and repletion of labile body nitrogen. Each plotted value is the mean ± SE of eight observations. Minimal and maximal urinary nitrogen excretion values were used to establish baselines for determining shaded areas $A_1$ and $A_2$. (From Biddle et al., 1975 and © J. Nutr. American Institute of Nutrition.)

increased to 9% by the third week of depletion. Realimentation with higher-nitrogen diets produced an immediate increase in retained nitrogen and reached a maximum value of 50% after 2 weeks of repletion.

Biddle et al. (1975) subjected Jersey steers having an average initial weight of 186 kg to a nitrogen depletion-repletion treatment. Low-fiber diets containing either corn gluten or urea as the only significant source of dietary nitrogen were fed. The experiment lasted for 17 weeks, and three periods were used to standardize, deplete, and replete body nitrogen. Plasma protein decreased from 6.05 g/100 ml during standardization to 5.44 g/100 ml after 5 weeks of depletion and did not approach predepletion levels until the sixth week of repletion. Labile nitrogen reserves were determined by integrating the areas from total urinary nitrogen excretion curves obtained during the depletion and repletion periods. The lag in urinary nitrogen excretion shown in Fig. 6.1 after abrupt shifts in nitrogen intake was interpreted as evidence for the existence of labile nitrogen reserves. The magnitude of labile nitrogen was reversibly depleted and repleted, as shown in the shaded areas ($A_1$ and $A_2$) of Fig. 6.1. The labile nitrogen was computed as being 6.5 ± 0.9% of the total body nitrogen on an empty body weight basis. The magnitude of labile nitrogen stores as a percentage of total body nitrogen was 44% greater in steers at 280 kg than in those at 144 kg body weight.

# 7

# Nonprotein Nitrogen Utilization

## I. INTRODUCTION

Generally, plant proteins are more efficient than nonprotein nitrogen (NPN) sources such as urea and biuret in stimulating animal growth and production. Ruminant response is generally quite satisfactory when NPN sources provide part of the nitrogen requirements, especially when animals are fed medium- and high-energy diets. After intensive research with NPN compounds, Hart *et al.* (1939) concluded that ruminants synthesize protein from simple nitrogen compounds through action of rumen microbes and that the tissue of steers fed urea in their diets contained normal protein. Growing and finishing cattle fed urea-containing diets were found to have normal livers and kidneys (Work and Henke, 1940). Harris and Mitchell (1941) showed that NPN could be used for maintenance and growth in diets deficient in protein but that when enough natural protein was in the diet, urea was poorly utilized.

The Association of American Feed Control Officials (AAFCO, 1955) approved the use of urea and ammonium bicarbonate, which were being used at the time in the feed industry, and recommended that not more than one-third of the total crude protein in the ruminant diet be from NPN products. Other NPN compounds that have been demonstrated to be utilized by rumen microbes are ammonium acetate, ammonium lactate, biuret, dicyanodiamine, glutamine, glycine, and melamine. Urea usage is a matter of economics (Oltjen, 1972); 1 kg NPN is equivalent to the nitrogen in 5–6 kg plant protein. Urea has the equivalent of 260–280% crude protein and biuret contains the equivalent of 250% crude protein, compared to 41–44% protein in cottonseed meal and 44–50% protein in soybean meal. However, plant protein sources supply energy, minerals, and other nutrients as well as nitrogen. In general, a mixture of 1.0 part urea or 1.1 parts of biuret with 7 parts of corn grain equals 8 parts of soybean meal in nitrogen and energy (Fonnesbeck *et al.*, 1975). Sulfur, potassium, and phosphorus may need to be added to the urea-corn mixture to make it equivalent to soybean meal. More biuret than urea may be fed to ruminants, since the low solubility of biuret allows it to remain longer in the rumen. It is hydrolyzed more slowly than urea to ammonia ($NH_3$) and $CO_2$ in the rumen and thereby decreases the accumulation of toxic levels of $NH_3$ in blood plasma.

Beef cattle can grow and reproduce when fed diets in which urea is the sole source of nitrogen (Oltjen, 1969). No ill effects were observed when cattle were kept on such diets over a 4-year period. Bond and Oltjen (1973a) investigated the effect of urea-containing diets on the reproductive performance of bulls and heifers by comparing purified diets with urea to those with isolated soy protein as the sole sources of dietary nitrogen. In heifers, puberty was reached at 634 days of age with urea compared to 364 days with soy protein diets; the heifers weighed less and gave less milk during the first lactation, and averaged 53 days longer between calving and first estrus.

Bond and Oltjen (1973b) provided monozygotic twin heifers with diets that contained two-thirds of the total nitrogen in a poor-quality 50% forage diet with (1) soybean meal, (2) equal parts of soybean meal and urea, and (3) urea. Diets had no effect on the length of the estrous cycle, services per conception, birth weight of calves, milk yield, average daily gain of calves, interval from calving to first estrus, and conception.

Thompson *et al.* (1973) studied the effect of dietary urea on reproduction in ruminants fed conventional diets consisting of corn silage or poor-quality hay supplemented with urea or soybean meal. Calving of cows and lambing rates of ewes fed urea-containing diets were approximately the same as for those fed soybean meal. During early and late pregnancy, urea fed to beef cows did not have adverse effects on embryonic development, parturition, or cow and calf weights. Urea had no effect on ovulation rate or subsequent fertility in ewes or on fertility in bulls or rams. Dairy cattle produced moderate levels of milk over an extended period when fed diets that contained urea or ammonium salts as the only source of dietary nitrogen (Virtanen, 1966). Sheep had normal growth, reproduction, and wool production when fed diets that contained urea or biuret equal to 75% of total nitrogen in diets fed over 593 days (Hatfield *et al.*, 1959).

Cows wintered on Argentine bahiagrass and provided Pangola digitgrass hay during December to March over 3 years were evaluated for their response to supplements of cottonseed meal cubes containing 41% crude protein, citrus pulp cubes with feed-grade biuret containing 38% crude protein, or citrus pulp containing 7% crude protein (Martin *et al.*, 1976). Mature cows averaged weight losses of 49, 60, and 70 kg with the cottonseed meal, citrus pulp-biuret, and citrus pulp cubes, respectively, during three winters. The conception rate, calf survival, and time required to conceive were not influenced by the supplements. Weaning weights of calves were 167, 166, and 162 kg from cows provided the cottonseed meal, citrus pulp, and citrus pulp-biuret cubes, respectively.

Horton and Nicholson (1981) compared diets composed of barley grain and wheat straw (35.7–40.5%) containing no protein supplement, 2.15% urea, or

19% soybean meal with diets containing 34% barley and 60% alfalfa meal fed to steers. Digestibilities of organic matter, crude protein, and acid detergent fiber were increased by urea and soybean meal but not by alfalfa meal. Propionic acid was highest and butyric acid lowest in the rumen of steers fed the unsupplemented diet. Ruminal ammonia and free amino acids were elevated in the three nitrogen-supplemented groups and were higher in steers fed urea or soybean meal than in those fed alfalfa meal. Daily gains and feed efficiency were not affected by urea, but soybean meal and alfalfa meal diets gave 8 and 24% more gain, respectively, than in steers fed the unsupplemented diet. Feedlot steers fed a diet containing 40% straw with a suboptimal level of crude protein improved when supplemented with soybean meal but not with urea.

Newton and Utley (1978) conducted trials with cattle to determine the value of melamine (2,4,6-triamino-1,3,5-triazine) as an NPN source, comparing it with urea and cottonseed meal. Melamine was found to be digested as well as cottonseed meal nitrogen but not as completely as urea nitrogen. The authors concluded that melamine may not be hydrolyzed in the rumen at a rate sufficient to promote maximal protein synthesis, and factions not completely hydrolyzed may be absorbed and excreted in the urine.

## II. NONPROTEIN NITROGEN ADAPTATION AND UTILIZATION

Cattle had lower gains during the initial phase of NPN feeding periods but later approached the gains obtained with plant protein (Kirk et al., 1958). Young et al. (1973) found that urea supplementation of ear corn finishing diets fed steers decreased 112-day gains by 11% compared to supplementation with soybean meal. Gains made by soybean meal-supplemeneted steers were similar whether the supplement was provided during the entire finishing phase (112 days) or only during the first 56 days. It was concluded that yearling steers need a preformed protein supplement during the early phase of the feeding period, and such requirements cannot be met entirely by microbial synthesis of protein from urea.

Mizwicki et al. (1980) studied the effect of timed ammonia release rates on digestibility and nitrogen retention in steers fed winter range grass. Digestibility of dry matter was increased by 5% with the addition of urea, regardless of the rate of urea administration. Nitrogen retention ranged from 11.4 to 12.4 g with urea, compared to −7.1 g with no urea.

A slow-release prilled urea compound (SRU) made by coating the material with a tungoil-linseed–oil–talc catalyst mixture was evaluated in comparison with untreated urea and natural protein under dormant winter range conditions with lactating cows in Oklahoma (Forero et al., 1980). Cows fed SRU tended to lose less weight than urea-fed cows, while both SRU- and urea-fed cows lost less weight than cows fed a 15% natural protein supplement. It was concluded that the cow and calf performance of the SRU-fed cows was still unacceptable even though it appeared to increase energy intake. In another study by Owens et al. (1980), the SRU was evaluated for ammonia release rate, animal acceptability, toxicity, diet intake, effects on dry matter digestibility, and nitrogen retention when added at a level of 1% urea in an 80% concentrate steer diet and fed twice daily. The SRU gave an ammonia peak of 32 mg per 100 ml rumen fluid at 1 hr postfeeding compared to a peak from SRU of 53 mg per 100 ml at 30 min and showed no toxic symptoms. Feed consumption was increased with SRU compared to urea, but dry matter digestibility and nitrogen retention were the same with the two types of urea.

Clemens and Johnson (1974) investigated the biuretolytic activity of rumen microbes in sheep as influenced by different levels of dietary biuret or urea and soybean meal. In one experiment, NPN sources were varied by replacing portions of the biuret with successive portions of equivalent urea so that 100, 75, 50, and 25% of the supplemental NPN was derived from biuret, with the remaining portion being from urea. In the other trial, supplementary biuret and soybean meal were varied so that soybean meal contributed 0, 33, 67, and 100% of the nitrogen. Neither urea nor soybean meal influenced the rate of biuret adaptation or the extent of *in vitro* hydrolysis. Biuret at the 25% level gave less biuretolytic activity than at higher feeding levels.

Mahadevan et al. (1976) studied the properties of bovine rumen urease. In the absence of high concentrations of dithiothreitol, urease activity was lost rapidly. Maximal urease activity was observed between pH 7 and 8.5. Only one type of urease was found in the rumen, and it had a much lower molecular weight than jack-bean urease. The enzyme was inhibited by *p*-chloromercuribenzene sulfonate, *N*-ethylmaleimide, and phosphate, but not by ammonium ions. Most divalent metal ions were inhibitory, and none were found to activate or be required for urease activity. Phenylurea and hydroxyurea inhibited urea hydrolysis.

G. Wilson et al. (1975) evaluated the factors responsible for reduced voluntary intake of urea-containing diets in ruminants. Urea at levels above 2% depressed the intake of diets whether it was supplied mixed in the diet or infused directly into fistulated cows. Increasing the levels of urea by either method elevated ammonia in the rumen, venous or arterial blood, and saliva. The data suggested that feed intake depression with diets containing more than 1% urea was due to a physiological parameter other than taste.

Oltjen et al. (1974) found that biuret in a mineral mixture improved the performance of steers fed hay during winter when the hay contained 4.5% or more crude protein, but was not as effective when the hay contained 6.0% or more crude protein. These investigators found that during a 112-day finishing trial, steers fed diets containing biuret-corn meal, urea-corn meal, and cottonseed meal as sources of supplementary nitrogen had gains of 1.11, 1.10, and 1.09 kg/day, respectively (Oltjen et al., 1974).

Perry et al. (1974) conducted metabolism and the feedlot studies on the effect of adding various levels of alfalfa solubles to complete mixed diets and liquid supplements for growing-finishing cattle in which the supplemental protein was predominantly urea. The 0.5% alfalfa solubles resulted in greater digestibility of dry matter and energy and lowered levels of ammonia in the rumen and of urea in plasma, while the 2% alfalfa solubles resulted only in an increase in energy digestibility. Both heifers and steers gained faster with 2.5% than with 5.0% alfalfa solubles in their diets.

Tiwari et al. (1973) studied degradation of biuret in the rumen. They concluded that urease participates in the total degradation of biuret and that urea is one intermediate compound. They demonstrated that minerals are not responsible for apparent complexing of 20–30% biuret immediately upon exposure to fermentation in rumen fluid. Brookes et al. (1972), utilizing urea-$^{14}$C, observed maximum recovery of 84.7% of expired $^{14}CO_2$ in rumen and blood of sheep within 6 hr and concluded that alternative nonhydrolytic pathways of urea metabolism do not appear to be quantitatively important in the rumen.

Hale and King (1955) administered ammonium carbonate, both orally and intravenously, to sheep and observed symptoms similar to those of urea toxicity. They suggested that this condition was probably not due to liberation of ammonia but to formation of ammonium carbonate from urea. While ureolytic enzymes appear to be always present in the rumen, microbes need to adapt in order to synthesize enzymes that hydrolyze biuret (Hatfield et al., 1955; Ewan et al., 1958). Urea, biuret, and uric and cyanuric acids have covalent bonds that must be hydrolyzed to ammonia before the ammonia can be utilized in the synthesis of amino acids. NPN substances that have ionic bonds, such as ammonium sulfate, ammonium phosphate, and ammonium chloride, will dissociate in water to ammonia without enzymatic activity.

Seven strains of *Pseudomonas aeruginosa* were found to hydrolyze biuret enzymatically, but cell growth indicated that biuret was not a rapidly utilized source of nitrogen, and the generation time of microbes was about one-half that obtained in urea media (Wheldon and McDonald, 1962). Biuret-hydrolyzing bacteria and fungi had a strong hydrolytic effect on urea and made it uncertain whether the agent is a specific "biuretase" or a "urease" that acts on biuret (Jensen and Kourmaran, 1956). Similar observations were made by Jensen and Schroder (1965) in studies with 80 fast-growing strains and 40 slow-growing strains of *Rhizobium* spp.; biuret was assimilated more slowly than urea with slower growth of the microbes. Nishihara et al. (1965) obtained an enzyme that decomposed biuret into ammonia, carbon dioxide, and urea. Urea of feed-grade biuret is hydrolyzed rapidly in the rumen to ammonia, while biuret, triuret, and cyanuric acid in the mixture are hydrolyzed gradually. In most studies, the feed-grade biuret has been used, while in others, purified biuret has been the only source of NPN; this would account for some inconsistencies among biuret experiments (Fonnesbeck et al., 1975).

An adaptation period of 30–40 days was reported with biuret for lambs (Welch et al., 1957) and dairy calves (Campbell et al., 1963). The time required to reach maximum biuretolytic activity was found to be about 15, 30, and 70 days when biuret was fed in diets that contained 3.5, 6.0, and 10.3% crude protein, respectively (Schroder and Gilchrist, 1969). Maximum biuretolytic activity was observed after 24 hr, but 100% breakdown of biuret did not occur until after 59 days (Schroder, 1970). Deadaptation to biuret may be rapid, since Clemens and Johnson (1973a,b) found that *in vitro* disappearance of biuret was greater in the rumen fluid of cattle and sheep that received purified biuret every day compared to that of animals consuming the biuret every 4 days.

## III. SULFUR REQUIREMENTS FOR NPN UTILIZATION

Microbial synthesis of methionine, cysteine, and cystine requires adequate sulfur. Thomas et al. (1951) observed that lambs fed urea-containing diets without added sulfur lost weight and were in negative nitrogen balance, while those supplemented with sulfur were in positive nitrogen balance. Starks et al. (1954) obtained similar results. The nitrogen : sulfur ratio of 15 : 1 found in animal tissue was considered sufficient for ruminant diets in the past, but Moir et al. (1968) found that the performance of sheep improved with a ratio of 9.5 : 1; this seems reasonable, since wool has a 5 : 1 nitrogen : sulfur ratio (Burns et al., 1964). Chalupa et al. (1971) observed that calves lost 0.1 kg/day when fed purified diets with an 86 : 1 ratio of nitrogen to sulfur but gained well with elemental sulfur ratios of 12 : 1 (adequate), 4 : 1 (excess), or 2 : 1 (excess).

When nitrogen in the diet exceeds the animal's requirement, there is no need to supply excess sulfur. Most feeds have adequate sulfur. However, when diets have high levels of NPN to make up a nitrogen deficiency in the diet or supplement, then additional sulfur should be provided. Diets low in sulfur (less than 0.1% dry matter) should be supplemented with 3 g inorganic sulfur per 100 g of urea, which is the equivalent of 1 part sulfur to 15 parts NPN (NRC, 1976b).

Bolsen et al. (1973) conducted trials with steers, heifers, and lambs to determine the effects of methionine and ammonium sulfate with soybean meal or urea-supplemented finishing diets on animal performance. Performance was not affected in the three trials by addition of sulfur to the soybean meal-supplemented diets. Steers and lambs fed urea plus ammonium sulfate gained more slowly, consumed less feed, and had less feed efficiency than animals fed diets with urea alone. The diets had no influence on carcass grade and dressing percentage. The performance of the animals indicated that the total sulfur content of the soybean meal- or urea-supplemented diets was adequate.

Pendlum *et al.* (1976) investigated the effect of varying levels of dietary sulfur and nitrogen sources on performance and on ruminal and plasma constituents of Holstein steers. Corn-cottonseed hull diets supplemented with soybean meal or urea (33.7% total crude protein) were fed ad libitum. Added sulfur was included with each nitrogen source equal to 0, 0.15, and 0.30%, which provided nitrogen : sulfur ratios of approximately 15.0, 7.0, and 4.0, respectively. Daily gains averaged 1.61 and 1.51 kg for the soybean meal- and urea-supplemented diets, respectively, and sulfur level had no effect on gain or efficiency of feed conversion.

## IV. QUALITY OF NITROGEN SOURCES FED RUMINANTS

Experiments with cattle and sheep have led to the conclusion that the quality of the dietary protein has relatively little importance due to microbial activity in the rumen. Most sources of nitrogen are degraded in the rumen and converted to microbial protein, which is the protein actually available to the host animal. When 12% protein from a number of protein sources was fed to lambs, a biological value of about 60 was obtained (Johnson *et al.*, 1942). A biological value of 71 was observed by Lofgreen *et al.* (1947) when diets providing 40% of the nitrogen from urea were fed to wethers; supplementary methionine increased the biological value to 74, linseed meal increased it to 76, and dried egg protein increased it to 80. Loosli and Harris (1945) found that methionine supplementation improved the performance of lambs fed urea-containing diets. Microbial proteins have a biological value of 60–70 (Loosli and McDonald, 1968). Very little protein value is obtained from NPN if it is added to diets that exceed about 13% crude protein unless the total digestible nutrients (TDN) is 80% or above (Satter and Roffler, 1975).

Virtanen (1966) found histidine to be very low in the plasma of lactating cows fed diets with urea as the sole source of nitrogen. Burris *et al.* (1975) observed the growth and plasma amino acids in steers fed diets of corn-cottonseed hulls containing no supplemental nitrogen (control) or diets supplemented with either urea or soybean meal. Daily gains were less with the control diets but equivalent with urea or soybean meal. On day 28, glycine and histidine were higher in plasma in the control group, arginine was lower in the controls than in the soybean meal group, and lysine was lower in steers fed urea than in those fed soybean meal. Little *et al.* (1969) reported reduced concentrations of leucine, isoleucine, valine, and lysine and increased concentrations of glycine in the blood plasma of steers fed urea-containing diets. Excess urea did not spare microbial utilization of natural protein, since diets with 21% fish meal had more than half of the protein degraded by microbes (Orskov and Frazer, 1970).

## V. RUMINANTS FED LOW-PROTEIN FORAGES WITH NPN

Most rangeland supports vegetation that can provide energy but is low in protein during the nongrowing season. Crop residues are a source of energy for ruminants. As grasses and residues mature, their protein and soluble carbohydrates decrease, and fibrous carbohydrates, lignin, and silica increase. Supplementation of such diets is necessary for satisfactory performance of the animals. Survival of sheep on low-quality roughage was used as a criterion for evaluating NPN supplementation by Briggs et al. (1960). When ruminant diets are low in nitrogen, the rumen microbes decline in number, resulting in poor utilization of fibrous materials and reduced intake (Moir and Harris, 1962). Under these conditions, ruminants may suffer from dietary protein and energy deficiencies, which will generally result in the mobilization of tissue protein and fat to meet the animal's energy and nitrogen requirements. Cattle fed 3.5% protein low-quality hay had heavy weight losses with or without molasses supplementation, but if a molasses-urea mixture was supplemented, body weight was maintained (Beames, 1959). Sheep lost 20% of their body weight over a 7-week period when fed oat straw with or without molasses, but if the oat straw was supplemented with molasses plus urea, body weight was maintained (Coombe, 1959). These observations indicate that insufficient nitrogen was present in the low-quality forages to utilize the extra energy in the molasses.

### A. Nonprotein Nitrogen Supplementation and Maintenance

South African cattle fed urea or biuret had increased body weight gains and increased intake of low-quality hay compared to cattle not supplemented (Kraft, 1963). MacKenzie and Altona (1964) found that beef steers lost weight (0.24 kg/day) when grazing forage containing 4.2% protein over 84 days, but those supplemented with urea or biuret made slight gains. Corn stalk pasture for pregnant beef cows over 84 days gave slight weight gains with biuret supplementation, but without supplementation average daily weight losses were $-0.42$ kg (Heath and Plumlee, 1971). Kearl et al. (1971) reported less body weight loss with cows grazing crested wheatgrass and increased growth of their nursing calves when animals were supplemented with cottonseed meal or biuret. Biuret was as effective as cottonseed meal on winter range in Texas (Templeton et al., 1970a) and was better than urea for wintering beef calves (Tollett et al., 1969).

Sheep fed a low-protein forage diet supplemented with urea, biuret, triuret, or cyanuric acid had equivalent nitrogen retention (Clark *et al.*, 1965). No differences in the weight gains of pregnant ewes or the weights of lambs weaned over 2 years were observed when animals were grazed during winters on range plants with supplements that replaced 0 to 50% of soybean meal protein with equivalent nitrogen from urea or biuret (van Horn *et al.*, 1969). Sheep fed an unsupplemented, low-quality pangola grass hay diet had a negative nitrogen balance, but when supplemented with soybean meal, urea, or biuret, they had increased feed intake and nitrogen balances of 1.4, 1.2, and 1.9 g per head daily, respectively (Ammerman *et al.*, 1972).

## B. Nonprotein Nitrogen for Body Weight Gains

To be efficient, animals fed for maximum growth or milk production require about two to three times the metabolizable energy (ME) and three to four times the protein required for maintenance (NRC, 1975, 1976a). Ruminants fed for rapid growth or lactation must be supplied with dietary nitrogen and energy in a more concentrated form than is acceptable for body maintenance. This requires replacing part of the roughage dry matter with cereal grains and nitrogen supplements such as oilseed meals or NPN. More dense energy and nitrogen feedstuffs take up less space in the digestive tract and allow the animal to consume more utilizable nutrients. Cereal grains generally do not meet the protein requirements of highly productive animals, and mature grass, corn, or cereal grain silages require too much intake due to their bulkiness for ruminants to meet either their protein or energy requirements for high production.

When high-energy diets containing less than 9% protein were fed to growing and finishing cattle, biuret produced growth equivalent to that obtained with urea or plant protein supplements (Campbell *et al.*, 1963; Owens *et al.*, 1967; Pickard and Lamming, 1968; Meiske *et al.*, 1969).

Lambs fed diets that contained either biuret, urea, or soybean meal nitrogen supplements gained faster and more efficiently than lambs fed a basal diet of 8.1% crude protein (Meiske *et al.*, 1955). Similar results were obtained by Karr *et al.* (1965b) utilizing a basal diet of 7.8% crude protein. With high-energy diets, both biuret and urea were as effective as soybean meal in nitrogen supplements when fed to lambs for weight gains (Karr *et al.*, 1965a). Templeton *et al.* (1970b) found that lambs maintained better body weight and ate more low-quality hay when consuming biuret in a liquid supplement than those fed a control supplement of molasses and corn meal. Their observations indicate that lamb growth will increase when biuret is fed in either low-energy or high-energy diets deficient in protein. A 3- to 5-week adaptation period to urea is usually required by lambs before they make equivalent weight gains to lambs fed plant protein-containing diets.

## C. Nonprotein Nitrogen for Milk Production

Using labeled $^{15}$N ammonium salts, Land and Virtanen (1959) found that 17–25% of the dietary nitrogen intake of cows appeared in the milk protein. Virtanen (1967) reported on a cow that produced over 4000 kg of milk during one lactation period with urea as the only source of dietary nitrogen, and concluded that approximately 39% of the urea nitrogen was excreted as milk protein. Cows fed purified diets containing urea and ammonium salts as the sole sources of nitrogen averaged 2750 kg of milk during standard 305-day lactations (Virtanen, 1966). Except for a slight increase in the percentage of fat and protein, the milk was similar to that produced by cows fed natural diets. Virtanen (1967) estimated that a cow can produce about 4000 kg of milk per year without dietary protein, about 5000 kg if 20% of the diet is protein, and 6000 kg if 30% of the diet is protein and the remainder of the dietary nitrogen is in NPN. Working with monozygotic twin cows, Flatt et al. (1967) found that one fed a purified urea-containing diet produced about 65% as much milk as her twin fed an equivalent but protein-containing diet. Oltjen (1969) reviewed studies on ruminants fed NPN-containing diets and concluded that NPN would support life and reproduction normally over at least several years of moderate milk production, milk similar in composition to the usual milk, normal ruminal synthesis of B vitamins, and the usual pattern of synthesized amino acids, but at a depressed level.

Van Horn et al. (1967) suggested that acceptability problems would limit urea to a maximum of 1% of the energy feed mixture offered lactating ruminants. Iwata et al. (1959) found that concentrate feeds containing 2–3% urea and 25–40% molasses and replacing soybean meal in diets fed to milking cows had no effect on milk production. Dairy cows, goats, and sheep were fed diets containing biuret, with no effect on their health, milk production, milk quality, or composition of blood. The usual protein-deficient diet fed dairy cows generally requires 1–3% additional crude protein to balance the protein requirement. Altona (1971) studied adaptation of dairy cows to biuret and found that heavy milkers produced as well on energy feeds containing biuret as on true protein over a 24-week lactation period, and that cows may be preconditioned by feeding 28 g biuret daily for a few weeks prior to replacing natural protein with a higher level of NPN.

## D. Supplementing Diets with NPN

Often when it is desirable to supplement diets with NPN, other dietary nutrients are also deficient and need to be supplemented. According to Fonnesbeck et al. (1975), these situations include (1) maintenance of ruminants on dormant forage or roughage (1.5–1.8 Mcal ME per kilogram of dry matter, 3–5% crude protein, 0.07–0.10% phosphorus, vitamin A and vitamin E deficiency, and a suspected trace mineral deficiency); (2) growing animals grazing or fed mature grass, forage or hay (1.8–2.2 Mcal ME per kilogram of dry matter, 5–10% crude protein, and 0.10–0.15% phosphorus); and (3) growing, finishing, or lactating ruminants fed diets of low-quality forage or roughage such as silages of corn or sorghum and cereal grains (8–11% crude protein). Supplementation of all deficient nutrients probably is necessary in order to have a profitable operation. Specially formulated self-fed supplements containing NPN and other nutrients lacking in the basal diet need to be prepared or are often available commercially as solids or liquids. In confinement, high-energy diets and supplements are best mixed, since this aids in preventing acceptability and toxicity problems as well as helping to provide a better-balanced ratio of calories to other nutrients.

### E. Nonprotein Nitrogen in Self-Feeding Supplements

Ruminants do not select a diet to meet their nutrient requirements, and it is nearly impossible to provide a balanced diet with a self-fed supplement. Salt in high concentrations restricts the intake of supplements but may be utilized at low concentrations to increase intake. Molasses and cereal grains are utilized to increase the intake of supplements. Fonnesbeck *et al.* (1971) prepared a cattle supplement for mature cows in intermountain areas of Utah that contained 30–35% feed grade biuret, 12–14% dicalcium phosphate, 5% salt, and 1% sulfur plus trace minerals and vitamin A, with the balance of the supplement being alfalfa meal, barley grain, corn grain, and cane molasses. This supplement could be self-fed and generally supplied the deficiencies of dormant grasses and cereal grain straws. Salt in excess of 10% restricted intake of the supplement. It took about 2 weeks of excessive consumption before intake declined to anticipated levels. If more than 35% biuret was added, sufficient intake was not consumed to balance cattle requirements. Consumption of the supplement was increased by locating feeders near watering areas, along trails and areas where cattle congregated.

### F. Liquid NPN Supplements

Molasses from the sugar industry is widely used as a source of energy supplementation and as a carrier of urea, biuret, and other NPN supplements as well as nutrients such as minerals and vitamins. The additive needs to be soluble in molasses or water in order to be held in suspension. Soluble carbohydrates of molasses are rapidly metabolized by rumen microbes to volatile fatty acids (VFA), and they neutralize the ammonia released from hydrolyzed NPN compounds. The carbohydrates of molasses also provide carbon sources for synthesis of amino acids by the microbes. Molasses and urea liquid protein supplements were reported to be effective by several investigators (Beames, 1960; Coombe, 1959; Tillman et al., 1951; Bohman et al., 1954) when they compared urea-molasses supplements to plant proteins and unsupplemented diets. Templeton et al. (1970b) compared liquid supplements containing feed-grade biuret and molasses, and urea in molasses for ruminants fed low-protein grass hay. More rapid and efficient weight gains were obtained with cattle and sheep fed the biuret supplement. Biuret must be finely ground in order to be kept in suspension in molasses, since it is less soluble in molasses and water than urea.

## G. Silage Supplementation with NPN

Corn silage contains an average of 8.2 ± 0.6% crude protein, and growing and finishing cattle require 10–13% crude protein for optimal growth. NPN substances have been added to silages to increase the nitrogen content at the time of ensilage or to the diet at the time of feeding. Ease of mixing at the time of ensilage is an advantage, but incorporation at this time may be a disadvantage, since it limits flexibility in balancing the nitrogen content of the total diet. Urea at a level of 0.5% added at the time of ensilage has generally increased growth and milk production. Goode et al. (1955) found that cattle performance was better when silage was fed without urea, and Conrad and Hibbs (1961) suggested that the readily available carbohydrate of corn silage was too low for efficient utilization of urea in ruminant diets. In metabolism trials with lambs, Karr et al. (1965b) fed whole-corn silage with urea, or biuret incorporated during ensilage or feeding times, and found that dry matter digestibility and nitrogen retention were greater with silage treated with biuret at ensilage. Biuret and urea produced similar dry matter digestibility and nitrogen retention when added at the time of feeding. No difference in weight gains was observed due to the time of adding NPN. Calves fed biuret or urea-supplemented silage gained faster with less feed per kilogram gained than those fed unsupplemented silage (Karr et al., 1965b).

Owens et al. (1970a,b) reported on the effect of biuret, urea, soybean meal, and sodium metabisulfite mixed in corn silage *in vitro*. Urea increased lactic and acetic acid levels and tripled total gas production. High-purity biuret or soybean meal resulted in fermentation similar to that produced by the control silage. Nitrogen loss was greater with soybean meal. Calves fed biuret or urea-supplemented silage gained faster and had better feed efficiency than when fed unsupplemented silage (Meiske et al., 1968). NPN added at ensilage time increases the nitrogen content but should not be expected to help the fermentation process or to preserve the silage. Urea may adversely affect silage preservation, and considerable urea nitrogen may be lost before feeding of the silage. Biuret does not affect the ensiling fermentation, and the silage is similar to untreated silage except that the nitrogen content is higher (Fonnesbeck et al., 1975).

## VI. RUMINAL AMMONIA CONCENTRATION AND NPN

A knowledge of the concentration of ruminal ammonia necessary for maximum microbial growth rates should aid in the utilization of NPN sources, since excess feeding would be a waste of nitrogen. Roffler *et al.* (1976) made an *in vivo* study with steers to determine if 5 mg $NH_3$-nitrogen per 100 ml of rumen fluid was sufficient to support maximum growth rates of rumen microbes. Once ammonia began to accumulate at this level, there was a linear relationship between urea intake and ruminal ammonia concentration, and no increase in plasma amino acids. Satter and Roffler (1976) studied the effect of ammonia concentration and microbial protein production in continuous culture fermentors charged with ruminal contents of steers fed either a protein-free purified diet, a corn-based all-concentrate diet, or a forage-concentrate diet infused with urea to maintain various concentrations of ammonia. The relationship between the crude protein of the mixture added to the fermentor on a dry matter basis and the tungstic acid precipitable protein and ammonia concentration is shown in Fig. 7.1. Protein synthesis in the fermentors increased with increased urea and then leveled off, showing that further increases in urea had no effect on protein production. The leveling off of protein production coincided with the point at which ammonia began to accumulate. An excess of 5 mg per 100 ml of fluid had no effect on protein synthesis. It appears that when ammonia starts to accumulate above 5 mg $NH_3$-nitrogen per 100 ml fluid in the rumen, further supplementation with NPN would be of no benefit. Working with 35 different diets that varied in crude protein from 8 to 24% and in TDN from 53 to 85%, Satter and Roffler (1976) found mean cattle ruminal nitrogen concentrations to vary from 0.8 to

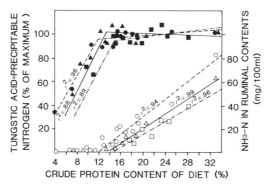

**Fig. 7.1.** Relationship between ammonia concentration of continuous culture fermentor contents and output of tungstic acid-precipitable nitrogen when either a purified, all-concentrate, or forage-concentrate (23:77) mixture was added to the fermentor. ○●, purified; □■, all-concentrate; △▲, forage-concentrate (From Satter and Roffler, 1974 with permission from *British J. Nutrition* and Cambridge University Press).

56.1 mg per 100 ml rumen fluid with increasing dietary protein levels. Mean ruminal ammonia and dietary protein percentages had the following relationship: $NH_3$-nitrogen (mg/100 ml) = 10.57 − 2.5% crude protein (CP) + 0.159 $CP^2$; $r^2$ = .88. Mean ammonia concentration in the rumen reached 5 mg/100 ml at approximately 13% dietary crude protein; above this level, ammonia increased rapidly with increasing dietary protein.

The amount of fermentable energy sources available influences the growth of rumen microbes and the quantity of ammonia converted to protein. Therefore, Satter and Roffler (1976) developed the following equation, including TDN, which should give a better prediction of the ammonia concentration in the rumen: $NH_3$-nitrogen (mg/100 ml) = 38.73 − 3.04% CP + 0.171% $CP^2$ − 49% TDN + 0.0024% $TDN^2$; $r$ = .92. Using this equation, the ruminal levels of ammonia shown in Table 7.1 were calculated for diets with varying TDN. Ruminal ammonia nitrogen reaches or exceeds 5 mg per 100 ml of fluid earlier with low-TDN diets than with high-energy ones with increasing protein. Ruminal ammonia nitrogen concentrations of 2 mg/100 ml or less should give NPN utiliza-

**TABLE 7.1.**

**Influence of Diet Composition on Mean Ruminal Ammonia Concentration and NPN Utilization**[a]

| Percent of crude protein in dietary dry matter | Total digestible nutrients (TDN) or digestible dry matter (DDM) (expressed as % of dietary dry matter)[b] | | | | | | | NPN utilization (%) |
|---|---|---|---|---|---|---|---|---|
| | TDN (DDM) | 55 (59) | 60 (63) | 65 (68) | 70 (72) | 75 (76) | 80 (81) | 85 (85) | |
| | | ---------------------------- (mg/100 ml) ---------------------------- | | | | | | | |
| 8  | | 6  | 5  | 4  | 3  | 2  | 2  | 1  |      |
| 9  | | 6  | 5  | 4  | 3  | 2  | 2  | 1  |      |
| 10 | | 6  | 5  | 4  | 3  | 2  | 2  | 1  | 90   |
| 11 | | 6  | 5  | 4  | 3  | 3  | 2  | 2  |      |
| 12 | | 7  | 6  | 5  | 4  | 4  | 3  | 3  | 0–90 |
| 13 | | 8  | 7  | 6  | 6  | 5  | 4  | 4  |      |
| 14 | | 10 | 9  | 8  | 7  | 6  | 6  | 5  |      |
| 15 | | 12 | 11 | 10 | 9  | 8  | 8  | 7  |      |
| 16 | | 14 | 13 | 12 | 11 | 10 | 10 | 10 |      |
| 17 | | 17 | 16 | 15 | 14 | 13 | 13 | 12 | 0    |
| 18 | | 20 | 19 | 18 | 17 | 16 | 16 | 15 |      |
| 19 | | 23 | 22 | 21 | 20 | 19 | 19 | 18 |      |
| 20 | | 27 | 26 | 25 | 24 | 23 | 23 | 22 |      |

[a]From Satter and Roffler (1976).
[b]TDN values are National Research Council (NRC) tabular values, and DDM values are the calculated equivalents.

## TABLE 7.2.

### Upper Limit for NPN Utilization[a]

| Percentage of crude protein in diet dry matter before NPN addition | TDN or DDM (expressed as a percentage of dietary dry matter)[b,c] | | | | | | |
|---|---|---|---|---|---|---|---|
| | TDN (DDM) | 55–60 (59–63) | 60–65 (63–68) | 65–70 (68–72) | 70–75 (72–76) | 75–80 (76–81) | 80–85 (81–85) |
| 8 | No[d] | 10.0 | 10.5 | 10.9 | 11.2 | 11.4 | |
| 9 | No[d] | 10.4 | 10.9 | 11.3 | 11.6 | 11.8 | |
| 10 | No[d] | 10.8 | 11.3 | 11.7 | 12.0 | 12.2 | |
| 11 | No[d] | 11.2 | 11.7 | 12.1 | 12.4 | 12.6 | |
| 12 | No[d] | No[d] | 12.1 | 12.5 | 12.8 | 13.0 | |
| — | — | 11.4[e] | 12.2[e] | 12.8[e] | 13.3[e] | 13.6[e] | |

[a]From Satter and Roffler (1976).

[b]TDN values are NRC tabular values and DDM values are the calculated equivalents.

[c]Percentage of crude protein after NPN adition.

[d]The upper limit for NPN utilization is below the crude protein content of the unsupplemented diet, and no benefit from NPN supplementation would be expected.

[e]Dietary crude protein where ruminal ammonia begins to accumulate when only plant protein is in the diet.

tion in excess of 90%. At levels of 3–5 mg/100 ml, utilization should range from 0 to 90%; above 5 mg/100 ml, utilization of added NPN would be nil.

When part of the dietary protein is replaced by NPN, the amount of true natural protein that escapes ruminal degradation is reduced and the fractional part of dietary nitrogen going through the ruminal ammonia pool is increased. Integration of the effects of dietary protein, diet digestibility, and allowable NPN substitution for true protein is shown in Table 7.2 for diets of varying natural protein and TDN contents. Natural protein levels of 12% or below would not utilize any NPN as urea with 55–60% TDN levels in the diet, but higher TDN levels would allow increasing levels of NPN to be added at given natural protein levels.

Slyter *et al.* (1979) conducted a study with fistulated steers to determine the minimum concentration of ruminal ammonia for maximum microbial growth and maximum nitrogen retention. Tungstic acid-precipitable nitrogen increased in ingesta as urea infusion increased to the point where the $NH_3$-N concentration was 2.2 mg per 100 ml of rumen fluid. Increases of $NH_3$-nitrogen above this level of rumen fluid had no effect on precipitable nitrogen. Increasing $NH_3$-nitrogen beyond 4.5 mg/100 ml had no effect on total VFA concentration. Nitrogen retention increased as urea increased until $NH_3$-nitrogen was 4.5

mg/100 ml, but no further increases in nitrogen retention were observed up to 22.5 mg $NH_3$-nitrogen per 100 ml rumen fluid.

Coppock et al. (1976) compared the rumen ammonia concentration in fistulated lactating cows fed urea twice daily in a concentrate mixture, urea fed once per day in a complete diet, and a gelatinized starch-urea product fed twice daily in the concentrate mixture. Sharp and similar ammonia peaks followed the feeding of urea and a gelatinized urea-starch product twice daily in the concentrate mixture, while a much lower ammonia peak following the once-daily feeding of the complete diet. It was concluded that complete diets offered ad libitum eliminate the need for products with sustained ammonia release.

## VII. AMMONIA LEVELS IN THE RUMEN OF FEEDLOT CATTLE

Burroughs et al. (1973) fed feedlot cattle the following diets: (1) low-protein basal; (2) basal plus low NPN; (3) basal plus high NPN; and (4) basal plus soybean protein isonitrogenous with diet 3. These diets were used by Satter and Roffler (1976) to evaluate their system of relating NPN utilization to rumen ammonia levels. In Fig. 7.2, the dotted line shows the point of ammonia accumulation and illustrates that the addition of NPN (urea) to appropriate low-protein diets results in increased growth rates of cattle but that further additions of NPN are without benefit. The animals, however, needed additional protein, as indicated by the higher gains achieved with the soybean protein-supplemented steers.

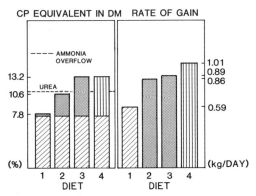

**Fig. 7.2** Growth response from plant protein or NPN supplementation of feedlot rations. ▨ basal; ▩ urea; ⊞ protein (From Satter and Roffler, 1976).

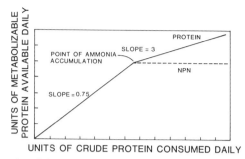

**Fig. 7.3.** Schematic relationship between MP, crude protein, and NPN (From Satter and Roffler, 1976).

## VIII. AMMONIA AND SYNTHESIS OF METABOLIZABLE PROTEIN

The absorbable protein in the gastrointestinal tract of ruminants was described as metabolizable protein (MP) by Burroughs *et al.* (1975a) and represents the dietary protein that is available in α-amino acid form for metabolism of the host. The amount of MP per unit of crude natural protein intake is greater if sufficient excess dietary energy is available for rumen microbes to utilize NPN sources. The amount of MP per unit of crude protein intake is higher when ruminal ammonia is utilized totally for microbial protein synthesis than when ammonia is in excess (Satter and Roffler, 1976). This point is illustrated in Fig. 7.3. Below the point of ammonia accumulation, all sources of nitrogen are approximately equal in providing MP. However, above this point, additional NPN contributes nothing to the amount of MP. Additional true protein in the diet adds to the amount of available protein for absorption to the extent that it escapes degradation in the rumen and is digested in the lower tract.

## IX. FACTORS IN MP FORMATION AND UTILIZATION

A number of factors involving the synthesis and utilization of MP have been considered, some of which have documented evidence and others which are tentative. Satter and Roffler (1976) itemized these factors as follows: (1) the amount of nitrogen recycled to the reticulorumen is equal to 12% of the dietary intake; (2) 85% of the crude protein intake with typical ruminant diets is in true protein form and 15% in natural NPN form; (3) 40% of the true dietary protein escapes degradation in the rumen and goes to the intestine, whereas all of the NPN and recycled nitrogen passes through the ruminal ammonia pool; (4) 90%

of the ruminal ammonia produced is incorporated into microbial nitrogen when the diet fed does not exceed the upper limit value for crude protein given in Table 7.1; (5) none of the ruminal ammonia derived from dietary crude protein fed in excess of the upper limit value in Table 7.2 is incorporated into microbial nitrogen; (6) 80% of microbial nitrogen is in true protein form and 20% in nonutilizable NPN form; (7) 80% of the microbial true protein will be absorbed as MP; (8) 87% of the dietary true protein that escapes degradation in the rumen will be absorbed as MP.

Quantitative information on the extent of degradation of different types of protein under different feeding conditions is quite limited. Values suggested by Miller (1973) for percentages of protein escaping degradation in the rumen are: barley, 10; cottonseed meal, 10; peanut meal, 10; sunflower meal, 25; soybean meal, 45; dried grass, 50; white fish meal, 50; and Peruvian fish meal, 70. Satter and Roffler (1976) concluded that their choice of 40% as an average for the amount of protein escaping degradation in the rumen is suitable, although a tentative value.

Approximately 80% of microbial crude protein is true protein and 20% is NPN, chiefly nucleic acids (Hungate, 1966; Smith, 1969). The amino acid composition of microbial protein appears to be little influenced by diet (Purser, 1970). Digestibilities of bacterial and protozoal crude protein fed to rats was 74 and 91%, respectively (McNaught et al., 1954). Digestibility data from Mason and Palmer (1971) and Bergen et al. (1968b) demonstrated that about 80% of bacterial nitrogen is absorbed by rats. While these data were obtained with rats, Satter and Roffler (1976) concluded that a value of 80% for true digestibility of microbial true protein in the ruminal gastrointestinal tract seemed reasonable for their equations. There is no direct measurement for the true digestibility of dietary protein that escapes degradation in the rumen. It might be assumed that digestibility in the intestine of that fraction of true protein that escapes degradation in the rumen would be slightly less than that of dietary true protein, since the more available proteins would have been degraded in the rumen. Egan et al. (1975) observed that 80–90% of the nitrogen entering the small intestine was truly digested. Based on these observations, Satter and Roffler (1976) assigned a value of 87% for digestibility of dietary protein not degraded in the rumen.

Satter and Roffler (1976), utilizing the facts and assumptions available, calculated MP from 1 kg of dietary crude protein, as illustrated in Fig. 7.4. When ammonia production in the rumen is not in excess of microbial need, 1 kg of dietary protein results in about 750 g of MP. Of this amount, 450 g is microbial and 300 g is of dietary origin. This indicates that if the crude protein content of a diet is equal to or less than the upper limit value in Table 7.2, the MP will be approximately 75% of the dietary protein intake. Dietary plant protein fed in excess of the upper limit will be utilized only at the 30% level of MP, and NPN above this limit would not be utilized at all for MP synthesis.

## X. Nitrogen Utilization by Dairy Cattle

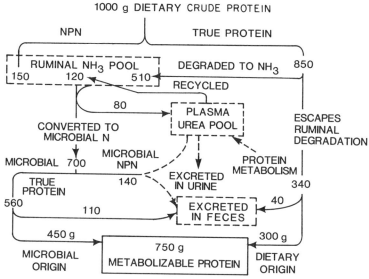

Fig. 7.4. Schematic summary of MP in calculations (From Satter and Roffler, 1976).

Huber and Kung (1981) reviewed protein and NPN utilization by dairy cattle. They concluded that there is still much to be learned about the symbiotic relationship between cows and rumen microbes needed to obtain maximum production of milk.

## X. NITROGEN UTILIZATION BY DAIRY CATTLE

Satter and Roffler (1975) proposed a number of factors to be taken into account in meeting the protein requirements of dairy cattle and ensuring optimal protein utilization. Lactating cows should not maintain ruminal ammonia nitrogen in excess of 5 mg per 100 ml rumen fluid, since increased levels have no effect on microbial protein synthesis; supplemental NPN is not utilized in typical diets containing more than 12–13% crude protein on a dry matter basis; NPN is approximately equal to true protein as a source of nitrogen in typical diets containing not more than 12–13% crude protein; and MP provides a better basis for calculating protein requirements and comparing protein sources than crude or digestible protein data. It has been found that 1 kg crude protein, regardless of the nitrogen source, is equal to about 0.75 kg MP in typical diets that contain no more than 12–13% crude protein, but 1 kg true plant protein fed in excess of this protein level equals only about 0.3 kg MP.

Protein supplementation of lactating cows should be related more to the stage of lactation than to milk production. Cows with above-average lactating ability may benefit from dietary protein as high as 16–17% on a dry matter basis during the first third of the lactation period. Cows in the latter two-thirds of lactation appear to require 12.5% or less dietary protein. During the first third of lactation, plant protein should be the source of nitrogen, with NPN providing most, if not all, of the supplemental nitrogen during the last two-thirds of lactation.

## XI. UREA FOR MP IN THE UREA FERMENTATION POTENTIAL (UFP) SYSTEM

Burroughs *et al.* (1973, 1975a,b) devised a system involving a UFP for the calculation of MP in ruminant diets. In utilizing this system, one must get acquainted with two values: the MP requirement of the animal and the UFP of the diet. MP and metabolizable amino acids (MAA) come from protein in the diet that escapes fermentation in the rumen and protein that arises from digestion of rumen microbes. Requirements for MP and feed values are determined from amino acid requirements for body maintenance, growth, and amounts deposited in milk. Figure 7.5 illustrates the MP requirements of growing feedlot steers making varying gains per day at given body weights.

An illustration of how to calculate the MP of a feed ingredient like alfalfa hay and the factors involved is given in the following equation: MP, of alfalfa hay per kilogram of dry matter = 193 (crude protein in hay, g/kg) × 0.05 (percentage of undegraded protein bypassing the rumen) × 0.90 (percentage of digestibility of undegraded protein) + [610 (g TDN/kg hay) × 0.104 (percentage factor for amount of microbial protein from TDN) − 15 (feed expense of forming metabolic fecal nitrogen)] × 0.80 (percentage of digestibility of microbial protein) = 47.6 (g/kg dry matter). The factor 0.104 is based on the fact that 52% of TDN is digested in the rumen and 25% of the digested TDN is transformed into microbial

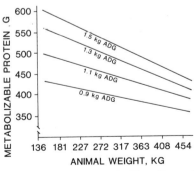

**Fig. 7.5.** MP requirements of feedlot cattle (From Burroughs *et al.*, 1973).

protein, 80% of which is digestible in the lower alimentary tract, i.e., 52 × 0.25 × 0.80 = 10.4.

A positive UFP value is an essential part of the UFP-MP system when urea is a part of the diet. Illustrations of how the UFP of corn grain or soybean meal can be calculated with the factors involved are given in the following equations:

$$\text{UFP for corn grain, g/kg dry matter} = \frac{(910 \times 0.104) - (100 \times 0.62)}{2.8} = +11.8$$

$$\text{UFP for soybean meal, g/kg dry matter} = \frac{(810 \times 0.104) - (515 \times 0.75)}{2.8} = -107.7$$

In the equation for corn, the value 910 is grams of TDN per kilogram of corn; 0.104 is the factor for the percentage of TDN converted to microbial protein; 100 is grams of crude protein in corn per kilogram; 0.62 is the factor for the percentage of crude protein degraded in the rumen; and 2.8 converts the nitrogen involved to equivalent grams of urea. Corresponding values are given in the soybean meal equation. The value of 11.8 in the equation for corn gives the grams of urea required per kilogram of corn grain dry matter to provide sufficient nitrogen to utilize the excess energy in the corn by rumen microbes. The negative value −107.7 indicates that excess nitrogen is present in the soybean meal, expressed as grams of urea per kilogram of soybean meal, and that for the excess nitrogen to be utilized, it needs to be fed with other ingredients that have equivalent +UFP value. In order for its excess nitrogen to be utilized, soybean meal with a negative value of −107.7 can be mixed with corn, which has a +11.8 UFP, since 107.7 ÷ +11.8 = 9.1 (kilograms of corn grain needed per kilogram of soybean meal required to utilize the excess nitrogen in the soybean meal).

In a mixed diet, it is necessary to summarize the positive and negative UFP values contributed by each ingredient. If the sum is negative, then one should forget about urea, since too much nitrogen is present. If the sum of the positive and negative UFP values of the various ingredients is positive, then one adds 1 g of urea per kilogram of diet per unit of positive UFP. That is, a UFP of +10 indicates that 10 g urea should be added per kilogram of diet (1% urea). Table 7.3 gives the crude protein, TDN, UFP, and natural MP values of many common feeds, as well as the amount of synthesized MP that can be produced by rumen microbes from feeds that have +UFP if sufficient urea (NPN) or plant proteins are present in the diets.

Tritschler et al. (1983) formulated five isocaloric diets with varying UFP values (+3.8, +1.2, −1.4, −3.9, and −6.9) with corn grain, corn silage, soybean meal, urea, and minerals, using values in Table 7.3, and fed them to beef steers in metabolism and comparative slaughter feedlot trials. Average daily gain, dry matter intake, ME intake, and energy balance were parabolically related to decreasing UFP levels in the diets. Maximal feedlot performance and

## TABLE 7.3.

### MP and UFP of Selected Feeds, Dry Weight Basis[a]

| Feed | Crude protein | TDN | Natural MP | UFP | MP from +UFP[b] |
|---|---|---|---|---|---|
| | % | % | g/kg | g/kg | g/kg |
| Alfalfa, hay | 19.3 | 61 | 47.6 | −42.8 | — |
| Alfalfa, hay | 16.3 | 57 | 43.0 | −34.0 | — |
| Barley, grain | 13.0 | 83 | 92.4 | −1.6 | — |
| Beet molasses | 8.7 | 89 | 58.0 | +3.7 | 8.2 |
| Beet pulp | 10.0 | 72 | 66.2 | −1.7 | — |
| Bermuda grass | 9.5 | 44 | 37.5 | −12.4 | — |
| Cattle milk, whole | 25.8 | 130 | 131.4 | −29.8 | — |
| Citrus pulp | 7.1 | 88 | 46.6 | +13.8 | 30.7 |
| Citrus molasses | 10.9 | 77 | 62.2 | −6.3 | — |
| Corn, silage | 8.1 | 70 | 55.4 | +6.4 | 14.2 |
| Corn, cobs | 2.8 | 47 | 11.1 | +10.0 | 22.2 |
| Corn, ears | 9.3 | 89 | 65.8 | +11.6 | 25.8 |
| Corn, gluten w/bran | 28.1 | 82 | 119.6 | −44.7 | — |
| Corn, grain, yellow | 10.0 | 91 | 71.8 | +11.8 | 26.3 |
| Cottonseed hulls | 4.3 | 41 | 23.5 | +3.8 | 8.5 |
| Cottonseed meal, solv. ext. | 44.8 | 75 | 151.4 | −92.0 | — |
| Fish, herring meal | 76.7 | 76 | 224.1 | −177.1 | — |
| Fish, menhadden | 66.6 | 75 | 200.5 | −150.4 | — |
| Grains, brewer's | 28.1 | 66 | 106.3 | −50.7 | — |
| Lespedeza, hay | 13.4 | 55 | 40.0 | −25.0 | — |
| Meat scraps | 57.1 | 76 | 180.0 | −124.6 | — |
| Oats, grain | 13.2 | 76 | 87.1 | −4.7 | — |
| Oats, aerial part | 9.2 | 61 | 51.4 | −5.2 | — |
| Peanut meal, solv. ext. | 51.5 | 77 | 168.2 | −109.2 | — |
| Rice grain, polished | 8.2 | 84 | 56.1 | +10.8 | 24.0 |
| Ryegrass, Italian, aerial part | 16.3 | 62 | 47.2 | −32.2 | — |
| Rye, grain | 13.4 | 85 | 95.1 | −1.8 | — |
| Rye, straw | 3.0 | 31 | 12.8 | +3.5 | 7.8 |
| Sorghum, milo grain | 12.4 | 80 | 93.2 | +6.8 | 15.1 |
| Soybean meal, solv. ext. | 51.5 | 81 | 171.6 | −107.7 | — |
| Sugarcane, mol., blackstrap | 4.3 | 91 | 22.8 | +20.1 | 44.7 |
| Urea, limited to +UFP values | 280.0 | 0 | 2225.0 | — | — |
| Wheat, bran | 18.0 | 70 | 95.1 | −18.9 | — |
| Wheat, grain | 14.3 | 88 | 100.1 | −2.9 | — |
| Yeast, brewers | 47.9 | 78 | 160.9 | −99.2 | — |

[a] From Burroughs et al. (1975a).
[b] Calculated as grams of +UFP per kilogram of diet × 2.225 = MP. The 2.225 value represents the grams of MP that may be synthesized for each unit of +UFP per kilogram of diet if sufficient urea is provided.

## XI. Urea for MP in the Urea Fermentation Potential (UFP) System 171

energy utilization were observed at UFP levels of $-1.4$ to $-3.8$, which corresponded to maximum MP concentrations. Empty body protein gain per day increased logarithmically with decreasing UFP levels. However, while predicted net protein for gain ($NP_g$) ranged from 196 to 283 g/day in the diets, the actual tissue protein gains ranged from 62 to 109 g/day.

Poos et al. (1979) fed dairy cows a basal diet that contained 8.6% crude protein (UFP = +5.2) and 16.2–17.0% crude protein (very negative UFP) after the addition of either urea or soybean meal to the basal diet. Changing the UFP from positive to negative increased the digestibilities of dry matter and acid detergent fiber and increased milk production. Thomas et al. (1984) conducted an experiment to test the metabolizable protein (MP) system in predicting the amount of urea useful in a corn-based diet fed to steers. They concluded that the MP system correctly predicted the amount of utilizable urea in their diets.

# 8

# Energy and Nitrogen Utilization in Feedstuffs

## I. INTRODUCTION

Corn grain and corn silage are the most extensively utilized sources of dietary energy for ruminants in the United States. While feedlot research with cattle has shown that whole shelled corn can be satisfactorily fed, grinding, flaking, crimping, and other processing of the grain is commonly practiced prior to feeding. Although corn grain, corn silage, and corn field wastes have considerable nitrogen, it is inadequate for efficient growth, reproduction, and lactation of animals. For this reason, extensive research has been conducted to find economical supplies of supplemental nitrogen from oil seed meals, slaughterhouse by-products, fish by-products, or nonprotein nitrogen (NPN) sources such as urea, biuret, and ammonia-containing compounds.

Soybean, cottonseed, peanut, flaxseed, sunflower, and other meals from plants have been widely investigated and utilized as supplementary sources of protein for corn-based high-energy diets fed ruminants. Since these sources of nitrogen, as well as those from slaughterhouses and marine products, are expensive, urea and biuret have been competitive with them as sources of nitrogen.

## II. NET ENERGY

Lofgreen and Garrett (1968) defined a net energy (NE) system for the evaluation of feedstuffs for cattle utilizing metabolism and comparative slaughter feedlot trials, and presented data on the NE for maintenance ($NE_m$) and NE for gain ($NE_g$) for many common feedstuffs. Corn grain exceeded the $NE_m$ and $NE_g$ values of all other grains. Fox et al. (1970) utilized the comparative slaughter technique with feedlot cattle to compare the relative NE values of corn grain and corn silage diets to those of similar diets containing bird-resistant (BR) sorghum grain and silage. The corn grain had an NE value of 1.29 Mcal/kg compared to 0.94 Mcal/kg for the BR sorghum grain.

Maxson et al. (1973) utilized the comparative slaughter technique with feedlot cattle to determine the NE values of corn, non-bird-resistant (NBR), and BR sorghum grains. In the metabolism phase, corn, NBR, and BR grains had 3.21, 2.63, and 2.49 Mcal of ME per kilogram of dry matter, respectively. Corresponding $NE_m$ values for the grains were 3.40, 1.98, and 1.78 Mcal/kg, and $NE_g$ values were 1.49, 1.15, and 0.73 Mcal/kg (dry weight) when diets were fed ad libitum. Jesse et al. (1976b) used a comparative slaughter technique with steers to determine the NE values of corn and corn silage fed in diet combinations of 30:70, 50:50, 70:30, and 80:20 of corn:corn silage on a dry basis. NE values for gain for corn and corn silage were 1.17 and 1.05 Mcal/kg, respectively.

## III. CORN

Vance et al. (1972b) conducted three experiments to evaluate the effects of feeding growing-finishing steers diets containing either dry, whole shelled, or dry crimped corn grain with varying proportions of corn silage. Whole shelled corn diets containing 0–4.5 kg corn silage per head daily gave the best gains and feed efficiency. When crimped corn diets were fed, better performance resulted if silage was provided at a level of 2.3–9.1 kg per head per day. Whole shelled corn and crimped corn gave equivalent performance of steers when 4.5 kg silage was fed per head daily. Carcass traits of steers were not affected by treatments. More extensive clumping of rumen papillae was found with the crimped corn diets than with the whole corn diets, but 2.3 kg of corn silage per head per day in diets alleviated this problem. Protozoa were reduced or eliminated completely after 14–21 days of feeding silage-free diets.

White et al. (1972) reported that grinding corn or replacing soybean meal with urea in diets fed steers had no effect on energy or crude protein digestibilities. The pH and the proportions of acetate and butyrate in rumen fluid tended to be higher, and the propionate and total volatile fatty acids (VFA) to be lower, on the whole, than with ground corn diets, which suggests that whole corn has some characteristics of roughage.

Corn may be rolled or flaked after exposure to steam, or steam under pressure, and studies have generally indicated that flaked corn has a higher feeding value than other forms of corn in cattle diets. Hentges et al. (1966) compared flaked, ground, and cracked corn in a 126-day feedlot trial with steers. They obtained daily gains of 3.6, 3.3, and 3.4 lb per day per head and a feed efficiency of 5.8, 6.5 and 6.7 lb per pound of liveweight gain, respectively. Matsushima and

Stenquist (1967) compared flaked and cracked corn with feedlot cattle and observed daily gains of 2.76 and 2.76 lb per head per day and feed efficiency values of 7.56 and 8.04 lb of feed per pound of gain, respectively. Johnson *et al.* (1968) compared flaked and cracked corn in a metabolism study and found flaked corn to have more digestible dry matter, protein, and ether extract; further, the estimated NE was 10.6% greater in the flaked corn.

In a metabolism trial, Cole *et al.* (1976b) found total starch digestibilities for flaked and cracked corn to be 99.0 and 93.6%, respectively. However, in a feedlot trial with cattle, Burkhardt *et al.* (1969) found average daily gains of 2.67, 2.56, and 2.49 lb/day for whole, cracked, and flaked corn and feed efficiencies of 8.74, 8.63, and 8.63, respectively.

## IV. GRAINS TREATED WITH ALKALI

Berger *et al.* (1981a) treated whole corn, sorghum grain, oats, barley, and wheat as follows: air-dried, reconstituted (19% moisture), 3% sodium hydroxide (NaOH), ammonium hydroxide 3% ($NH_4OH$), and 6% $NH_4OH$. The nylon bag technique was used to measure the digestibilities of the treated samples. The greatest response to 3% NaOH was with wheat, which increased 64 percentage units over that of reconstituted wheat. Sorghum grain digestibility showed a 11.4 percentage unit increase, and corn, oats, and barley averaged about 20 percentage unit increases in digestibility with the 3% NaOH treatment. Similar effects were observed by Orskov and Greenhaugh (1977) when they processed whole grain with alkali and fed it to cattle.

## V. HIGH-MOISTURE CORN

Preston (1975) summarized three experiments with cattle fed diets with dry whole, dry rolled, or high-moisture (26%) rolled corn grain. The high-moisture rolled corn grain diets had 4% higher $NE_g$ values than those with either dry rolled or whole grain. Significant linear decreases in $NE_m$ and $NE_g$ values were observed when corn silage was substituted for grain in the diets, and dietary NE was an additive function of the ingredients.

Prigge *et al.* (1976) examined the influence of 80% ground high-moisture corn and 80% dry ground corn diets on the utilization of soybean meal or urea nitrogen supplements with steers. They observed that peak rumen fluid ammonia levels at 0.5–1.0 hr postfeeding were highest with the ground corn-urea diets. Urea was higher in the plasma of steers fed the dry corn diets. Lambs fed the high-moisture corn produced less urinary nitrogen and retained more nitrogen than those fed dried corn. It was concluded that high-moisture corn in both trials utilized nitrogen more efficiently than dried corn and that urea can be used satisfactorily as a high-moisture corn supplement.

Bothast et al. (1975) treated high-moisture (27%) corn with ammonia, ammonium isobutyrate, isobutyric acid, and propionic-acetic acid at concentrations of 0.5, 1.75, 1.5, and 1.2%, respectively, and stored the materials in partially open wooden bins. All of the chemicals immediately reduced bacterial and actinomycete counts and eliminated yeasts and molds. However, after about 30 days of storage, all treatments had secondary fungal growth.

Anderson et al. (1981) processed whole high-moisture corn (1) in an oxygen-limiting structure only; (2) the same as (1), except that it was rolled just prior to feeding; (3) treated fresh upon harvesting with a final concentration of 3% NaOH on a dry matter basis; and (4) treated fresh upon harvesting with a final concentration of 4% $NH_4OH$ on a dry matter basis. The NaOH-treated corn produced the highest dry matter and fiber digestibilities and rumen and fecal pH. The $NH_4OH$-treated corn produced little feeding response in the digestibility trial. In a feedlot trial, cattle fed the NaOH-treated corn had lower gains and poorer feed efficiency than controls.

Mowat et al. (1981) conducted trials with steers fed high-moisture shelled corn in diets containing about 60% corn silage on a dry basis, with treatments of the grain as follows: (1) ensiled, ground, (2) a 1.5% mixture of acetic and propionic acids with whole grain, and (3) 2% anhydrous ammonia with whole grain. Urea was included to supply sufficient nitrogen in all diets. The ammonia treatment resulted in similar starch and energy digestibilities compared to the ensiled ground corn, but the acid treatment gave a large decrease in starch digestibility. Both ammoniation and acid treatments improved the digestibility of acid detergent fiber. In a feedlot trial with steers, the ensiled ground high-moisture corn was compared in diets with 3.7% urea (dry basis), with and without monensin. The urea tended to reduce gains and feed efficiency slightly, while monensin decreased feed intake and slightly improved feed efficiency.

Beeson and Perry (1958) fed 2-year-old finishing steers and heifers diets that contained high-moisture (32%) ensiled ground ear corn and found that it was utilized 10–15% more efficiently than regular ground ear corn (dry basis). Tonroy et al. (1974) compared the nutritive value of dry corn, ensiled high-moisture corn, ensiled reconstituted high-moisture corn, and VFA-treated high-moisture corn for growing-finishing beef cattle. Equal weight gains were made with 3–13% less dry matter when cattle were fed high-moisture corn than when fed dry corn.

Horton and Holmes (1975) found that rolling propionic acid-treated, high-moisture corn improved the daily gains of beef cattle over controls. Similar amounts of corn were voided in feces by both 10- and 20-month-old steers when propionic acid-treated, high-moisture whole corn was fed. Utley et al. (1977)

evaluated diets fed feedlot heifers that contained rolled dry corn or propionic acid-treated, high-moisture corn, with and without the addition of monensin. Treating unground high-moisture corn with propionic acid preserved it by preventing fungal growth and aflatoxin contamination. Feedlot performance and carcass characteristics of heifers were similar when they were fed the two types of corn. Greater nitrogen-free extract digestibility was observed with diets that contained acid-treated, high-moisture corn, but greater crude fiber digestibility occurred in diets containing dry corn.

## VI. CORN SILAGE

Ensiled whole corn (*Zea mays*) is widely utilized in both feedlot and dairy cattle diets. It is considered a moderate-energy, low-protein feedstuff, but with small additions of protein and minerals, as well as added energy, it forms the basis of very acceptable diets for high-performing ruminants. In practice, the whole plant with grain is stored with dry matter ranging from 22 to 55%. Total digestible nutrients (TDN) in three silages stored at 25, 30, and 33% dry matter were found to be equivalent when fed to cows by Huber *et al.* (1965), even though voluntary intake of dry matter increased as dry matter in the silage increased.

When corn was harvested for silage at 32 and 55% dry matter, it was observed that the more mature silage had 11% less dry matter yield per acre and 10% lower dry matter digestibility by cows (Byers and Ormiston, 1964). Colovos *et al.* (1970) ensiled corn plant at soft dough, medium-hard dough, early dent, and glazed-frosted stages of ear maturity and fed it in diets to Holstein steers and adult wether sheep. Digestibility and nutritive value in steers were greatest for silage harvested at the medium-hard dough stage. The two more mature silages were digested and utilized with greater efficiency by wethers than by steers. It was concluded that the feeding value of corn silage determined with sheep was not applicable to cattle.

Jesse *et al.* (1976a) fed Hereford steers diets that contained the following ratios of corn to corn silage: 30:70, 50:50, 70:30, and 80:20. Daily gains for these four dietary groups were 0.90, 1.06, 1.13, and 1.11 kg, respectively. Gross energetic efficiency (megacalories retained per megacalories of gross energy consumed) increased as slaughter weight increased, with maximum conversions of 16.5 and 11.9% for empty body and carcass gains, respectively. This indicated that finishing appears to be a more efficient process (calorically) than growth (protein deposition). Krause *et al.* (1980) fed steers drought or normal corn silages supplemented with either 0 or 2.7 kg corn grain and 0.57 kg supplement per head per day. They observed that those fed the drought silage gained as rapidly and efficiently as those fed the normal silage-containing diets.

Steers fed corn grain gained faster and converted feed more efficiently than those fed no corn grain. Urea supplementation benefited normal silage diets but not those containing drought silage.

Keith et al. (1981) compared the performance of feedlot cattle fed either brown midrib-three (bm3) corn silage or its normal genetic counterpart, with and without various levels of corn grain. The average daily gains for bm3 and normal silages were 0.90 and 0.82 kg when no additional grain was fed, and 1.03 and 1.01 kg when grain was fed at 2% of body weight. The lower lignin content of bm3 corn silage resulted in greater fiber utilization by feedlot cattle when little and no added corn grain was fed, but when corn grain was fed at 2% of body weight, the bm3 silage had no advantage. More dependence is placed on corn silage to supply the energy needs of ruminants as grain is decreased in their diets. The fiber in the corn plant that must be utilized by rumen microbes for energy is bound to lignin, and lignin inhibits microbial digestion of much of the cellulose and hemicellulose of the plant cell wall (Muller et al., 1972). The bm3 gene reduces the percentage of lignin in the corn plant (Kuc and Nelson, 1964; Barnes et al., 1971; Lechtenberg et al., 1972).

Utley et al. (1973) conducted a 3-year study to compare corn silage and oat or rye pasture used separately and in combination for finishing steers. Those fed corn silage ad libitum plus small amounts of grain pasture consumed one-third as much silage as steers finished on corn silage and cottonseed meal in drylot when allowed two-thirds as much grazing as steers that grazed oat or rye pasture without supplementation. Bertrand et al. (1975) compared corn silage and sorghum silages in beef cattle diets and concluded that when the cattle were fed at 50% on an as fed basis, (1) there was no tendency toward digestive problems, (2) in many cases the most profitable feedlot diet occurred when just enough silage was utilized with reconstituted grain to furnish the "roughage properties" necessary for finishing steers, (3) corn silage was utilized approximately 13% more efficiently than sorghum silage, and (4) rolling the silages to crimp the grain for feeding in high-energy diets was not worthwhile.

Young and Kauffman (1978) compared Hereford steers fed grain, corn silage, or 50% corn silage (dry basis) in finishing diets, and observed daily gains (in kilograms) and feed : gain ratios of 1.44, 8.4; 1.09, 11.5; and 0.91, 12.1, respectively. Steers fed the corn silage diet had to be fed to 8.2% heavier weights to attain ultrasonic fat thickness readings similar to those of the grain-fed steers. Steers fed grain diets had higher dressing percentages and more fat over the longissimus muscle, but organoleptic evaluations of steaks and roasts from all three dietary groups were equivalent.

Joanning et al. (1981) fed steers diets that contained immature corn silage, mature silage, immature silage plus grain, and mature silage plus grain, with a 10% soybean meal-mineral mixture in all diets on a dry matter basis. Observed digestibilities of dry matter, starch, neutral detergent fiber, and protein averaged

## VI. Corn Silage

68, 98, 63, and 63% for immature silage; 67, 98, 49, and 60% for mature silage; 70, 86, 53, and 61% for a 1 : 2 ratio of immature silage to grain; 70, 89, 37, and 62% for a 1 : 2 ratio of mature silage to grain; and 84, 98, 54, and 74% for an all-grain diet. Incomplete starch digestibility was the principal reason for the decreased efficiency observed when corn silage and grain were fed as mixtures. Pendlum et al. (1978) conducted experiments to study the effect of feeding different quantities of protein, either throughout a growth trial or during the first third of a growth trial, on the performance of steer calves fed corn silage. Steers were fed the following levels of total protein and supplemental protein (44% soybean meal) in kilograms per head per day, respectively, in treatment groups: (1) 0.27, 0; (2) 0.53, 0.18; (3) 0.62, 0.27; and (4) 0.71, 0.36, with steers in group (1) offered corn silage ad libitum and others fed equal levels of corn silage. Steers fed soybean meal gained faster than those fed no supplemental protein, consumed more dry matter, and had greater feed efficiency.

Bolsen et al. (1976) reported that steers fed corn silage had better performance (7.8% faster gains) than those fed wheat silage. Oltjen and Bolsen (1980) found that steers fed corn silage from corn harvested at the hard-dent stage gained faster and consumed more dry matter than steers fed barley silage from barley harvested at the dough stage; feed efficiency was equivalent for the two silages. Corresponding wheat silage supported poorer gains than corn or barley silages. Silages supplemented with soybean meal or urea gave equivalent feedlot performances. Proximate analyses of corn, barley, wheat, and oat silage and the composition of acid detergent fiber accounted for the variations in daily gains of steers.

Byers (1980a) conducted a study to determine the effects of limestone and monensin on the NE value of corn silage with Hereford steers in a comparative slaughter trial. Limestone (1%) was added at the time of ensiling, and monensin was provided in the mineral supplement. Neither limestone nor monensin altered daily gains, but dry matter intake was decreased by 7.8, 12.2, and 18.7% with monensin, limestone, and in combination, respectively, and corresponding NE values were increased by 8.8, 9.6, and 15.4%.

Wilkinson and Penning (1976) conducted studies with 3-month-old bull calves to determine if cattle could meet a target growth rate of 1 kg per head per day from 100 kg liveweight to slaughter at 450 kg and 15 months of age on a diet consisting predominantly of corn silage with urea as a sole source of supplemental nitrogen. Their studies showed that bulls could be grown from 100 to 450 kg body weight during a 12-month feeding period when fed corn silage ad libitum plus 288 kg protein concentrate (peanut meal, rape meal, corn gluten, hydrolyzed feather meal, meat and bone meal) plus 243 kg rolled barley per head but not with urea as the sole nitrogen supplement. Reduced energy intake during the first 9 months of the study and the absence of significant compensatory growth later in bulls fed urea resulted in decreased slaughter weights. More efficient use was made of the protein supplement when it was offered from the start of the feeding period.

## A. Anhydrous Ammonia-Treated Corn Silage

Cook and Fox (1977) conducted trials with Hereford and Charolais crossbred cattle allotted to diets that contained (1) untreated corn silage plus minerals and vitamins, (2) one-half treated anhydrous $NH_3$ corn silage, (3) three-fourths treated anhydrous $NH_3$ silage, (4) fully treated anhydrous $NH_3$ silage, (5) one-half treated $NH_3$ silage plus soybean meal until steers reached about 600 lb, (6) three-fourths $NH_3$ silage plus soybean meal, as in (5), and (7) untreated silage plus soybean meal gradually decreased from 12.5% to 10.5% crude protein as cattle increased in weight. The decreasing soybean meal diet was started at 12.5% crude protein and was decreased 0.5% for each 100-lb gain until the final level of 10.5% was reached. Steers fed decreasing soybean meal-containing diets had higher daily gains and better feed efficiencies than steers fed the full $NH_3$-treated silage, and the decreasing soybean meal system gave better performance than the continuous high-soybean meal level. However, the daily gains between the full-treatment anhydrous $NH_3$, decreasing soybean meal, and continuous high-soybean level were not significant, and all had significantly greater daily gains than controls, i.e., 2.4–2.5 versus 0.73 lb, respectively. In another trial, when anhydrous $NH_3$ was added to corn silage at levels of 5.1, 7.0, or 9.0 lb per ton of dry matter, to obtain one-half, three-fourths, and full treatments of $NH_3$, average daily gains were 1.50, 2.18, and 2.38 lb during the first 92 days, respectively, but all treatments ranged in gains from 2.40 to 2.49 lb per day after 180 days of feeding the diets.

Owens et al. (1970b) investigated the effects of urea, biuret, and soybean meal as nitrogen sources and sodium metabisulfite, an antibacterial agent, on nitrogenous fractions of corn silage and corn stalklage during fermentation in laboratory silos. Urea caused the greatest increase in ammonia, and only small amounts of urea remained after fermentation. Nitrate was greater when the bacterial agent was added but appeared to be lowered by the extended fermentation that occurred with urea. Owens et al. (1970a), in a similar study with laboratory silos utilizing urea, biuret, soybean meal, and sodium metabisulfite added to corn silage and corn stalklage, observed that nitrogen additions were associated with increased wet and dry losses.

Britt and Huber (1975) treated corn silage (35% dry matter) with urea, aqua ammonia, or an ammonia-mineral suspension at an added nitrogen level of 0, 0.23, 0.45, or 0.9% and placed the materials in polyethylene bags that were evacuated and sealed. On day 86 of fermentation, portions were placed in open containers at 25°C, and on days 0, 7, 21, and 28 of refermentation, samples were taken. Upon refermentation, all samples with added nitrogen had large pH increases; the 0.23% nitrogen stimulated lactic acid production, but higher levels depressed it until it disappeared. Silages with the ammonia-mineral suspension required the longest time for visible fungi to appear, but total fungi fell more rapidly with the ammonia treatments. All NPN treatments depressed fungal counts, with the higher concentrations having the most effect. Yeasts dominated throughout refermentation, but some *Pencillium* and *Aspergillus* species were identified.

Glewen and Young (1982) reported that the application of anhydrous ammonia at 1% of dry matter at the time of ensiling increased the stability of the silage following aeration. The dry matter loss was less for treated silage. Neither treated nor untreated silage showed large changes in ammonia, but an increase in acid detergent insoluble nitrogen occurred in the untreated silage near the end of the refermentation period.

## B. Urea Supplementation of Corn Silage

Coleman and Barth (1974) conducted digestibility trials with steers fed corn silage and urea-containing diets. Greater percentages of urea (5.4–45.4% of dietary nitrogen) resulted in decreases in the percentage of intake nitrogen retained, the percentage of absorbable nitrogen retained, and net protein utilization. Barth *et al.* (1974) reported that increasing the percentage of urea in supplements of corn silage diets fed steers to supply 24, 36, and 48% of the total nitrogen resulted in an increase in crude protein digestibility and TDN, but a decrease in retention of absorbed nitrogen and net protein utilization.

Aston and Tayler (1980) fed dairy bulls initially weighing 432 kg corn or grass silage ad libitum plus 0, 5, or 10 g barley on a dry matter basis per kilogram of liveweight for 80 days. Daily liveweight gains were 1.00, 1.32, and 1.46 kg for the 0-, 5-, and 10-g levels of barley supplement with corn silage, respectively. Corresponding daily gains with steers fed grass silage with these levels of barley supplement were 0.65, 0.98, and 1.22 kg, respectively. In a similar trial, these workers found that bulls fed corn silage ad libitum with either urea or aqueous ammonia mixed at the time of feeding, plus 0 or 5 g barley per kilogram of liveweight daily for 90 days, had gains ranging from 0.63 to 1.09 kg/day. Ammonia treatment had no effect on feed intake or feed efficiency. Barley improved both liveweight and carcass gains.

## C. Effect of Microbial Additives on Silage Quality

Nutrient preservation of silages is largely due to fermentation by lactobacilli or other lactic acid-producing microbes. In order for these microbes to thrive, there must be adequate fermentable material, anaerobic conditions, and sufficient numbers of lactobacilli present per 100 g (Stirling, 1953; Anderson, 1956).

Burghardi et al. (1980) treated corn silage in laboratory silos with $22.5 \times 10^7$ live *Lactobacillus bulgaricus, L. acidophilus, L. brevis, Streptococcus lactis,* and *Strep. cremoris.* NPN levels in the silage tended to be increased by the organisms, but recoveries of dry matter and crude protein were not affected by the inoculations. Lactic acid, acetic acid, ethanol, and acid detergent fiber concentrations were not affected by the addition of microbes. *Lactobacillus acidophilus, Aspergillus oryzae,* and mixed lactic ferment enzymes in silages failed to improve the feedlot performance of steers or the digestibility by lambs fed diets containing treated silage.

## D. Formic Acid Treatment of Silage

Breirem and Ulvesli (1960) concluded that preservation of cut grass silage by formic acid was partially due to reduction in pH and to selective antibacterial effects. Waldo et al. (1971b) reported that alfalfa and sorghum hybrid silages preserved with formic acid improved the dry matter intake and increased the weight gains of dairy heifers fed the treated materials. Formic acid added to corn silage at two stages of maturity at harvest (28 versus 44% dry matter) increased silage intakes and milk production (Huber, 1970).

McCullough (1972) prepared corn silage without additives, with 0.5% formic acid, or with 5% dried steep liquor concentrate (DSLC), which is a dried, condensed, fermented corn extractive containing 7.8% lactic acid, and fed the silages to wethers and steers. Both of the treated silages had improved dry matter, crude protein, crude fiber, and nitrogen-free extract digestibilities. A short feeding trial of 40 days with Hereford steers gave improved growth, reflecting the differences in the digestibilities of the silage-containing diets.

## E. Monensin Added to Corn Silage

Gill et al. (1976) fed steers feedlot diets that contained 14, 30, or 75% of their dry matter from corn silage containing either 0 or 300 mg of monensin per day. The monensin improved feed efficiency by 6% across all roughage levels with 5–14% less feed intake but had no effect on gain or carcass parameters. Pendlum et al. (1980) compared corn silage-containing diets fed steers with soybean meal or urea supplements providing 0 or 200 mg monensin per head daily in a 106-day feedlot trial. Steers fed monensin in diets gained slightly faster with both sources of nitrogen, and had a lower feed intake with improved feed efficiency compared to steers not fed monensin. The total and essential amino acid levels in plasma were higher in steers fed monensin at the 106-day sampling time.

Byers (1980b) conducted an energy balance study with steers fed corn silage diets with and without monensin supplementation and found that weight gains were similar with the two diets (920 versus 923 g/day), while dry matter consumption was reduced 5.91% with monensin. Feed conversion was improved 5.84% with monensin, but terminal body composition and total energy retained were not altered. Limiting feed intake to 70% of the ad libitum level reduced daily rates of fat deposition more ($-47.3\%$) than it reduced protein ($-25.1\%$). Monensin increased $NE_m$ by 5.7% but had no effect on $NE_g$.

## F. New Endosperm Types of Corn

An improved amino acid balance in corn protein occurred with the introduction of the *opaque-2* gene in corn breeding systems. The most important improvement was a higher lysine content of opaque-2 corn than was present in normal hybrid corn (Mertz *et al.*, 1964). Improved growth rates were reported for rats (Mertz *et al.*, 1965), chicks and pigs (Drews *et al.*, 1969), and swine (Cromwell *et al.*, 1967). Since ruminants can synthesize microbial protein from NPN, it was not expected that they would benefit as much from the improved quality of protein in corn as nonruminants.

Thomas *et al.* (1975) compared diets fed feedlot cattle that contained opaque-2 corn, opaque-2 corn silage, or normal corn grain or silage. There were no advantages for opaque-2 corn grain or silage over their normal counterparts in metabolism studies. More digestible energy (DE) and higher levels of ammonia in the rumen were found with normal corn silage diets than with those containing opaque-2 corn silage. Steers fed normal corn grain or silage gained faster than those fed opaque-2 corn grain or silage-containing diets. Steers fed normal corn silage also had higher carcass grades than those fed opaque-2 corn silage.

Redd *et al.* (1975) utilized abomasal cannulated Hereford steers to determine the effects of feeding normal compared to opaque-2 corn on abomasal amino acids and plasma nitrogen constituents. Equivalent amounts of trichloroacetic acid-precipitable nitrogen and individual amino acids reached the abomasum on the normal and opaque-2 corn-containing diets.

Thomas *et al.* (1976a) conducted metabolism trials with a number of new endosperm types of corn in growing steers. In one trial, waxy corn had 3.5 percentage units lower DE than normal, opaque-2, or waxy opaque corn. In another trial, sugary-2 opaque-2 corn had 17 percentage units higher absorbed nitrogen retained and 15 percentage units higher dietary nitrogen retained than normal and sugary-2 corn. All other parameters studied indicated that the genotypes were equivalent.

## VII. GRASS SILAGES

An excess of herbage in one season and a shortage in another has always been a problem with stockmen depending on nonconcentrate sources of nutrients. In the conservation of any forage crop, it is important to harvest it when vegetation is at the most nutritive stage of growth and preserve it. Due to rainfall and other problems of hay making, ensiling may be the most satisfactory means of preservation when forage quality is highest. In silage making, the ensiled mass should attain a concentration of lactic acid sufficient to inhibit certain forms of microbial activity in order to preserve the material in its most palatable and nutritious form until it is required for feeding (Barnett, 1954).

Silage of good appearance, odor, and palatability was made from Pangola grass containing 65–75% moisture without the use of additives or preservatives when thorough packing removed air pockets in the ensiled mass (Peacock et al. 1961). Long yearling steers gained 1.29 lb/day when fed an average daily diet of 59 lb Pangola grass silage and 2 lb cottonseed meal. Sometimes grass silage has insufficient energy for good lactic acid fermentation. Wing and Becker (1963) reported that citrus pulp or ground snapped corn could be used satisfactorily as aids in preserving forage silage when used at the rate of 150 lb/ton.

Baldwin et al. (1975) compared the feeding value of ensiled Pangola grass and ensiled water hyacinths in diets fed sheep. The Pangola grass was harvested without wilting and chopped to 1.6 cm in length, while the hyacinths were harvested from a freshwater lake, chopped, and pressed in a Vincent press (1.27 kg/cm$^2$) to remove the water. Then 4% dried citrus pulp and 0.5% sugarcane molasses were mixed with the material prior to ensiling. Dry matter intake by sheep was greater for the Pangola grass silage than for the hyacinth silage. Digestibility of organic matter and crude protein was also higher for the Pangola grass silage. Linn et al. (1975b) fed lambs ensiled aquatic plants, aquatic plants plus corn, or aquatic plants plus alfalfa silage and observed dry matter digestibilities of 41.4, 32.0, and 38.5%, respectively. Energy and nitrogen digestibilities were lower in diets that contained aquatic plants than in those that contained alfalfa silage or alfalfa silage plus corn. Lambs fed ensiled diets of alfalfa or alfalfa plus corn had dry matter digestibilities of 61.9 and 66.2%, respectively.

Petchey and Broadbent (1980) compared steers fed diets that ranged from grass silage alone to rolled barley plus 1 kg long hay per day. The barley was offered alone once daily or in a homogeneous mixture. Steers fed the mixture had 8.9% greater feed intake than those that consumed the barley once daily, but there were no increases in daily gains.

## Additives Added to Grass Silage

Shultz et al. (1974) evaluated several additives to ryegrass straw silage and determined their effect on *in vitro* dry matter digestion (IVDMD). Material ensiled with 4.5% sodium hydroxide (NaOH)·potassium hydroxide (KOH) after

2 days had 20% greater IVDMD than untreated native straw and more than 13% improvement after 21 days ensiling with 0.5% limestone. Adding 1% urea improved digestibility, and 20% molasses maintained a pH of 5 and 4 for hydroxide-treated and untreated material ensiled for 64 days, respectively. Organic acids formed in hydroxide-treated silage changed from predominantly acetate-lactate to acetate-butyrate after ensiling for 16 days. Untreated silage has an acetate-lactate fermentation during 64 days of ensiling. Shultz and Ralston (1974) reported that lambs consumed more dry matter and nitrogen and had higher digestibility of organic matter from ryegrass straw silage with 4.5% NaOH·KOH, 20% molasses, 1% urea, and 0.5% limestone than from ensilage not treated with alkali. Body weight gains were equivalent in heifers fed ryegrass silage or straw ensilage with urea and limestone, with and without hydroxide, when each heifer consumed 0.9 kg barley-cottonseed supplement daily.

Hinks and Henderson (1977) ensiled wilted Italian ryegrass with either formic acid, formic acid plus formaldehyde, or a mixture of formic acid plus formaldehyde plus propionic acid, and fed the ensilage to steers (350 kg initial weight). The additives inhibited fermentation, resulting in higher levels of water-soluble carbohydrates and lower levels of organic acids than untreated silage. Higher levels of true protein and lower levels of ammonia with the formaldehyde treatment indicated some protein protection by the additive. Formic acid alone had little effect on digestibility by steers but increased dry matter intake and liveweight gains compared to untreated silage. The propionic acid was added to increase the stability of the silage on exposure to air.

Hidiroglou *et al.* (1977) ensiled barley and oat forage grown on low-selenium soils and fed the silages to cows that had been injected 2 months prior to calving with selenium plus vitamin E. Three calves born from untreated cows on barley silage died from nutritional muscular dystrophy by 50 days of age, whereas no calves died from any of the selenium-vitamin E–treated cows or from the untreated cows fed oat silage.

## VIII. SORGHUM GRAIN

Sorghum grain is widely grown, especially in semiarid regions of the world, and ranks fifth in acreage of crops, being exceeded by wheat, rice, corn, and barley. Grain sorghums were bred for the quality and quantity of their kernels; sweet sorghums or sorgos were chosen for optimum sugar contents in their stems and acceptability as forages; and grassy sorghums were cultivated for forage. The grains are usually white, yellow, red, or salmon pink, but brown seeds predominate among the BR types. The brown seeds are generally more bitter due to tannins or tannin-like pigments. The sweet sorghums have in common tall, juicy, sweet stalks and generally smaller kernels than grain types. Grass sorghums have slender stems, narrow leaves, numerous tillers, and small seeds compared to sweet and grain sorghums.

## A. Proximate Analysis and Total Digestible Nutrients (TDN)

According to Hale (1970), the protein content of a sorghum grain from 24 sources ranged from 8.67 to 11.73% (mean, 10.2%) on a dry matter basis. Breur et al. (1967) reported a range in crude protein from 7.9 to 15.2% for Texas varieties of sorghum grain. Grain grown in a particular area had similar protein concentrations. Ether extract values of over 2% are rare, and values as low as 1% have been observed compared to values as high as 4% with corn. Since the energy value of ether extract is high, this may partially explain a generally lower feeding value of sorghum grain compared to corn. Sorghum grain, corn, barley, and wheat averaged 75.8, 92.9, 87.2, and 88.9% TDN, respectively. Differences among the grains in TDN are not indicated by proximate analysis, but are attributed mostly to variations in the digestibility of certain components. Processing of sorghum grain generally increases its digestibility. Sheep masticate sorghum grain more thoroughly than cattle, and processing is not as important with them as with cattle.

## B. Effect of Variety

Grain sorghums have been estimated to have 90–95% of the feeding value of corn grain even though they are chemically similar (Morrison, 1959). However, Hall et al. (1968) found very little difference in the NE of sorghum grain and corn for finishing cattle in a comparative slaughter trial utilizing finely ground grains. They suggested that other varieties of sorghum grain may give different results. Maxson et al. (1973) compared ground BR and NBR types of sorghum grains with corn grain in metabolism and comparative slaughter trials with steers. The corn, NBR, and BR sorghum grains had 3.21, 2.63, and 2.49 Mcal of metabolizable energy (ME) per kilogram on a dry matter basis, respectively. $NE_m$ values for the grains were 2.40, 1.98, and 1.78 Mcal per kilogram and $NE_g$ values of 1.49, 1.15, and 0.73 Mcal per kilogram of dry weight of diets fed ad libitum, respectively.

White and Hembry (1978) reported that incorporating 20% rice straw in diets improved the DE values of two BR sorghum grain concentrates by 2–11%. Energy, nitrogen-free extract, and crude protein digestibilities and nitrogen retention were lower when diets contained BR instead of NBR sorghum grain. Brommelsiek et al. (1979) found that methionine hydroxy analog (MHA), when supplemented at a level of 7 g per day per steer in BR sorghum grain diets, slightly improved TDN, ME, $NE_m$, and nitrogen balance values but had no effect on $NE_g$. Daily gains were not affected by MHA, but protein deposition in tissues was increased.

## C. Processing

Sorghum grain has a dense, hard endosperm and a waxy bran cover that makes it difficult for rumen microbial fermentation. Apparently, cattle masticate whole sorghum grain only for sufficient salivation to allow swallowing, and very little of the grain is regurgitated to be remasticated. When sorghum grain makes up 60–85% of the concentrate in finishing diets for cattle, any improvement in energy utilization, as reflected by increased gains and feed efficiency, needs to be utilized. Various processing methods for sorghum grain for cattle include ground, dry-rolled, soaked, pelleted, steam-rolled, steam-processed flaked, pressure cooker flaked, reconstituted, and ensiled.

Grinding is probably the least expensive method for preparing sorghum grain for cattle diets. The grain should not be ground so fine that it is dusty or floury, since this is more expensive and may make it less palatable. If molasses or silage is used in the diet, some of the dusty effect may be overcome. Diets containing 78–85% sorghum grain that was fine ground, coarse ground, or dry rolled had equivalent digestibilities (Husted et al., 1968; Buchanan-Smith et al., 1968). Digestibility of dry-rolled sorghum grain by steers was found to be much higher in a diet that contained 50% grain and 50% alfalfa than in a diet containing 98% grain. This may have been due to a slower rate of passage through the rumen with the high-roughage diet or to a better balance of energy and nitrogen in the diets.

Husted et al. (1968) reported a slight improvement in TDN for sorghum grain soaked prior to rolling, but Ely and Duitsman (1967) showed no advantage of water-soaked grain compared to dry-rolled grain fed finishing steers. Soaking of grains can lead to fermentation or souring and to mechanical problems during milling of wet material. Steam treatment of sorghum grain for 3–5 min at about 180°F prior to rolling gave no improvement in steer gains and feed efficiency compared to dry-rolled sorghum grain (Hale, 1963).

Flaked sorghum grain is produced by subjecting the grain to sufficient low-pressure steam to raise the moisture content to 18–20% and passing the material through a roller mill having a clearance of about 0.002 inch. If the flaked grain is to be stored for more than 1 day, it must be dried. Hale et al. (1966a) found that steam processing and flaking of milo increased daily gain and feed intake of steers by 0.12 and 0.45 kg, respectively, over a dry-rolled milo diet. Processing reduced feed requirements by 38 kg per 100 kg of gain. Steam processing increased TDN from 71 to 79% but did not affect protein digestibility.

Popping of sorghum grain disrupts the organization of the kernel, and the finished product is similar to popped corn. Feeding trials generally indicate that popped sorghum grain has very low density, and cattle fail to consume enough to gain at a high rate (Hale, 1970).

## D. Reconstituted Grain

Sorghum grain may be reconstituted to 25–30% moisture and stored in an airtight silo for approximately 21 days prior to feeding. It is necessary to roll or grind the grain upon removal from storage before feeding. Buchanan-Smith *et al.* (1968) used cattle and sheep to determine the digestibilities of organic matter, starch, and reducing sugars, as well as the ME and nitrogen retention of reconstituted-rolled, and coarsely and finely ground sorghum grain. Digestibilities of organic matter and energy by cattle were greater with diets that contained the reconstituted grain than in finely or coarsely ground grain diets. No differences were observed in digestibilities of the nutrients by sheep due to processing, except that steam processing decreased nitrogen digestibility. Steam processing and reconstitution resulted in the same digestibility of diets by cattle and sheep.

## E. Early-Harvested Grain

At maturity, moisture should be in the range of 12–15%, but due to weather conditions and the threat of birds, grain may be harvested at 20–35% moisture with maximum dry matter obtained. The grain alone may be harvested or the entire head may be taken to produce "sorghum grain head silage." With either method, the grain may or may not be ground before storage. Grinding permits better packing and exclusion of air, which aids in preventing spoilage. Riggs (1965) reported that early-harvested sorghum grain, cracked or ground prior to feeding steers, improved feed efficiency compared to grain harvested at a moisture level of 12–15%. Generally, weight gains are not improved by early harvesting of grain compared to dry ground sorghum grain.

## IX. CITRUS BY-PRODUCTS

The frozen citrus concentrate industry produces large amounts of citrus pulp consisting of peel, rag, and seeds. The dried pulp, citrus molasses, and citrus meal are all valuable feedstuffs for both dairy and beef cattle. Their protein content is low, but all of the products are rich in available carbohydrates and are considered to be energy feeds. While it is considered a concentrate, citrus pulp has a roughage-like action in the rumen because it maintains an acetate : propionate ratio of approximately 4 : 1, similar to that of roughages. However, it is not a substitute for roughage such as hay or pasture. Citrus molasses is palatable to all classes of cattle, and much of it is added to citrus pulp in the final drying process. Urea may be readily mixed with citrus molasses.

Kirk and Davis (1954) recommended using citrus by-product containing diets for finishing feedlot cattle, supplementation of calves after weaning, creep feeding, and for cows on pasture during periods of feed shortage. Peacock and Kirk (1959) compared the feeding value of dried citrus pulp, corn feed meal, and ground snapped corn for feedlot steers. Their data showed no differences in daily gains, TDN of diets, or dressing percentages and carcass grades of steers fed the three ingredients in diets. Chapman et al. (1953) compared cane molasses and citrus molasses in diets containing ground snapped corn, dried citrus pulp, and cottonseed meal fed steers grazing grass pasture. The citrus molasses-containing diets had less feed efficiency.

Chapman et al. (1972) compared the feeding value of dried citrus pulp, citrus meal, citrus molasses, and wet citrus pulp in diets fed beef cattle. All of the citrus products were excellent sources of energy. The dried citrus pulp was found to be high in DE and calcium but low in protein and phosphorus. With proper supplementation with protein and phosphrous, citrus pulp could provide 40% of feedlot diets with excellent results. Pelleted citrus pulp is well utilized by feedlot cattle, but it loses some of its roughage property. Citrus seed meal is variable in protein (averaging about 23%), low in fiber, and variable in fat content, crude fiber, and nitrogen-free extract. It can be used as a protein supplement in beef cattle diets. Citrus molasses was used successfully at a level of 10–20% in feedlot diets and at a maximum level of 6 lb per head daily on pasture. Wet citrus pulp is not widely fed, due to the cost of transporting the 75–85% moisture present, and needs equivalent supplementation as the dried pulp.

Hentges et al. (1966) replaced corn meal in pelleted high-concentrate diets fed finishing steers with steam-dried citrus meal plus sources of protein and phosphorus. Diets containing 0, 15.8, and 31.6% steam-dried citrus meal had equivalent weight gains and feed efficiency. Diets that contained 47 and 63% citrus meal resulted in reduced weight gains, feed efficiency, and propionic acid production in the rumen. Increased increments of citrus meal caused smaller and darker papillae in the rumen mucosa. Dressing percentage and carcass grade decreased in value with increased increments of citrus meal. All diets produced steaks of equal tenderness and juiciness. The writers (Hentges et al., 1966) pointed out that superior dried citrus meal is golden yellow; little or none of it is charred or dark.

Hadjipanayiotou and Louca (1976) compared dried citrus pulp and grape marc as substitutes for part of the barley in diets fed bull calves. The diets were supplemented with crude protein to compensate for the less digestible crude protein in the citrus pulp and grape marc. Including dried citrus pulp at 60% of the concentrate diet resulted in no effect on the feed intake and performance of calves. Citrus pulp appeared to compare favorably with barley as an energy source. Grape marc in diets at levels of 15 and 30% tended to increase feed intake and decrease feed efficiency. At the 30% level of grape marc in diets, slightly less weight gains and dressing percentages were observed, and its value as an energy source appeared to be about half that of barley.

## A. Citrus Condensed Molasses Solubles

Citrus condensed molasses solubles (CCMS) is the residue obtained from the fermentation of citrus molasses to alcohol. Chen et al. (1981) evaluated CCMS as an energy source for steers and lambs. The CCMS contained 10.6% crude protein, 74.6% nitrogen-free extract, and 14.6% ash on a dry matter basis. In a steer feedlot trial, the CCMS was substituted at levels of 0, 3.15, 6.30, and 9.45% of the dry matter of either corn or citrus pulp in the diets. Weight gains, feed efficiency, and carcass characteristics of steers were equivalent with the various dietary treatments. When lambs were fed diets that contained 0, 10, or 20% (dry matter) of CCMS as a replacement for dry corn and soybean meal, there were no effects on organic matter digestibility. However, crude fiber and ether extract digestibilities increased as the dietary CCMS increased from 0 to 20%. Crude protein digestibility was decreased by CCMS at 20%, but there were no differences in nitrogen retention. The feeding of CCMS in sheep diets increased propionate and valerate but decreased acetate and isovalerate in rumen fluid. The CCMS was palatable in diets, with little, if any, laxative effect when fed at levels of 20% or less on a dry matter basis.

Ammerman et al. (1963) studied the nutrient composition of whole dried citrus seeds, citrus seed hulls, and citrus seed kernels and their value as a source of protein for lambs. Average values for nutrients in seeds expressed as a percentage of dry matter were: protein, 16.2; ether extract, 45.1; crude fiber, 13.2; nitrogen-free extract, 22.1; and ash, 3.4. When supplying 88% of the total protein in lamb diets, protein in the dried citrus seed meal was equal in digestibility and biological value to protein in soybean meal. Ammerman et al. (1965b) fed three samples of dried citrus pulp, citrus meal, and one sample each of light- and dark-colored citrus meal pellets to wethers in a metabolism trial. The citrus products made up 72.5% of the total diets, and approximately 33% of the total protein was supplied by citrus pulp or citrus meal. Digestibility of protein in the diets varied from 84.1% for one of the citrus pulps to 56.6% for the dark-colored citrus meal pellets. Energy digestibility of the samples varied from 82.4 to 64.9%.

Ammerman et al. (1966) utilized dried citrus pulp as an absorbent for droppings of broiler chicks and determined the digestibility of nutrients in the resulting litter with lambs. The poultry litter diet had greater apparent digestibility of crude protein and less digestibility of ether extract, but other nutrients had equivalent digestibilities in the citrus pulp and poultry litter plus citrus pulp diets.

The data indicate that citrus pulp can be used as poultry litter and subsequently fed to ruminants.

## B. Citrus By-Products for Dairy Cattle

Dried citrus pulp has been shown to be an excellent feedstuff for dairy cattle (Becker and Arnold, 1951; Keener *et al.,* 1957). Citrus pulp with corn silage diets increased fat-corrected milk output by lactating cows and maintained rumen acetic:propionic acid ratios (Drude *et al.,* 1971). Schaibly and Wing (1974) replaced 0, 33, 67, and 100% of corn silage with dried citrus pulp in diets and determined their digestibility by the nylon bag technique in fistulated Jersey steers. Digestibility of dry matter and energy increased at the 67% level of citrus pulp. Protein digestiblity decreased only when all of the corn silage was replaced by citrus pulp.

Wing (1974) conducted three experiments with dairy cattle. In the first experiment, fistulated steers were fed diets consisting of 33 and 67% concentrates with 0–60% dried citrus pulp. Digestibilities of energy, dry matter, and protein were not affected by the level of citrus pulp. Acetate:propionate ratios in rumen fluid were over 4:1 and were equivalent with all diets. In the second trial, the proportion of pellets in a 33% citrus pulp diet was varied with fistulated steers. The pellets did not influence the acetate:propionate ratios in rumen fluid. In the third experiment, pelleted and conventional dried citrus pulp-containing diets were compared when fed to lactating cows. The two forms of citrus pulp resulted in equivalent milk production and percentages of fat, protein, and solids-not-fat, as well as equivalent titratable acidity and chloride content of milk.

Pinzon and Wing (1976) studied the effects of pelleted and conventional dried citrus pulp as a replacement for corn in diets fed fistulated dairy steers. Soybean meal was added to keep protein comparable, and diets contained 5% urea and 33% sugarcane bagasse. Citrus pulp was varied in the diets from 0, 19, and 38 to 55%. There were no differences between pelleted and nonpelleted citrus pulp in terms of the amount of urea in the blood and the pH and VFA of rumen fluid; however, with both forms, rumen ammonia was lower with the two highest percentages of citrus pulp.

## X. SUNFLOWER MEAL

Studies with growing and lactating ruminants indicate that sunflower meal is equivalent to soybean and cottonseed meals as a source of high-quality protein. Richardson *et al.* (1981) reported the crude protein, crude fiber, ether extract, nitrogen-free extract, ash, and gross energy of sunflower meal to be 33.7%, 22.0%, 2.7%, 27.9%, 6.3%, and 4.10 Mcal/kg, respectively, on a dry weight basis. The nutritional value of sunflower meal was studied with cattle by Pearson *et al.* (1954). Sunflower meal in diets fed finishing steers gave gains, feed efficiency, dressing percentages, and carcass grades equivalent to those of cottonseed meal on the basis of the crude protein level. Studies with growing calves showed that diets containing sunflower meal, rapeseed meal, or soybean meal were equivalent for animal performance (Stake *et al.*, 1973). Schingoethe *et al.* (1977) found that sunflower meal and soybean meal were equivalent as protein supplements for lactating cows. VFA proportions were unchanged, but the pH of the rumen fluid was lower in the cows fed the sunflower meal diet.

Richardson *et al.* (1981) studied the nutritional value of sunflower meal as a protein supplement for growing cattle and sheep. Sunflower meal and cottonseed meal were equivalent in value on a protein basis in diets fed growing-finishing steers. No differences in digestibility or nitrogen retention were observed at levels of 0, 5.5, and 11% sunflower meal in diets when substituted for equal amounts of cottonseed meal on a protein basis. Lambs were fed sunflower meal, cottonseed meal, or a combination of the two protein sources in 8 and 12% crude protein growing-finishing diets. Lambs fed the diets that contained 12% crude protein had similar gains and feed efficiencies. Lambs fed formaldehyde-treated sunflower meal retained a higher percentage of dietary nitrogen than lambs fed soybean meal treated with formaldehyde (Amos *et al.*, 1974).

## XI. FERMENTED AMMONIATED CONDENSED WHEY (FACW)

Whey is commonly an environmental pollutant that contains livestock nutrients. Clark (1979) pointed out that $5.7 \times 10^6$ kg lactose and $73 \times 10^6$ kg lactalbumin are discarded annually in waste whey in the United States. A fermented, ammoniated whey product was reported to lack palatability for lactating cows and growing heifers by Hazzard *et al.* (1958). A similar product was reported to support satisfactory growth of 6-month-old calves by McCullough and Neville (1972).

An improved whey fermentation process was described by Reddy *et al.* (1976). In this process, sweet whey was fermented to lactic acid by *L. bulgaricus* bacteria, with the lactic acid produced continually neutralized with anhydrous ammonia. The fermentation broth was evaporated to 70% solids to yield a liquid called "fermented ammoniated condensed whey (FACW)." The FACW contained 55% crude protein, of which 75–77% was from ammonium lactate, 17% from whey proteins, and 6–8% from bacteria.

Crickenberger *et al.* (1981) conducted trials to evaluate FACW as a crude protein supplement for feedlot cattle fed corn silage or corn–corn silage diets. In one trial, the addition of FACW, soybean meal, or urea as crude protein supplements for corn silage or 40% corn plus 60% corn silage diets, rates of gain and feed efficiency were significantly increased over those of steers fed similar diets not supplemented with crude protein. Supplementation of similar diets with two sources of sulfur (sodium sulfate and calcium sulfate) had no effect on the feedlot performance or carcass traits of steers. Feed efficiencies tended to be better for FACW and soybean meal supplements than with urea in the diets. The data obtained in these trials indicated that FACW was equivalent to soybean meal as a source of crude protein for cattle fed corn silage or corn plus corn silage finishing diets.

Erdman *et al.* (1981) studied the effects of FACW on the acid-base metabolism of lactating Jersey cows. The treatments included replacement of 0, 33, 66, and 100% of soybean meal with FACW in a 60% corn silage plus 40% concentrate diet (dry basis). Increasing FACW to 100% replacement reduced feed intake and milk production 27 and 32% compared to cows fed 0 and 33% FACW, respectively.

## XII. FEATHER MEAL AND HAIR MEAL

Feathers, hair, and horn contain keratins and, on a dry weight basis, are 85–99% protein. They are by-products of the poultry and meat-processing industries. Such products in their natural state are very resistant to digestion by proteolytic enzymes of the alimentary tract. The products are first autoclaved or cooked at 140–150°C with live steam (2.8–3.5 kg/cm) in a closed cooker. After 30–45 min at optimum heat and pressure, the pressure is released and the resulting slurry is cooked for an additional hour to drive off excess water. The slurry is then transferred to a hot air dryer or steam tube, where it is dried to 6–8% moisture, followed by grinding to yield free-flowing meals (Thomas and Beeson, 1977).

Feather meal was found to replace half of the soybean meal supplement in lamb finishing diets (Jordan and Croom, 1957). In a metabolism study, Thomas and Beeson (1977) compared feather meal, hair meal, and soybean meal in diets fed growing Hereford steer calves. Digestible energy (DE) values of 74.9, 72.0, and 70.9% were obtained with soybean meal, feather meal, and hair meal diets, respectively. The authors concluded that if soybean protein is 90% digestible, then only 62.8 and 49.7% of the protein in feather meal and hair meal are digestible when calculated by difference.

Daugherty and Church (1982) fed a fistulated steer grass hay ad libitum plus 1.5 kg protein concentrate (20% crude protein) twice daily with soybean meal, urea, feather meal, feather meal plus urea, hair meal, or hair meal plus urea supplying the supplemental nitrogen. Increased levels of ruminal ammonia-nitrogen and total VFA were found when urea was fed in combination with feather meal or hair meal compared to feather meal or hair meal alone. Ammonia-nitrogen and VFA levels were lower with the feather meal and hair meal diets than they were with the diets containing soybean meal.

Church *et al.* (1982) conducted digestion trials with wethers fed diets that contained 15% crude protein, of which 12, 22, or 32% was provided by hydrolyzed feather meal or hair meal. The control diet had soybean meal for supplemental protein. Crude protein digestibility decreased with increasing levels of feather meal or hair meal, but there were no differences in dry matter or energy digestibilities in the diets. However, in a second trial, dry matter digestibility decreased at the higher levels of the feather and hair meals. In a feedlot trial with steer calves fed a 40% roughage diet in which feather meal or feather meal plus urea was substituted for soybean meal on a 40/60 crude protein basis, there were no differences in weight gains or carcass quality, but feed efficiency was improved with the feather meal plus urea diet. In a second feedlot trial with 27.5% roughage diets, steers fed feather meal plus urea or hair meal plus urea diets (50/50 crude protein basis) showed greater feed efficiency than steers fed dried poultry waste plus feather meal (35/65 crude protein basis) diets, but there were no effects on weight gains or carcass quality (Church *et al.*, 1982).

## XIII. MOLASSES AND OTHER LIQUID FEEDS

Molasses is obtained most commonly from sugarcane, beets, citrus, corn, and wood. These various sources provide molasses of variable nutrient composition, as shown in Table 8.1 (NRC, 1971b). Sugarcane, beet, and corn molasses have

**TABLE 8.1.**

**Composition of Molasses from Various Sources**[a]

| Item | Sugarcane | Beet | Corn | Citrus | Wood |
|---|---|---|---|---|---|
| Dry matter, % | 77.0 | 77.5 | 72.5 | 67.7 | 62.4 |
| Brix | 79.5 | 79.5 | 78.0 | 71.0 | 65.0 |
| Nitrogen-free extract, % | 64.7 | 61.9 | 64.2 | 56.4 | 58.5 |
| Crude protein, % | 4.5 | 6.6 | 0.3 | 5.7 | 0.6 |
| Digestible protein, %[b] | 2.6 | 3.8 | −2.5 | 2.5 | −1.8 |
| TDN[b] | 57.2 | 61.1 | 60.5 | 52.5 | 53.4 |
| Calcium, % | 0.81 | 0.12 | — | 1.20 | 1.45 |
| Phosphorus, % | 0.08 | 0.03 | — | 0.12 | 0.03 |
| Ash, % | 7.8 | 8.9 | 8.0 | 5.4 | 3.1 |
| ME, Mcal/kg[b] | 2.67 | 2.20 | 2.19 | 1.90 | 1.93 |

[a]From NRC (1971b).
[b]As fed to cattle.

more dry matter and nitrogen-free extract (source of sugars), as well as more TDN and ME available for cattle than citrus and wood molasses.

Plain molasses can be fed in open troughs to ruminants. It may also be fed as a liquid feed containing urea (or other NPN sources) as a high-protein supplement in beef and dairy concentrates, a medium protein and mineral supplement fed in a lick-wheel device under range conditions, a complex mixture of natural proteins and NPN (urea, etc.,), or a mixture of molasses, NPN, phosphorus, and vitamin A. Many applications can be made of molasses-liquid feeds that fit the need of different situations. The increased use of urea in cattle feeds, relative ease of handling liquids, improved stability of vitamins and minerals, increased cost of natural protein, and the knowledge that liquid and dry feeds have comparable nutritive value have encouraged greater utilization of liquid feeds. Other advantages of liquid over dry feeds are generally less labor required in handling, decreased dustiness, little wastage, more uniformity, good appetite, and possibly better availability of nutrients.

## A. Liquid Feeds in Cow and Calf Operations

In recent years, there has been increased utilization of liquid feeds in range areas. Liquid feeds aid in providing energy, protein, and minerals to cattle, especially under poor grazing conditions. There has been an increased use of a 32% crude protein feed utilizing urea in molasses that limits intake to 0.5–1.0 kg per cow. Where greater energy supplementation is desirable, a 16% crude protein liquid feed is often fed in open troughs, where 2–3 kg per cow is consumed. Plain mill-run cane molasses has been fed to cattle on medium-quality pasture with good results (Chapman et al., 1965). Over a period of 3 years in Florida, it was shown that beef cows consuming an average of 2.3 kg of mill-run cane molasses per head daily had a higher conception rate and produced heavier calves than controls.

Bond and Rumsey (1973) compared the value of liquid molasses plus either urea or biuret when timothy hay was fed beef cows and yearlings in wintering trials and also when beef calves were fed an alfalfa-timothy mixed hay diet. The hays and supplements were offered free choice. In general, hay intake was decreased when molasses was fed, and molasses consumption tended to be greater when urea and biuret were not included in the supplement. Greater energy and protein consumption, when associated with molasses feeding, did not affect body weight changes. The cattle preferred the molasses-urea supplement over molasses-biuret. Yearling animals had mean ruminal ammonia values in descending order: hay, hay and molasses plus biuret, hay and molasses plus urea, hay and molasses. It was concluded that molasses supplements with a low level of NPN fed free choice with poor-quality hay may not consistently improve cattle performance.

Webb et al. (1972) reported that simultaneously administering molasses with urea or ammonium acetate to fistulated cows reduced the ammonia and pH of the rumen, as well as ammonia and urea in the blood. The molasses apparently increased nitrogen utilization by rumen microbes and decreased ruminal pH, thus reducing the rate at which ammonia was absorbed from the rumen.

White et al. (1973) conducted metabolism trials with steers to determine the influence of 0, 1, 2, and 4% urea and 0, 5, 10, and 20% cane molasses on nutrient digestibility and nitrogen balance. The 1% urea, which was equivalent to 5% urea in a molasses mixture fed 20% of the diet, improved the digestibility of crude fiber and slightly improved the digestibilities of dry matter and energy. Adding 2 or 4% urea had no additional effects. Digestibility of crude protein was increased with each urea addition, but the 1% levels gave the greatest nitrogen retention. Dry matter, energy and nitrogen-free extract digestibilities were increased by including 5 or 10% molasses in the diet and were further improved by the 20% level of molasses.

## B. Molasses in Feedlot Diets

Cane molasses as well as molasses from beets, corn, and wood are primarily sources of energy and need nitrogen from other feed ingredients in balanced diets. According to Burroughs et al. (1975a), sugarcane and beet molasses have urea fermentation potentials (UFP) of +20.1 and +3.7, respectively, while citrus molasses has a UFP of −6.3. These values indicate the lack (+) of or excess (−) of nitrogen present, expressed as grams of urea equivalent per kilogram of dry matter.

Hatch and Beeson (1972) conducted metabolism trials with steers to determine the effects of replacing corn with 5, 10, and 15% cane molasses. At 10 and 15% levels of molasses in the diets, digestibilities of dry matter and energy were improved, as well as nitrogen retention. White et al. (1975) reported that feeding a 6% urea-94% cane molasses mixture free choice to feedlot cattle provided with either whole or ground shelled corn diets resulted in greater feed intake, daily gain, and carcass weight than feeding 6% urea in a dry supplement as 25% of the diet. Average daily gain and carcass weight of steers fed soybean meal or urea-molasses supplements were equivalent.

Preston et al. (1970) fed Brahman or Brown Swiss × Brahman crossbred cattle finishing diets with final molasses levels of 68–76% of the ME and 10% Napier grass. The diets had supplementary protein (fish meal or poultry waste) replacing urea at low, medium, or high levels. Mean daily gains were better for the crossbreds than for the Brahman cattle (0.89 versus 0.77 kg). Fish meal improved gains compared to poultry waste by 44% and feed efficiency by 39%. Carcass composition was not affected by dietary source of protein.

Wahlberg and Cash (1979) compared the performance of growing-finishing steers fed diets varying only in liquid by-products. The diets consisted of 55%

corn silage, 35% cracked corn, and 10% liquid supplement. The liquid supplement consisted of 5% cane molasses and 5% liquid by-product. Four by-products (condensed whey solubles, condensed cane solubles, condensed beet solubles, and cane molasses) contained urea, while ammonium lignin sulfonate contained ammonium salts as sources of crude protein. All diets gave equivalent daily gains, but steers fed the lignin sulfonate had poorer feed efficiency and those fed the condensed beet solubles or condensed cane solubles had the best feed efficiency.

## C. Masonex (Hemicellulose Extract)

Masonex, a hemicellulose extract that contains approximately 60% solids with 10% simple sugars, 25% simple sugars after hydrolysis, 0.5% crude protein, 0.5% fat, 1.0% fiber, and 6.0% ash (Masonite Corporation, 1977), has been added to feedlot diets as a substitute for sugarcane molasses. Perry (1964) reported that hemicellulose extract from hardwood was equivalent to cane molasses when either was incorporated at a 10% level in diets fed feedlot cattle. Diets that contained either 7% Masonex or 7% cane molasses were equivalent for finishing lambs (Pfander et al., 1964).

Galloway (1975) partitioned Masonex into carbohydrate and polyphenolic fractions. It has been suggested that the polyphenolic fraction binds with certain proteins, decreases their microbial degradation, and increases the amount of dietary protein that escapes fermentation in the rumen (Hartnell and Satter, 1975). Chalupa and Montgomery (1979) compared the fermentability of Masonex with that of cane molasses in an *in vitro* batch culture of rumen fluid. Fermentability of the two substrates were similar. The investigators concluded that xylose and mannose of the Masonex fermented in a similar manner to the sucrose in molasses. Their data indicated that the phenolic fraction of Masonex does not interfere with utilization of the carbohydrates, and that Masonex and cane molasses have equivalent energy values for ruminants.

Hartnell and Satter (1978) mixed soybean meal with Masonex (10%) to determine if the phenolic constituents of Masonex would react with protein and reduce its degradation by rumen microbes. Soybean and cane molasses received a similar treatment. The fermentors were fed twice daily for 10 days with a diet consisting of alfalfa meal and treated soybean meal, and protein synthesis was determined by the amount of tungstic acid-precipitable nitrogen produced. About 15% more soybean protein escaped rumen microbial degradation when the soybean meal was mixed with Masonex than with sugarcane molasses. When soybean meal was extruded with 10% Masonex or sugarcane molasses and fed in diets to sheep, there were no differences between the dietary groups regarding body growth, feed efficiency, wool growth, and ruminal VFA or ammonia concentrations.

Crawford et al. (1978) compared Masonex and cane molasses in diets fed cattle in metabolism, feedlot, and palatability trials. The hemicellulose extract enhanced digestibility. Average daily gains (in kilograms) and feed : gain ratios were: Masonex, 1.35 and 6.80; cane molasses, 1.36 and 6.83. Carcass characteristics were not affected by the type of molasses. In a palatability trial, there were no differences in free choice consumption of the two molasses products when fed with a high-fiber diet to meet maintenance requirements.

Bartley *et al.* (1968) compared grain diets containing 10% cane molasses, Masonex, or dry hemicellulose extract fed lactating dairy cows. The cows responded similarly to the three diets in milk production, milk composition, grain intake, hay intake, and body weight. Vernlund *et al.* (1980) compared diets that contained either 8% Masonex or cane molasses with pelleted cottonseed hulls or pelleted, undelinted cottonseed hulls with and without added fat. The liquid treatments produced no effects on feed intake, milk yield, or fat percentage in milk.

## XIV. ALCOHOL PRODUCTION BY-PRODUCTS

Distillers grains and other products resulting from alcohol production have been available for many years. The initial material after fermentation will generally range from 7 to 10% dry matter and is referred to as *whole stillage*. Whole stillage may be screened or centrifuged to produce wet distillers grains and thin stillage, which is called *distillers solubles*. The wet distillers grains generally range from 30 to 50% dry matter and the thin stillage from 2–5% dry matter. Wet grains or mash is that portion of the grain remaining after fermentation. The solubles or thin stillage contains yeast cells in suspension, sugar alcohols, and organic acids (Klopfenstein and Abrams, 1981).

The two primary by-products of alcohol fermentation can be utilized in feeding programs for ruminants in a number of ways. They can be fed as whole stillage, wet distillers grains (WDG), dried distillers grains (DDG), thin stillage, condensed distillers grains (CDG), dried distillers solubles (DDS), or dried distillers grains plus solubles (DDGS). The feeding of DDG and DDGS have been most investigated for utilization by ruminants.

The National Research Council (NRC, 1971b) gave the nutrient composition for distillers grains from a variety of grain sources, as shown in Table 8.2. Since starch in the grain is fermented to alcohol and carbon dioxide, the protein and other nutrients are concentrated, resulting in an energy feedstuff with a residue after fermentation that is a protein source. However, distillers grains are generally poor sources of protein for nonruminants due to their limited levels of lysine, methionine, and tryptophan. These amino acids are typically low in the initial whole grains. Ruminants may upgrade the distillers grain protein by degradation and resynthesis of the nitrogen into microbial protein. But the resistance of the

## TABLE 8.2.

### Nutrient Composition of DDG from Various Grain Sources[a,b]

| Source | Corn | Milo | Wheat | Rye | Potatoes |
|---|---|---|---|---|---|
| Moisture, % | 6.2 | 6.2 | 6.6 | 7.8 | 4.3 |
| Ash | 2.4 | 4.2 | 3.3 | 2.5 | 7.0 |
| Crude fiber | 13.4 | 13.1 | 13.6 | 14.4 | 21.5 |
| Ether extract | 9.9 | 9.0 | 6.4 | 8.1 | 3.2 |
| Nitrogen-free extract | 44.6 | 40.5 | 43.3 | 52.8 | 44.3 |
| Crude protein | 29.7 | 33.3 | 33.5 | 22.1 | 23.9 |
| TDN (cattle) | 84.3 | 81.2 | 78.9 | 47.7 | 63.7 |
| Calcium | 0.10 | 0.15 | 0.08 | 0.14 | — |
| Phosphorus | 0.43 | 0.63 | 0.56 | 0.45 | — |
| Magnesium | 0.07 | — | — | 0.18 | — |
| Potassium | 0.18 | — | — | 0.08 | — |
| Sulfur | 0.46 | — | — | 0.47 | — |

[a]From NRC (1971b).
[b]Dry matter basis.

protein to microbial degradation may be utilized to advantage by bypassing the rumen and being available for absorption in the small intestine.

## A. Bypassing the Rumen

Klopfenstein and Abrams (1981) presented data shown in Table 8.3 that demonstrated that various distillers by-products were degraded less to ammonia by rumen microbes than protein in soybean meal. It was reasoned that if this were true, then some distillers grain protein could be replaced by urea without

## TABLE 8.3.

### In Vitro Ammonia Release (mg/100 g) of Soybean Meal and Distillers By-Products[a]

| Hours post-inoculation | SBM[b] | Condensed solubles | Dry DG[b] | Wet DG | Ensiled wet DG | |
|---|---|---|---|---|---|---|
| | | | | | No additive | Ca(OH)$_2$ |
| 0 | 2.0 | 3.7 | 2.0 | 2.0 | 2.2 | 3.1 |
| 12 | 24.7 | 13.8 | 2.4 | 2.9 | 3.6 | 8.0 |
| 24 | 37.4 | 21.9 | 10.1 | 11.7 | 12.3 | 15.3 |

[a]From Klopfenstein and Abrams (1981).
[b]SBM = soybean meal; DG = distillers grains.

TABLE 8.4.

Brewers Dried Grains Compared with Soybean Meal for Growing Calves[a,b]

| Supplement | Daily gain (lb) | Feed/ gain | Gain/ protein[c] | Comparative value (%)[d] |
|---|---|---|---|---|
| Urea | 1.45 | 7.74 | — | — |
| Soybean meal | 1.92 | 6.83 | 0.67 | 100 |
| Brewers dried grain-urea | 1.93 | 7.09 | 1.26 | 188 |

[a]From Klopfenstein and Abrams (1981).
[b]Diet for 106 days based on corn silage and 4% NaOH-treated corn cobs (50 : 50); average weight of cattle was 260 kg.
[c]Gain and protein in excess of urea control.
[d]Expressed as a percentage of soybean meal.

decreasing the performance of the ruminant. Brewers dried grains plus urea were compared to soybean meal in diets fed growing calves and beef cattle to test the possible advantage of utilizing bypassed protein from the dried brewers grains (Tables 8.4 and 8.5). Greater weight gains were obtained with all diets containing distillers grains plus urea than with those containing urea or soybean meal. It was estimated that, on the average, distillers grain protein had 173% of the value of soybean meal for growing cattle and DDGS had 137% of the value of soybean meal.

Waller et al. (1980) compared the performance of steers fed corn silage supplemented with DDGS or soybean meal and anhydrous ammonia-treated corn silage plus DDGS. The DDGS, when provided to supply the same quantity of supplemental protein as soybean meal, resulted in 10% faster gains and better

TABLE 8.5.

Efficiency of Protein Utilization of DDG and DDGS by Beef Cattle[a]

| Source | Average daily gain (lb) | Gain (lb)[b] | Supplemental protein (lb)[c] | Gain/ protein | Comparative value (%)[d] |
|---|---|---|---|---|---|
| Urea | 1.39 | — | — | — | — |
| SBM[e] | 1.54 | 0.15 | 0.29 | 0.51 | 100 |
| DDG | 1.69 | 0.30 | 0.29 | 1.03 | 200 |
| DDGS | 1.66 | 0.27 | 0.29 | 0.93 | 180 |

[a]From Klopfenstein and Abrams (1981).
[b]Average daily gain above that of urea control.
[c]Protein from natural supplemental sources per day.
[d]Expressed as a percentage of soybean meal.
[e]SBM = soybean meal.

feed efficiency than soybean meal supplements. It was concluded that the improved performance was not enough to justify the added cost of DDGS compared to soybean meal, but when DDGS was used with NPN-treated silage, the value of DDGS was optimized.

### B. Condensed Distillers Solubles (CDS)

Beeson (1975) fed steers a liquid supplement alone or with 2.5 or 5.0% CDS and obtained improved weight gain and feed efficiency. Beeson and Chen (1976) reported the CDS increased cellulose digestion and protein synthesis in the rumen.

### C. Wet Distillers Grains (WDG)

Feeding distillers grains in the wet form eliminates the cost of drying. The pH of by-products of grain fermentation is in the range 3.7–4.0 due to lowering of the pH of fermentation media to exclude undesirable microbes. This low pH prohibits ensiling by preventing further fermentation. The use of limestone or ammonia to increase the pH to 6.0 resulted in a butyric acid type of fermentation and an increase in soluble nitrogen. The dry matter intake of lambs was increased by feeding ensiled wet grains compared to dry grains (Klopfenstein and Abrams, 1981). Abrams et al. (1983) reported steers fed wet distillers grain–urea diets had better protein efficiency values than those fed soybean meal–urea or calcium hydroxide-ensiled wet distillers grain–urea diets.

### D. Whole Stillage

In a reveiw on the utilization of ethanol production by-products, Poos (1981) concluded that the principal problems expected with whole-stillage feeding include freezing during cold weather, mold and digestive problems, need for immediate consumption, and a continuous fresh supply. She indicated that a minimum of 2–5 lb of dry hay or hay equivalent should be provided for growing animals fed only stillage or stillage and grain for normal rumen function. Large-breed dairy cattle may consume 15 gallons of stillage daily without digestive problems or decreased total dry matter intake.

### E. Sheep

Satter et al. (1979) fed sheep fitted with duodenal reentry cannulae diets that contained DDG or DDGS, and observed that reduced levels of ammonia and increased levels of protein reached the small intestine compared to soybean meal diets. Diets that contained 0, 15, 30, and 45% of DDGS (dry basis) had an increased flow of amino acids to the small intestines with increasing levels of DDGS.

## F. Dairy Cattle

Satter *et al.* (1979), utilizing reentry cannulated Holstein heifers, found that as DDGS was increased in the diet from 0 to 45%, a corresponding increase in nonammonia nitrogen flowed to the small intestine. There was also an increase in apparent nitrogen digestion. Warner (1970) reported that distillers grains supported higher levels of milk production than soybean meal, cottonseed meal, linseed meal, and urea. Cows fed DDGS had a higher acetate:propionate ratio than cows fed control diets and a higher milk fat percentage (Palmquist, 1978).

# 9

# Protein and Energy in Forages and Roughages

## I. INTRODUCTION

Feeds that contain 18% or more crude fiber on a dry matter basis are classified as forages or roughages (NRC, 1971b). The level of ruminant nutrition related to the roughage : concentrate ratio in diets has been investigated extensively. Replacing part of the concentrate in a diet by an equal weight of roughage will reduce its energy content. A small decrease in energy intake will decrease the amount of energy in weight gain while producing little or no effect on rate of gain, and this improves feed efficiency. But if the energy intake is restricted further, daily gains will decrease until a point is reached where the energy requirement for maintenance will nullify this effect. Roughage at some level is generally essential for microbes in the rumen and overall performance of ruminants, but many roughages, when fed alone, will give only small gains or provide only maintenance requirements, or may be inadequate to maintain body weight.

The nutritive value of roughages is generally inversely related to their fiber content. The degree to which ruminants adapt to high-fiber diets varies with the proportions of structural carbohydrates contained in the plant cell walls. The quality of the fiber and its influence on utilization of nonfiber components of the diet are important factors in ruminant performance. Sometimes nonplant sources of roughages, such as oyster shell and polyethylene pellets, have been used. They provide stimulus for rumination and for prevention of bloat in cattle fed concentrates.

Performance of animals may be related more to diet intake than to diet digestibility. In Fig. 9.1, van Soest (1965) presents data on the feed intake of legumes and grasses as related to the percentage of cell wall content. A cell wall content above 60% is strongly associated with declining voluntary feed intake. Grasses that have a high cell wall content have relatively low lignin levels, which

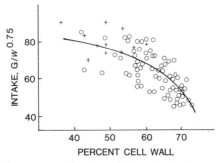

**Fig. 9.1.** Relationship between cell wall and feed intake. +, legumes; ○, grasses (From van Soest, 1965).

promotes higher digestibility. Legumes with high lignin and low cell wall content aid higher feed intake. Grinding and pelleting of forages usually result in greater intake, but the effect is greater with high cell wall-containing forages (Moore, 1964).

The relationship of lignin to digestibility varies with different plants, but the general relationship is shown in Fig. 9.2 (van Soest, 1964). As lignin increases in grasses and legumes, digestibility of dry matter by ruminants decreases in a linear manner. Van Soest (1967a) concluded that the effects of lignin are restricted to the hemicellulose and cellulose of the plant cell wall and stated that the conclusion of Drapala *et al.* (1947) that availability of soluble cellular contents is decreased by entrapment in lignified cells should be rejected.

Thiago and Kellaway (1982) separated wheat straw, oat straw, and paspalum hay manually into botanical fractions. They then (1) analyzed the three largest fractions for cell wall constituents, silica and nitrogen, and (2) determined the digestibility of dry matter, cellulose, and hemicellulose with nylon bags in the

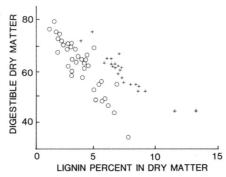

**Fig. 9.2.** Relationship between lignin content and digestible dry matter. +, legumes; ○, grasses (From van Soest, 1964).

## I. Introduction

rumen of sheep. The proportions of digested dry matter, cellulose, and hemicellulose after 72 hr in nylon bags were negatively correlated with the lignin content of dry matter.

Xylan is generally less digestible than other cell wall fractions (Gaillard, 1962; Lyford et al., 1963). If the true digestibility of cellular contents is complete, the fecal residues should represent the true indigestible fraction and should consist only of plant cell walls, which should divide the dry matter of forages according to their availability (van Soest, 1969).

Jahn and Chandler (1976) studied the effect of diet protein and fiber levels on feed intake, growth, and digestibility in 8- to 20-week-old calves. Protein intakes for maximum gains with diets that contained 11, 18, and 25% acid detergent fiber (ADF) diets were 0.47, 0.66, and 0.58 kg/day. These levels of protein and ADF gave gains of 0.97, 0.91, and 0.76 kg/day, respectively. Dry matter intake increased at all three fiber levels as protein was increased from 9 to 14.5%, but decreased as protein was further elevated with the 11 and 25% ADF diets.

Jahn et al. (1976) reported the effects of four crude protein (9.0, 11.5, 14.5, and 17.5%) concentrations on protein and energy utilization by calves in diets that had three different ADF levels. Tissue protein accumulation increased sharply at all fiber percentages as dietary protein was increased from 9 to 11.5%. Energy accumulation increased in tissues as protein was raised from 9 to 17.5% for the 18% ADF diet, and maximum energy accumulation was reached at about 14.5% crude protein for the 11 and 25% ADF dietary levels.

Golding et al. (1976a) related crude protein and Van Soest fiber fractions to the *in vivo* dry matter intake and/or dry matter digestibility of 32 bermudagrass (*Cyanodon dactylon* (L) Pers.) hays with sheep. Observed and predicted digestible dry matter intake values ranked the hays in the same order. The investigators emphasized that multiple regression equations based on chemical analyses of forages may not be consistently accurate predictors of forage quality even when they are rational and exhibit high $R^2$ and low $S_{y.x}$ values. Nonforage factors may lower the accuracy of quality predictions.

Other factors also affect the nutritive value of roughages such as tannins (Burns, 1963), silica (van Soest, 1967b), along with maturity, light intensity, temperature, and rainfall patterns. Nitrogen fertilization of Sudan grass grown in West Virginia increased the protein and lignin content and tended to decrease the cell wall content during intermediate stages of growth, so that digestibility was not affected consistently.

Hunter and Seibert (1980) conducted digestion studies with steers fed low-quality spear grass alone, supplemented with cottonseed meal, or urea plus sulfur. The cottonseed meal was added to the rumen directly once each day, while the urea plus sulfur was administered by continuous infusion into the rumen. Digestible organic matter intakes were 1.48, 1.88, and 2.28 kg/day for spear grass, spear grass plus urea and sulfur, and spear grass plus cottonseed meal, respectively. Limited benefit was derived from urea on the low-quality forage.

Coleman and Barth (1977) evaluated the effect of supplemental energy on utilization of urea and biuret supplements added to fescue and broomsedge hay (6% crude protein). In the first trial, corn meal or molasses with biuret increased the digestibility of dry matter and energy but had no effect on nitrogen utilization. In a subsequent trial, biuret increased Pangola digitgrass hay intake after the first week of adaptation. In metabolism trials, steers receiving biuret were in positive nitrogen balance (12–17 g/day), while those given no biuret were in negative nitrogen balance. Fick *et al.* (1973) reported that biuret increased feed intake and apparent digestibilities of cellulose and nitrogen in low-quality forage-fed sheep.

Grasses with adequate water, fertilizer, and management may vary in productivity per hectare from a high of 20 animal unit months (AUM) for cool-season grasses to a low of about 2.5 AUM for warm-season grasses. The potential AUM of fall and winter grazing of crop residues approaches that of warm-season grasses, and may be more if crop residues are harvested for feeding (Ward, 1978). Moore and Mott (1973) reviewed inhibitors of quality in tropical grasses and summarized current knowledge.

## II. UTILIZATION OF CORN AND SORGHUM RESIDUES

Research has demonstrated that nonlactating beef cows can be maintained on diets consisting primarily of crop residues. Producers of beef cows in areas of corn and grain sorghum production have a valuable feed resource. Because of the relatively low availability of metabolizable energy (ME) in crop residues, they are most effectively utilized in maintenance diets for gestating cows when energy requirements are less than those during lactation. Low digestibility of dry matter in crop residues indicates that such diets should be fed ad libitum for maintenance. Residues of corn and sorghum crops may be deficient in protein for gestating cows and in both energy and protein for lactating cattle. Certain minerals, such as phosphorus, as well as vitamin Α, would be expected to be quite low in crop residues and would need to be provided in diets.

Generally, in areas where corn and grain sorghum crop residues are available, summer pasture is utilized during the spring, summer, and fall. Utilization of such pastures, along with winter grazing of crop residues, can provide year-round grazing except during periods of excessive snow during winter months. Ward (1978) illustrated such a system of grazing in Fig. 9.3, and ME requirements of spring calving cows according to the NRC (1976a) are shown in Fig. 9.4.

## II. Utilization of Corn and Sorghum Residues

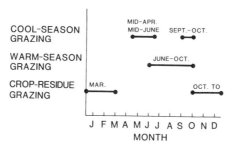

**Fig. 9.3.** Conventional grazing calender (From Ward, 1978).

Energy intake of cows grazing crop residues may at times be deficient, but dry cows should maintain their body weight if they have optimal intake of corn or sorghum grain crop residues. Dry cows grazing an entire field of crop residue may select grain, leaves, and husks early in the grazing season and have no need of supplemental protein. However, as the grazing season progresses and forage quality decreases, the crop residue consumed will tend to be low in protein, and digestibility of protein may be as low as 20–30% (Schmitz, 1976).

Lamm et al. (1977) conducted two trials with spring-calving cows to evaluate the source and level of nitrogen supplementation while they were grazing dryland corn crop residues. Cows self-fed supplemental soybean meal or biuret at a level of 0.11 kg crude protein per day gained less weight or lost more weight than cows fed 0.23 kg crude protein daily. Cows grazing irrigated corn crop residues had greater gains when supplemented with soybean meal than when fed no supplement, corn cubes, dehydrated poultry waste, or liquid urea supplements. Subsequent calf performance or reproductive performance of cows did not appear to be influenced by cow winter treatments.

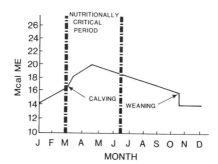

**Fig. 9.4.** Metabolizable energy needs of spring calving cow on a calender basis (From Ward, 1978; NRC, 1976a).

## III. GRAZING INTENSITY

As grazing intensity increases, ruminants have less chance to graze selectively due to greater removal of preferred species of plants and parts of plants. Bement (1969) observed daily gains of cattle on blue grama grass and reported that as grazing intensity increased, total kilograms of beef produced per hectare increased but individual animal gains decreased. The yield of herbage and weight gain per animal and per hectare were decreased with increased grazing pressure (Bryant *et al.*, 1970). If grazing pressures was sufficiently intense, availability and quality of forage declined. This resulted in the coarser, more mature portions of plants being eaten, with decreased digestibility and nutrient content of the diet.

Vavra *et al.* (1973) reported that heavy grazing over 2 years in Wyoming resulted in somewhat lower values for dry matter intake and digestibility, with differences in intake greater later in the season when total forage available was limited on heavily grazed pasture. Individual cattle gains on lightly stocked pasture were greater and reflected more digestibility and intake of forage. However, total gains per hectare were obtained on the heavily stocked pasture.

Hodges *et al.* (1964) established Pangola grass pastures with either superphosphate, superphosphate plus lime, rock phosphate, colloidal phosphate, triple superphosphate, or basic slag sources of phosphorus fertilizers and grazed beef cows over 11 years with very limited supplementation during winter. Omission of phosphorus resulted in a 40–50% reduction in carrying capacity. Application of basic slag produced the greatest average annual beef gain (252 lb/acre), superphosphate was second (220 lb/acre), and other treatments averaged downward to 194 lb/acre for rock phosphate compared to 113 lb/acre for no phosphate fertilizer. Horton and Holmes (1974) conducted a study on the effect of two levels of nitrogen fertilizer (50 and 504 kg/ha) on the productivity of pastures grazed by young Friesian cattle. Responses per kilogram of nitrogen fertilizer were about 1 kg gain, 19–24 Mcal ME, and 8–9 kg dry matter of forage. Productivity of set-stocked pasture was equivalent to rotational grazing at high nitrogen fertilization levels but lower with low nitrogen levels.

F. P. Horn *et al.* (1979) determined the effect of three levels of nitrogen fertilizer (67, 202, and 326 kg/ha) on Midland Bermuda grass (*Cynodon dactylon*) grazed by cows and calves on a "put and take" basis of intensity. Overall dry matter intake values were 12.3 $g/W^{0.75}$ for cows in May and 97.7 $g/W^{0.75}$ in July, and 42.3 $g/W^{0.75}$ for calves in July. Increasing levels of nitrogen fertilization had no effect on the quantity of forage consumed by cows or calves. Intake was positively correlated with *in vitro* dry matter digestibility and negatively correlated with lignin content. Calves tended to select forage with greater levels of crude protein and lower levels of ADF and cellulose than cows in two trials.

Burton (1970) fertilized Coastal Bermuda grass annually over a 5-year period with 90 kg nitrogen, plus adequate phosphorus and potassium, and stocked the pasture with three, four, or five steers per 0.9 ha. During this 5-year period, daily gains averaged 0.68, 0.59, and 0.50 kg, respectively, while corresponding annual per-hectare gains were 390, 470, and 500 kg.

Variety of grass has been demonstrated to have a marked influence on steer gains per hectare. Utley *et al.* (1974) grazed yearling beef steers on Pensacola bahia grass, Coastal Bermuda grass, and Coastcross Bermuda grass over a 4-year period and obtained average daily gains of 0.43, 0.49, and 0.68 kg, respectively. The calculated average beef production per hectare per year was 249, 372, and 527 kg, respectively, for the three grasses. Bertrand and Dunavin (1975) grazed growing beef calves on pure stands of three small grain crops (triticale, wheat, and rye) and two mixtures composed of a small grain (triticale and rye), ryegrass, and crimson clover. Dring the first 84 days of the study, under conditions of unrestricted intake of high-quality forage, animal potential rather than forage quality was the limiting factor influencing daily gains, while from 84 to 127 days (end of trial), highly significant correlations existed between gains and *in vitro* organic matter digestibility and crude protein content of the forages.

## IV. SUPPLEMENTATION OF GRAZING RUMINANTS

Providing protein and energy supplements to grazing cattle during drought or winter is a widespread practice. Usually there is considerable year-to-year variation in the response to supplementation, but cattle in Montana responded most strongly during severe winters when forage was limited (Black *et al.*, 1938). Rittenhouse *et al.* (1970) reported little effect of protein supplementation on intake and digestibility of range forage, but energy supplementation depressed forage intake. Protein supplements in Utah tended to increase winter forage intake and digestibility by cattle and sheep, but energy or energy plus protein supplementation generally had no effect or tended to decrease intake and digestion (Cook and Harris, 1968). As a result of supplementing cows during winter grazing of browse-shrub forage in which energy was limited with 0.45 kg barley or 0.45 kg protein cake, both treatment groups of cattle maintained body weight or had slight gains, while cows receiving no supplement lost weight (Speth *et al.*, 1962).

Kartchner (1981) compared cottonseed meal (0.75 kg) fed individually at 2- or 3-day intervals to dry, pregnant, mature cows grazing native fall-winter range in Montana to a cracked barley isocaloric supplement and observed no significant differences in dry matter intake, digestibility, and daily gains. During the following winter, soybean meal was substituted for an equivalent amount of cottonseed meal and daily forage dry matter intake was increased from 6305 g for the controls to 6813 and 8009 g for the barley- and soybean meal-supplemented cattle, respectively. Body weight and condition scores were not affected by the supplementation. The difference observed was attributed to differences in the quantity and quality of the pasture available between years. Protein supplementation of cattle in winter has been reported to increase weight gains (Bohman *et al.*, 1962; Lewis *et al.*, 1964; Clanton *et al.*, 1966), while increasing energy without sufficient protein has not always given such increases (Clanton *et al.*, 1966). Adding readily available carbohydrate supplements to predominantly roughage diets have decreased voluntary intake of the roughage (Bines and Davey, 1970; Rittenhouse *et al.*, 1970). Addition of protein to low-quality roughages has increased voluntary intake (Elliot, 1967; Crabtree and Williams, 1971).

Pinney *et al.* (1972) studied the effect of winter supplementation of beef cows on longevity and productivity. The cows grazed native grasses year long, with a stocking rate of about one cow per 10 acres in Oklahoma. The treatments were: (1) 0.45 kg 43% cottonseed meal, (2) 1.13 kg cottonseed meal, and (3) 1.13 kg cottonseed meal plus 1.31 kg whole oats daily from early November to mid-April. Hereford heifers supplemented at the various levels were bred to calve at either 2 or 3 years of age. The level of winter supplementation significantly affected winter and summer weight gains during the first few years, but mature cow size and body weight were not affected by either winter feed level or age at first parturition. However, the later-calving cows receiving the high level of winter supplemental feed had the greatest body weight and height. The average calving date was delayed 7–12 days by low winter supplementation, but there were no differences in birth or weaning weights.

Rush and Totusek (1976) compared supplements for cows grazing low-quality forage that contained approximately 62% sorghum grain, 19% soybean meal, 5% alfalfa, and either 5.31% urea or 6.46% biuret to liquid molasses supplements. In the first trial, a comparison was made between molasses and urea (30% crude protein) in a lick tank and a cottonseed meal plus alfalfa (25% crude protein) supplement hand-fed (1.35 kg per cow). The cows fed the liquid supplement consumed more supplemental nitrogen (91.4 versus 54.5 g per cow per day), but there were no differences in winter weight loss, summer gain, and calf weaning weights. In a second trial, cows that consumed natural protein gained 3 kg during the winter compared to cows that lost weight when fed urea-molasses ($-8$ kg) or biuret-molasses ($-26$ kg). Comsumption of the urea and biuret supplements were similar, but the amount of salt required to control the intake of the natural protein supplement was considerably higher than that required for the nonprotein nitrogen (NPN) supplements. In a third trial, cows fed dry urea supplement lost less weight ($-39$ kg) during the winter than cows fed the soybean meal ($-68$ kg), dry biuret ($-58$ kg), or molasses-urea lick tank ($-89$ kg) supplements. Calf weaning weights for the four treatment groups averaged 209, 206, 200, and 195 kg, respectively.

## IV. Supplementation of Grazing Ruminants

Martin et al. (1976) compared the effects of isocaloric supplements that contained nitrogen in the form of biuret, or nitrogen from cottonseed meal on cows and heifers wintered on pasture near central Florida. Supplemental cubes were formulated with cottonseed meal or citrus pulp, molasses, and minerals, with and without biuret. They were fed on the ground from December through April each year at a level to provide 0.9 kg cottonseed meal per day per head, 1.0 kg of supplement with biuret and 0.86 kg supplement without biuret. Nitrogen from biuret was utilized during 2 of the 3 years to reduce the weight loss of mature cows. Both the cottonseed meal and biuret supplements were sufficient to give some weight gains with heifers during the winter.

In southern Australia, Allden (1968, 1980) observed that grazing sheep were unable to maintain their body weight during the dry season, although abundant feed was available. During 4–6 months of the year, intake and digestibility of useful energy were suboptimal and the animals needed to be given energy-rich supplements. Increasing maturity of forage leads not only to a decrease in digestibility but also to a reduction in nitrogen or protein. Forages low in available protein may have inadequate nitrogen to meet the needs of rumen microbes, and may thereby induce energy deficiency and decreased nutrition in the host animal (Williams et al, 1953; Moir and Harris, 1962; Egan and Moir, 1965).

Correction of a dietary nitrogen deficiency in forage is dependent on knowing what animals are eating and how much nitrogen is needed for an active population of rumen microbes. When the nitrogen content of pasture is below 1.3 g per 100 g digestible organic matter (DOM), it is likely that a protein shortage will be imposed on an energy shortage (Egan, 1976). Maturing tropical and subtropical grasses commonly have less than 1% nitrogen on a whole-plant basis for long periods each year (Butterworth, 1967; Minson, 1976). Pastures in temperate regions are more likely to have higher nitrogen content than those in the tropics, especially when nitrogen fertilizers are applied or when legumes are present (Minson, 1976). The nitrogen content of the leaves is generally greater than that in the stems, and the pasture selected by grazing animals is commonly higher in average protein values than the samples analyzed (Hardison et al., 1954).

Grazing animals may not be able to obtain sufficient energy when pastures are sparse, even though the forage has high digestibility (Riddett et al., 1941; Johnstone-Wallace and Kennedy, 1944). Concentrates or hay may then be utilized to supplement the deficient pastures. Grazing animals fitted with esophageal fistulae have been utilized to obtain reliable estimates of forage actually eaten during grazing (Torell, 1954; van Dyne and Torell, 1964). Forage collected by esophageal fistulae can be used with the *in vitro* digestibility method of Tilley and Terry (1963) to estimate the composition and digestibility of the diet with reasonable accuracy (Langlands, 1966), but the technique is expensive.

Thompson et al. (1973) studied the influence of dietary urea on reproduction in ruminants in four experiments involving 143 Angus and 97 Hereford cows, 70 Dorset and 106 crossbred ewes, 4 Holstein bulls, and 4 Suffolk and 4 Dorset rams. The effects of urea diets on cow and calf weight gains were not significant. The experiments demonstrated that high-roughage diets can be supplemented with urea without reducing reproductive performance. Clanton (1978) conducted experiments in Nebraska during six consecutive winters with growing calves on native range and observed that the average daily gains per head of those fed 0.68 kg of a 40% protein supplement containing all natural protein, 3% urea, or 6% urea were 0.26, 0.25, and 0.19 kg, respectively. Feeding urea in extruded starch-urea or clay-urea carriers did not improve its utilization.

Hennessy et al. (1981) fed young steers in Australia protein supplements or sorghum grain for 140 days during their first winter. The protein-supplemented steers were heavier (265 versus 250 kg) at the end of the first summer. At this time, the supplemented steers fed the protein supplement during the winter had more essential amino acids in their plasma, which correlated with their live-weight performance over controls. The authors postulated that the increased availability of essential amino acids may have caused an increased appetite for pasture dry matter. Over a 3-year period in Indiana, Lemenager et al. (1980) conducted three trials with heifers that involved feeding low-quality fescue hay ad libitum plus 0, 1.22, or 2.45 kg ground ear corn (GEC) daily during the winter feeding phase. Final winter and summer condition scores increased as the winter level of GEC was increased, but summer gains were in reverse order. Conception rate and weaning weight of progeny increased linearly with increased winter level of GEC.

## V. CEREAL RESIDUE

Wheat, barley, and oat residues remaining after harvesting of grain have been widely utilized in wintering beef cattle. These residues are generally low in protein and phosphorus, marginal in calcium, and high in fiber and lignin. The digestibility of cereal residues is low, passage through the rumen is slow, and voluntary intake is low. Urea and/or readily available carbohydrate supplements have given variable responses in utilization. Cattle in feedlots cannot consume sufficient pelleted feed containing high levels of residues to support satisfactory gains, but the residues can provide part (25–50%) of the maintenance diets for dry, pregnant cows. Common supplements for residues fed ruminants are alfalfa hay, protein blocks, molasses supplements, and corn silage.

The digestible (DE), metabolizable (ME) and net (NE) energies of wheat straw fed lambs were reported to be 1.65, 1.42, and 0.84 Mcal/kg, respectively (Wilson et al., 1976). Morrison (1959) gave estimated NE values for barley straw and alfalfa hay of 0.49 and 0.89 Mcal/kg, respectively. In a comparative slaughter trial with beef cattle, Lofgreen and Christiansen (1962) found the NE value of barley straw and alfalfa to be 0.50 and 0.97 Mcal/kg, respectively. Such low energy values for straws indicate that they cannot be utilized in large quantities in diets for high rates of weight gain.

Campling et al. (1961) fed long oat straw ad libitum to dry Shorthorn cows and found that the low intake relative to hay was due to less digestibility and longer retention time. Blaxter and Wilson (1962) concluded that a 43% lower intake of oat straw than hay by steers fed ad libitum was due to the less apparent digestibility of energy. Andrews et al. (1972) compared barley and oat straw alone, and with two levels of protein and two levels of energy supplements fed to cattle. There were no differences in the performance of cattle fed oat or barley straws. The low-protein, high-energy combination (6.4% crude protein in the whole diet) diet was frequently refused, and gains on this treatment were low. Mulholland et al. (1976) reported feed intakes of wheat and barley straws to be 53% and 33% less, respectively, than that of oat straw.

Horton and Holmes (1976) fed 350-kg liveweight steers barley straw ad libitum either alone or with 1.5, 3.0, 4.5, 6.0, or 7.5 kg of a barley : dried alfalfa (1 : 2) concentrate. Straw intake decreased and total organic matter intake increased linearly with increasing levels of concentrate. The crude protein intake level of the whole diet did not affect straw consumption. Mathison et al. (1981) fed diets that contained 94% barley straw for 83 days during the winter in Canada. Eight cows died and three had abomasal compactions. The cows consumed 47 MJ of DE daily, which was only 70% of their estimated requirements. When protein in the total diet was increased from 5.2 to 6.2%, the cows increased their feed consumption by 17% and gained more weight. Magnesium supplementation enhanced feed intake and weight gains in the low-protein dietary group.

Bass et al. (1981) fed Hereford cross cattle oat straw ad libitum with (1) no supplement, (2) 100 g/day of a fully soluble liquid supplement containing urea (1090 g crude protein per kilogram equivalent) in 1.5 kg cubed barley per day, (3) the same amount of urea as in (2) in drinking water with 1.5 kg cubed barley per day, or (4) the same amount of urea in a molasses lick (3 parts molasses : 2 parts liquid supplement : 1 part water). Voluntary oat straw dry matter intake increased from approximately 2.9 to 3.25 kg/day with each method of supplementation.

Swingle and Waymack (1977) conducted trials with growing steers to determine the digestibility and NE values of wheat straw and grain sorghum stover fed with a molasses-urea supplement. Wheat straw had DE and total digestible nutrient (TDN) values of 2000 kcal/kg and 45%, compared to corresponding values of 2130 kcal/kg and 51% for the sorghum stover. White et al. (1974) compared diets fed steers that contained 0, 20, 40, 60, 80, or 100% rice straw or dehydrated Coastal Bermuda grass pellets and found that as the level of roughage increased, the digestibilities of energy, dry matter, organic matter, and nitrogen-free extract decreased.

## VI. MECHANICAL PROCESSING OF CEREAL STRAWS

Burt (1966) concluded that grinding and pelleting barley straw changed it from an unpalatable feedstuff incapable of supporting maintenance of heifers into one that was utilizable in highly productive diets. Campling and Freer (1966) observed that the voluntary intake of ground, pelleted oat straw was 26% greater than that of long straw when fed to cows, but the ground straw was lower in digestibility than the long form. Minson (1963) found greater gross feed efficiency with ground straw, since the greater intake produced more weight gain. Pelleted diets containing 15% wheat straw fed bull calves resulted in greater feed intake and organic matter digestibility than diets with 30% pelleted wheat straw (Levy et al., 1972).

## VII. SUGARCANE AND BAGASSE

There are times when sugarcane can be utilized in ruminant diets rather than processed for sugar. Bagasse, the fibrous material remaining after processing of sugarcane for sugar, is a potential roughage for ruminants. Henke (1952) observed that steers fed a bagasse-containing diet for 164 days had an average daily gain of 1.07 lb, while those fed an alkali-treated bagasse diet gained 1.32 lb/day. Steer gains increased as bagasse was reduced from 30 to 10% in the diets and cane molasses was increased from 50 to 70% (Wayman et al., 1953).

Fonseca (1957) observed that Hereford heifers fed a low-protein diet plus Pangola grass hay, citrus pulp, and sugarcane molasses gained 1.24 lb/day, but when the hay was replaced with Camola (4 parts bagasse and 10 parts molasses), only an 0.86-lb gain per day was observed. Kirk et al. (1969) fed a diet of cottonseed hulls, cottonseed meal, citrus pulp, and alfalfa pellets to steers and obtained 2.34-lb daily gains. When ammoniated bagasse replaced 50% of the hulls and 14% of the cottonseed meal, daily gains were 2.38 lb. Replacing all of the hulls and 28% of the cottonseed meal with ammoniated bagasse reduced gains to 2.07 lb/day.

Marshall and van Horn (1975) fed lactating dairy cows complete diets ad libitum containing either 25% sugarcane bagasse pellets, 25% cottonseed hulls, or 12.5% of each roughage. The bagasse pellets were as satisfactory as the cottonseed hulls as a roughage. Roman-Ponce et al. (1975) fed sugarcane bagasse (30 and 40%) in the form of pellets (pelleted right after processing or after ensiling) with soybean meal, urea, or Starea as sources of protein in complete dairy diets. They observed feed intake to be lower for regular than for ensiled pellets and lower for urea than for soybean and Starea diets. Milk fat percentages were greater for 40% than for 30% bagasse diets. However, the lower bagasse level gave increased solids-corrected milk yields.

Chapman and Palmer (1972) found that sugarcane bagasse pellets containing 6–8% cane molasses were equivalent to cottonseed hulls when both roughages were fed in beef cattle finishing diets at a level of 7.5–10%. Rangnekar et al. (1982) evaluated the effect of high-pressure steam (5, 7 and 9 kg/cm$^2$) for 30 and 60 min on the chemical composition and in vitro digestibilities of sugarcane bagasse, paddy straw, and sorghum straw. The treatments increased soluble carbohydrates and volatile fatty acids (VFA), but ADF remained constant. Increments brought about by treatments in in vitro dry matter digestibility compared to untreated material were 23–64% for sugarcane bagasse, 19–36% for paddy straw, and 31–42% for sorghum straw. Digestibilities of cell wall constituents improved by treatments of all three roughages.

Pate (1981) fed diets to feedlot steers that contained 30% freshly chopped sugarcane (a low-fiber and a high-fiber variety) and observed the digestibilities of dry matter, organic matter, neutral detergent fiber (NDF), and ADF to be 63.8, 62.3, 39.4, and 32.6%, respectively, for diets containing high-fiber material and 63.0, 62.4, 34.7, and 27.2% for diets containing the low-fiber variety. In a feedlot trial with steers fed 30% dry matter as sugarcane, there were no differences in performance or carcass measurements due to variety. Steers fed at the 30% level of sugarcane had higher daily gains, dressing percentages, quality grade, and yield grade than steers fed at the 60% level.

Kung and Stanley (1982) determined the nutritive value of sugarcane silage harvested at various stages of maturity using a 2-year maturing variety currently used for sugar production in Hawaii. The fresh, whole-plant sugarcane harvested at 6, 9, 12, 15, or 24 months of age had a significant linear increase in in vitro dry matter digestibility. However, it was found that ensiled sugarcane harvested at comparable maturities decreased in in vivo dry matter digestibility with increasing maturity when fed to sheep. The sugarcane silage was lower in TDN than corn silage. Sugarcane, because of large losses of dry matter and high levels of ethanol, did not lend itself to ensiling.

## VIII. AQUATIC PLANTS

Aquatic plants contain sufficient nutrients to be considered a potential livestock feed. In many areas of the world, they are abundant and may be removed from streams, lakes, and ditches rather than chemically controlled. However, the water content of fresh aquatic plants is generally about 90% or more, which causes handling and feeding problems. Boyd (1968) reviewed the limited chemical composition data on aquatic plants. On a dry matter basis, these plants generally contain as much or more crude protein and nutrient minerals as conventional forage crops and usually have less crude fiber.

Easley and Shirley (1974) collected six species of aquatic plants from lakes or rivers in Florida at monthly intervals over a 1-year period and reported the concentrations of 10 nutrient elements. On a dry weight basis, phosphorus, potassium, magnesium, copper, zinc, and manganese were in the range of concentrations of land forages in the United States, sodium was 10–100 times greater, and iron exceeded the range by 4–19 times. The concentration of calcium was higher and that of phosphorus was generally lower, except in water hyacinth, which contained about 2% calcium. The calcium : phosphorus ratio in hyacinths was suitable for cattle. Hydrilla, naiad, hornwort, pondweed, and eelgrass had average calcium : phosphorus ratios of 30 to 90, and 1 kg of dry plant of these species would provide, at average levels, three to six times the daily calcium requirement of a cow on a maintenance diet.

Stephens et al. (1972) compared hydrilla (*Hydrilla verticillata* Casp.) and water hyacinth (*Eichhornia crassipes* Mart.) and a land forage, Coastal Bermuda grass (*Cynodon dactylon* L.) Pres), at 33% of the dietary organic matter in pelleted diets fed steers for net retention of calcium, phosphorus, magnesium, sodium, sulfur, potassium, manganese, iron, copper, and zinc, and the apparent absorption of ash and sand-silica. The mean voluntary intakes of dry matter per day for the hydrilla, water hyacinth, and Coastal Bermuda grass were 17, 21, and 24 g per kilogram of body weight, respectively. The net retention and apparent absorption values for the various elements, ash and sand-silica of hydrilla, and water hyacinth-containing diets were comparable to those of the Coastal Bermuda grass diet. There were no differences in the apparent absorption of nitrates, oxalates, and tannins in the various diets.

Baldwin et al. (1975) ensiled water hyacinth (*E. crassipes,* Mart) and Pangola grass (*Digitaria decumbens,* Stent) and fed them to sheep in order to compare voluntary intake and nutrient digestibilities. Intake of dry matter was less for the hyacinth silage than for the Pangola grass silage. Digestibilities of dry matter, organic matter, and crude protein were greater for the Pangola grass silage than for the hyacinth silage.

Linn et al. (1975b) reported on the chemical analyses of 21 species of dried aquatic plants grown in Minnesota. Fourteen species contained more than 10% protein, and all species contained less than 30% crude fiber. NDF and ADF averaged 42.3% and 32.6%, respectively. Calcium and phosphorus averaged 1.62 and 0.27%, respectively. When mixed aquatic plants were ensiled with organic acids (acetic, formic, propionic), corn, or alfalfa for 47 days, the silages had pH values above 4.5 and lactic acid values below 0.4% of dry matter. Ensiling mixtures of aquatic plants and alfalfa resulted in silages that were similar to those of aquatic plant. However, addition of alfalfa to sterilized aquatic plants at ensiling resulted in silage similar in composition to alfalfa silages.

Linn *et al.* (1975a) conducted studies on digestibility by lambs with two species of dried aquatic plants (*Myriophyllum exalbescens* and *Potamogeton pectinatus*) and an ensiled mixture of aquatic plants. The dried *M. exalbescens* and *P. pectinatus* were unpalatable (less than 600 g dry matter consumed daily) to the lambs. When equal proportions of dehydrated alfalfa were mixed with either of these plants, the digestibilities of dry matter and crude protein rose to 43.8 and 46.0% for *Myriophyllum* and 43.4 and 44.1% for *Potamogeton,* respectively, as determined by difference. Digestibility of energy was higher for *Myriophyllum* (53.7%) than for *Potamogeton* (47.4%). Aquatic plants, aquatic plants plus corn, or aquatic plants plus alfalfa silages had dry matter digestibilities of 41.4, 32.0, and 38.5%, respectively, compared to 61.9 and 66.2% for diets of ensiled alfalfa or ensiled alfalfa plus corn diets, respectively. Energy and nitrogen digestibilities were lower for lambs fed diets that contained ensiled aquatic plants than for lambs fed alfalfa silage or alfalfa plus corn silage.

Chavez *et al.* (1975) observed the performance of rats fed diets that contained 0, 10, 20, or 30% dehydrated water hyacinths over three generations. The average 21-day weaning weights decreased with increasing levels of hyacinths in the diets and were markedly lower in the second generation with the 20 and 30% levels of hyacinths. The third-generation weaning weights were generally lower on all diets, but especially with the 30% hyacinth-containing diets. There was essentially no difference in the number of dams that littered on the four diets in the first two generations. During the third generation, only 75 and 38% of the dams littered on the 20 and 30% hyacinth diets, respectively.

## IX. PINEAPPLE PLANT FORAGE

Postharvest plant material is a forage resource available for ruminants in Hawaii and other tropical areas where pineapples are grown. Kellems *et al.* (1979) reported composite mean values (in percentages) on a dry matter basis of stumps, ratoon stems, and green and dry leaves of postharvest pineapple plants to be: crude protein, 4.85; ether extract, 1.82; ADF, 33.4; NDF, 58.42; lignin, 7.95; starch, 13.84; glucose, 1.33; ash, 5.18; silica, 0.32; calcium, 0.36; and phosphorus, 0.07. The crude protein in unsupplemented pineapple silage was found to be 33.9% digestible, with 21.7% nitrogen retention when fed wethers.

Dehydrated pineapple hay was found to have 53.1% TDN and 1.47% digestible crude protein when fed lactating dairy cows on an as-fed basis (Otagaki *et al.*, 1961). Stanley and Morita (1966) reported that fat-corrected milk (FCM) production was maintained when up to 5.44 kg of dry matter from pineapple bran was replaced with an equivalent amount from pineapple silage. Dronawat *et al.* (1966) reported similar observations for loose versus pelleted dehydrated pineapple hay fed at 6.35 kg/day.

## X. APPLE POMACE

Apple pomace, a by-product of apple processing, is a source of nutrients for livestock. Ensiled apple pomace has the following composition (in percentages) on a dry weight basis: crude protein, 5.6; ether extract, 7.3; nitrogen-free extract, 55.1; crude fiber, 28.1; and ash, 1.8 (Fontenot et al., 1977). Burris and Priode (1957) found that apple pomace, when supplemented with natural protein, was similar to grass silage for wintering beef cattle.

Fontenot et al. (1977) conducted trials with beef cows to determine the feasibility of supplementing apple pomace with NPN for wintering from approximately 12 weeks prepartum until 12 weeks postpartum. Feeding the pomace with urea, biuret, or a combination of the two NPN sources lowered feed consumption and increased body weight losses in cows compared to the results gained by supplementing corn silage with NPN or supplementing the pomace with protein. Feeding pomace with NPN to beef cows resulted in the birth of small, deformed, dead, or weak calves. Increasing the dietary energy did not prevent these harmful effects, but a small amount of coarse hay appeared to mitigate them.

Oltjen et al. (1977) compared apple pomace and corn silage with supplemental NPN or cottonseed meal protein providing 50–70% of the total nitrogen in diets fed gestating cows or steers. The apple pomace-containing diets resulted in greater molar percentages of acetic and valeric acids and less propionic, butyric, isobutyric, and isovaleric acids and lower pH in rumen fluid. Cows and steers fed the urea supplement had lowered plasma concentrations of several essential amino acids. Rumsey (1978) compared an apple pomace-cottonseed meal diet, an apple pomace-urea diet, an apple pomace-urea plus cornstarch diet, and an apple pomace-urea plus straw diet with a corn silage diet fed fistulated steers. Feeding pomace in diets was associated with slightly reduced pH in rumen fluid, a higher acetic acid: propionic acid ratio, lower proportions of ruminal butyric, isobutyric, and isovaleric acids, and more ethanol. The addition of starch to pomace-urea diets tended to lower the acetic acid: propionic acid ratio and to increase butyric acid.

Rumsey et al. (1979) studied the effects of apple pomace diets on rumen microbial population, movement of ingesta from the rumen, and water intake with fistulated beef steers fed either corn silage plus cottonseed meal, apple pomace plus cottonseed meal, apple pomace plus urea, pomace plus urea and starch, or pomace plus urea and straw. Apple pomace dyed with Sudan IV was used as a measure of pomace particle depletion in the rumen, and polyethylene glycol was used to measure the ruminal liquid washout rate. Total ingesta volume of the rumen and polyethylene glycol liquid volume were not affected by the diet. Total bacterial numbers and tungstic acid-precipitable nitrogen were similar when steers were fed either corn silage plus cottonseed meal or apple pomace plus cottonseed meal diets, but were higher with these diets than for pomace diets that contained urea. The liquid washout rate from the rumen tended to be greater when apple pomace was in the diets.

## XI. DRIED CELERY TOPS

Celery is a major vegetable crop in some areas, and it has been estimated that 500 lb of dry feed equivalent per acre is left in the field after the harvest. The high moisture content of celery waste suggests that it should be dehydrated to aid handling and to prevent spoilage prior to feeding ruminants. Haines et al. (1959) reported that dried celery tops contained 25.3% crude protein, 15.3% crude fiber, 3.0% ether extract, and 11.6% ash. Four diets were fed steers for 123 days that contained 0, 10, 20, or 30% dehydrated celery tops and were formulated to contain 12.8% crude protein. Dehydrated Bermuda grass was provided free choice. Average daily gains of the steers were 2.23, 2.28, 1.88, and 2.46 lb for those consuming concentrates that contained 0, 10, 20, and 30% dried celery tops, respectively. Neither carcass nor meat characteristics appeared to be affected by inclusion of the celery waste in diets. In a digestibility trial with steers, the celery waste had 77.2, 84.0, 86.1, and 63.4% digestibilities for crude protein, crude fiber, nitrogen-free extract, and ether extract, respectively, and an average TDN value of 79.5%.

## XII. TOMATO PULP

The disposal of cull tomatoes in tomato-growing regions is a serious problem. The dehydration of cull tomatoes provides a means of disposal as well as a by-product feedstuff for livestock. Tomhave (1932) reported that dried tomato pomace was a suitable feed for dairy cows when fed at a level of 15% in concentrate diets. Chapman et al. (1958) observed satisfactory gains by steers on pasture that consumed 16–19 lb daily of a concentrate in which dried tomato pulp replaced 10–30% of dried citrus pulp. Carcasses of the steers fed the tomato pulp were equivalent in grade and acceptability to those fed no pulp in their diets.

Ammerman et al. (1965a) evaluated dried tomato pulp for nutrient utilization by steers and protein utilization by lambs. The dried pulp of cull tomatoes (flakes plus fines) had an average composition (in percentages) on a dry matter basis as follows: protein, 24.0; crude fiber, 17.8; nitrogen-free extract, 43.4; and ash, 10.6. Steers had average coefficients of digestibility for protein, ether extract, crude fiber, and nitrogen-free extract of 56.3, 90.2, 46.2, and 79.0%, respectively. When supplying 58% of the dietary protein for lambs, nitrogen in the tomato pulp had a lower digestibility coefficient and a lower biological value than nitrogen in soybean meal. The addition of an enzyme supplement containing protease, amylase, and gummase did not improve the digestibility of nutrients in a diet containing 33% tomato pulp-fed lambs.

## XIII. POTATO PROCESSING RESIDUE

Potatoes (*Solanum tuberosum*) are processed commercially in increasing amounts. Approximately 35% of the processed potatoes are discarded as waste and should be utilized in livestock feed. Filter cake is a high-energy material containing 18–23% dry matter, of which 60–75% is starch (Howes and Sauter, 1974). Heinemann and Dyer (1972) indicated that potato-processing slurry had a dry matter digestibility of 73.5% when fed steers at 19.2–37.5% of the dietary dry matter, with DE values of 3.5 and 3.1 Mcal per kilogram of dry matter at the two levels of intake, respectively.

Stanhope *et al.* (1980) conducted metabolism trials with steers fed filter cakes of potato-processing residue. The residue had the following composition (in percentages): dry matter, 14.1; crude protein, 4.8; crude fiber, 2.3; ether extract, 7.7; ash, 3.1; calcium, 0.14; phosphorus, 0.19; and potassium, 0.24. In the first trial, steers were fed diets that contained 0, 15, 30, 45, or 60% potato residue. At 15% of the dietary dry matter, the residue had a DE value of 3.68 Mcal/kg, which corresponds to 121% of the DE of barley. However, at levels of 30, 45, and 60% of the dietary dry matter, the mean DE was 3.10 Mcal/kg, or 102% that of barley. In another experiment with steers having ruminal and abomasal cannulae, inclusion of potato residue had no effect on the location or extent of digestion of dry matter or starch in the diets. Dry matter and starch digestibilities in the entire tract of the steers were 86.6 and 99.1%, respectively. The site and extent of digestion of the residue were comparable to those of barley, and it was concluded that potato-processing residue can replace barley as an energy source for feedlot cattle.

## XIV. COFFEE GROUNDS

Sizable quantities of coffee grounds or meal have resulted from the soluble or instant coffee industry. Coffee grounds have considerable protein, lipids, nitrogen-free extract, and fiber, which are potential nutrients for ruminants. Mather and Apgar (1956) fed dried coffee grounds in diets to dairy cattle and found that they did not affect milk production, the milk fat test, or milk flavor significantly, but did reduce the body weight of cows. The coffee grounds also reduced the feed intake and growth rate of calves on starter diets.

Campbell *et al.* (1976) fed Holstein steers coffee grounds in diets in which they replaced grain at levels of 0, 5, 10, or 20%. The grounds were analyzed to have 11.8% protein, 23.1% fat, 42.5% crude fiber, 0.7% ash, and 0.13% caffeine. Digestibilities of dry matter, crude protein, and energy decreased, and ether extract digestibility increased progressively as coffee ground percentages increased in the diets. Nitrogen retention decreased with diets that contained 10 or 20% coffee grounds, Bartley *et al.* (1978) fed grain diets to Holstein cows that contained 0 or 5% coffee grounds in one study and 0, 5, or 10% in a second study. In the first trial, 5% grounds did not have a detrimental effect on feed intake and production, but in the second study, grain intake, milk production, and weight gain progressively declined with increasing concentration of coffee grounds in the diets. In a feedlot trial with Hereford heifers fed sorghum grain diets in which incremental additions of grounds (5 or 10%) were added to the control diets, there were incremental decreases in grain intake and body weight gain (Bartley *et al.*, 1978).

## XV. PAPER

Many tons of newspaper, magazines, boxes, and miscellaneous paper are available in many areas of the country. Their cellulose content could be utilized as a roughage and energy source for ruminants. The nutritional value of paper varies with the pulping treatment, source of wood, and additives such as ink, glue, clay, and plastic (Mertens *et al.*, 1971b; van Soest and Mertens, 1974).

Steers fed diets that contained 8% paper had normal weight gains, but those consuming 16 and 24% levels of paper had marked decreases in gains (Dinius and Oltjen, 1971b). Steers preferred newsprint paper without ink (Dinius and Oltjen, 1972). Sheep fed office wastepaper up to 45% of the total diet had increased digestibility up to 30% paper, after which it remained constant. Peavy *et al.* (1980) fed complete diets to lactating Holstein cows ad libitum that contained either 30% cottonseed hulls or 10, 20, or 30% ground corrugated paper boxes as sources of roughage. Within each roughage treatment, a citrus molasses solubles-soybean millfeed product (SuperFerm) was compared with 0, 15, or 30% citrus pulp. Ground corrugated boxes at the 20% level were a satisfactory roughage replacement for cottonseed hulls.

## XVI. PECAN HULLS

Utley and McCormick (1972) reported that steers fed finishing diets containing 0, 20, or 30% pecan hulls had improved daily gains at both the 20 and 30% levels, with similar intakes of concentrate at all three levels. Molar percentages of acetate increased and those of propionate decreased with increasing levels of pecan hulls in the diets. The all-concentrate diet exceeded the 20 and 30% hull diets in dry matter and crude fiber digestibilities. In a comparative slaughter trial with steers, Cullison et al. (1973) compared ground pecan shells to ground hay as a roughage in concentrate diets. The pecan shells reduced liver abscesses and caused no reduction in feed intake, but were inferior to ground hay on the basis of liveweight gains, energy gains, and feed efficiency.

## XVII. OYSTER SHELLS AND PLASTIC POLYMERS

Efforts have been made to provide the "roughage factor" in concentrate feedlot diets with inorganic or synthetic substances that may be incorporated readily in diets with mixing equipment. Williams et al. (1970) reported that about 226 g daily intake of "hen-sized" oyster shells with all-concentrate diets fed steers resulted in decreased feed intake, weight gains, and feed efficiency. However, the steers consuming oyster shells had fewer liver abscesses than the controls.

Bartley et al. (1981) fed an elastomeric copolymer at a level of 90 g per head per day to cattle consuming only grain. The polymer resulted in healthier rumen papillae and epithelia of the abomasum and small intestines than control animals. A $^{14}$C-labeled copolymer fed to milking cows produced no activity in milk, blood, or urine. Fordyce and Kay (1974) fed 1.36 kg of "Ruff Tabs" (polyethylene particles) to steers consuming an all-concentrate diet and observed that it reduced the flow of fluid from the rumen. It had no effect on the total time spent daily in eating and ruminating or on VFA production with ad libitum feeding.

## XVIII. WOOD AND WOOD BY-PRODUCTS

Wasted cellulosic materials are a source of feed, since ruminant microbes and associated enzymes can utilize cellulose as energy. However, only a limited fraction of the carbohydrates in whole wood are accessible to rumen microbes (Millet et al., 1970).

### A. Sawdust

Anthony et al. (1969) fed steers feedlot diets with 0, 2.5, and 10% oak sawdust as roughage and obtained 1.10, 1.12, and 1.08 kg/day gains, with feed efficiencies of 7.5, 7.7, and 8.0 kg feed per kilogram of gain, respectively. More normal livers occurred with the 10% sawdust level in the diets. Since the low-bulk density of sawdust and poor flow characteristics made sawdust unsuitable for mill additions using conveying and mixing equipment, the authors recommended that hardwood sawdust be pelleted with a protein, cane molasses, and vitamin and mineral supplements and fed in a 85% corn grain plus 15% pellet diet.

Cody *et al.* (1972) fed kiln-dried screened sawdust from shortleaf southern pine at levels of 0, 10, 15, 25, 35, and 45% with concentrates to young calves for periods of up to 20 months. The results indicated that screened sawdust-containing diets did not physically injure the gastrointestinal lining, nor was any toxic effect apparent. The 25% level of sawdust appeared to be the most desirable level for roughage substitution. Higher levels occasionally caused impaction of digesta. Cody *et al.* (1968) fed sawdust to dairy cattle as a diluent to limit voluntary consumption of concentrate.

Experiments were conducted by Dinius and Williams (1975) to evaluate sawdust as a diluent for adapting steers and heifers to concentrate diets. In one trial, fistulated animals were abruptly switched from a forage diet to concentrate diets containing 20, 35, or 50% sawdust for 5 or 10 days and then provided a full-concentrate diet for another 30 days. There were no differences in concentrate diet intake related to level of sawdust or to interval of feeding sawdust diets. In other trials, growing steers were group-fed and abruptly switched from forage to concentrate diets diluted with varying levels of sawdust and fed for 5 or 10 days prior to full feeding of concentrates. There was less fluctuation in daily grain intake when the dietary sawdust was reduced from 50 to 0% in three steps than when the sawdust was abruptly withdrawn.

Fritschel *et al.* (1976) conducted experiments to determine the feeding value of aspen bark and aspen pulp mill fines (by-product of an ammonia-base sulfite tissue mill) in cattle and sheep diets. Daily gains (kilograms per day), dry matter intake (kilograms per day), and feed efficiency (kilograms of feed per kilogram of gain) were 0.45, 7.68, and 17.1 for steers fed diets that contained 75% pulp fines, respectively, compared to corresponding values of 1.09, 8.50, and 7.8 for control steers fed alfalfa-based diets. In another trial, pelleted feedlot starter diets containing either oat hulls or pulp fines were compared to corn silage as aids in switching cattle to high-grain finishing diets during the first 3 weeks of the feeding period. Average daily gains were 1.38, 0.76, and 0.63 kg/day for the corn silage, oat hull, and pulp fines treatment groups, respectively. Ewes and lambs fed diets containing alfalfa hay, 72.5% pulp fines, or 72.5% aspen bark had similar and satisfactory performance.

Singh and Kamstra (1981) utilized the entire mature aspen tree, including branches, leaves, and bark, in diets of growing yearling Hereford steers. The wood was chipped, dried, and ground prior to incorporation in diets at levels of 0, 12, 24, 36, 48, and 48%. The 48% levels were the same except that 4% NaOH was added to one prior to pelleting of the diets. The ground wood largely replaced alfalfa in diets, and soybean meal was used to replace protein as alfalfa was decreased in diets. While all diets containing wood except the 12% level resulted in improved weight gains and feed efficiency over the alfalfa control, the investigators concluded that it should not be assumed that aspen wood is an equivalent replacement for alfalfa. The diet that contained 48% wood treated with 4% NaOH resulted in improved feed efficiency but no increase in daily gain.

## B. Wood Pulp By-Products

There are approximately 1.7 million tons of pulping fiber residues produced annually that ruminants might use (Millet *et al.*, 1973). Clarke and Dyer (1973) fed chemically degraded wood in diets to finishing cattle and estimated that hexose energy in a 70% treated wood diet fed steers was converted to VFA energy with 9% less efficiency than the hexose energy of the control barley diet.

Baker (1973) investigated the effect of lignin on the *in vitro* digestibility of wood pulps made by the sulfate process from paper, birch, red oak, red pine, and Douglas-fir wood. At high lignin levels, pulps from hardwood species were more digestible than those from softwood species, but at lignin levels below 7%, the pulps had about equal digestibility. The extent of digestibility depended on how much of the original lignin was removed. In general, softwoods increased more slowly in digestibility than hardwoods as lignin was removed. Kamstra *et al.* (1980) treated Ponderosa pine wood chips with peroxyacetic acid to improve *in vitro* rumen digestibility. While almost complete delignification was achieved, it appeared that 60% delignification would be most practical for ruminant feeds.

A. J. Baker *et al.* (1973) evaluated 10 sulfite pulps and two mechanical pulps and fines for *in vitro* rumen dry matter digestibility. Digestibility values ranged from 67 to 98% for the chemical pulps and pulp fines and from 0 to 7% for the mechanical pulps and fines. Dinius and Bond (1975) fed heifers a diet containing 50% wood pulp fines for 99 days and found that they gained weight more rapidly than heifers fed a control hay diet (0.74 versus 0.47 kg/day). Pregnant heifers were fed a diet of 75% pulp fines for 209 days, including parturition, with no differences in weight gain, calf birth weight, or calving problems compared to heifers fed a control hay diet.

Williams *et al.* (1979) isolated lignin–hemicellulose complexes from cell-free rumen fluid and determined their ability to stimulate *in vitro* microbial protein synthesis. The lignin–hemicellulose fraction isolated from rumen liquor from a sawdust-containing diet fed a cow improved *in vitro* microbial protein synthesis.

Hemlock clarifier sludge from an ammonia-base sulfite pulp mill was used in growth and metabolism studies with steers by Fifield and Johnson (1978). Steers were fed the clarified sludge at levels varying from 55 to 75% of the diet for 168 days (growing) and were then fed an 85% barley diet for 92 days (finishing). Average daily gains of all steers were 0.74 and 1.71 kg for the growing and finishing periods, respectively. Carcass characteristics were similar. In another growing and finishing trial in which 31 and 50% sludge dry matter was fed, the daily gains were 1.14 and 0.95 kg/day during the growing phase. There were no pretreatment effects when the steers were finished on an 85% barley diet, since they averaged 1.44 kg/day gains. Mean values for TDN, DE, and ME were 65.9%, 2.90 kcal/g, and 2.59 kcal/g, respectively, for clarifier sludge fed steers at maintenance levels.

## C. Solid Cellulose Waste

Solid cellulose waste makes up about $8 \times 10^6$ tons of printing and writing papers and $15 \times 10^6$ tons of packing and other grades of paper in the United States each year (U.S. Environmental Protection Agency [USEPA], 1975). Waste paper and other solid waste products have been shown to be generally low in energy content and to have *in vitro* dry matter digestibilities similar to those of low-quality forage (van Soest and Robertson, 1974; Mertens *et al.*, 1971a). The protein content of such materials is generally quite low. However, solid cellulose waste should provide a source of roughage in diets in holding pens at stockyards, a fiber source for high-grain diets where milk fat depression is a problem, and a roughage source in areas where hay is in short supply.

Belyea *et al.* (1979), working with St. Louis city solid cellulosic waste, found it to contain large amounts of fiber and ash, and to be low in protein and *in vitro* digestibility. The corrugated fraction was the most digestible and least contaminated with toxic elements. Lead and arsenic occurred in near-toxic concentrations, and mercury, cadmium, and chromium were marginal. Pesticides in some samples were above allowable levels, and polychlorinated biphenyls were often close to or greater than the tolerance limits set by the U.S. Food and Drug Administration.

Daniels *et al.* (1970) fed Holstein steers diets that contained either 0, 8, or 12% newspaper-containing diets for 120 days. There were no differences in rates of gain and feed efficiencies among steers consuming the three diets. The specific gravity values of steers fed the 0 and 8% newspaper-containing diets were equivalent, but the values of those fed the 12% diet were higher. Carcass grade was not affected by the diets, although steers fed the paper diets had less fat in their carcasses. There were no differences observed by a sensory panel in tenderness and flavor of meat among the dietary groups, or accumulation of toxic minerals or ink in the livers due to the various treatments.

When 16 or 24% newsprint (Dinius and Oltjen, 1971a) or up to 50% boxboard paper (Thomas *et al.*, 1970) was fed as energy constituents in diets to cattle, voluntary feed intake decreased even though molasses was incorporated to increase palatability. Dinius and Oltjen (1971b) observed cattle to consume the grain but leave the newsprint in diets, indicating that newsprint is unpalatable. Dinius and Oltjen (1972) conducted experiments to determine the basis for rejection by beef steers of diets that contain moderate levels of newsprint. Diets containing 24% newsprint with and without ink were fed cafeteria style. The difference in consumption was not significant, although steers tended to select the nonscript newsprint-containing diet. However, it was concluded that 24% newsprint in the total diet impaired feed intake because of the ruminoreticular fill. When 20% newsprint in diets was fed to dairy cows, there was a significant decrease in yield of milk; 15% newsprint appeared to be more acceptable and more efficiently utilized (Mertens et al., 1971a).

Becker et al. (1975) absorbed whey (a by-product of cheese manufacturing) on 11 types of paper, i.e., telephone directory yellow pages, white directory pages, feedsacks, glossy magazine, brown bags, telephone book covers, daily newsprint, cardboard boxes, computer punch cards, computer printout sheets, and coasters. Each type of paper was chopped into 1.2-cm squares and ground in a Wiley mill using a 20-mesh screen prior to absorption of added whey. After soaking with whey, the material was drained and dried at 65–70°C before *in vitro* rumen digestion trials. Digestibilities of dry matter in samples containing no whey varied from 37.2 for white pages to 81.7% for computer cards. Soaking in whey made paper more digestible.

## XIX. BY-PRODUCTS OF THE ESSENTIAL OIL INDUSTRY

Essential oils and oleoresins are extracted from plants, and considerable residues remain after extraction and must be disposed of. These residues have not been investigated extensively in feeding trials with ruminants. Wohlt et al. (1981) reported concentrations of crude protein, soluble nitrogen, nitrogen bound to ADF, ether extract, NDF, ADF, lignin, ash, calcium, magnesium, phosphorus, and silica in the residues from nutmeg hulls, cinnamon bark, black pepper, vanilla bean, ginger root, coffee bean, cocoa bean, and celery seed (Table 9.1).

As shown in Table 9.2, appreciable amounts of the plant residues were degraded in Dacron bags placed in the rumen of a heifer and by *in vitro* rumen microbe fermentation, indicating that the residues have potential as feed ingredients for ruminants. Digestibility of dry matter *in vitro* of residues from black pepper, vanilla bean, ginger root, coffee bean, cocoa bean, and celery seed ranged from 43 to 65%. However, only 25 and 29% of the residues from nutmeg hulls and cinnamon bark were degraded *in vitro* and 21 and 13% *in vivo*, respectively. The greater value for degradation of cocoa bean residue by the Dacron bag technique than by fermentation *in vitro* was believed to be due to its fine powder form, which allowed it to seep through the cloth into the rumen.

The 15% or greater levels of crude protein in black pepper, coffee bean, cocoa bean, and celery seed residues contained 51.5, 39.5, 39.7, and 22.2% of the

**TABLE 9.1.**

**Chemical Composition of a Hay and By-Products of the Essential Oil Industry (%)[a]**

| Ingredient[c] | Hay[d] | Plant residues of the essential oil industry[b] | | | | | | |
|---|---|---|---|---|---|---|---|---|
| | | Nutmeg hulls | Cinnamon bark | Black pepper | Vanilla bean | Ginger root | Coffee bean | Cocoa bean | Celery seed |
| Dry matter, as received | 88.0 | 65.2 | 72.5 | 65.1 | 37.4 | 58.0 | 95.7 | 25.4 | 97.2 |
| Crude protein | 13.5 | 8.8 | 4.0 | 16.5 | 7.1 | 8.0 | 16.1 | 27.0 | 19.9 |
| Soluble nitrogen | 18.4 | 2.0 | — | 15.0 | — | 17.1 | 30.4 | 31.6 | 21.3 |
| Nitrogen bound to ADF | 14.4 | 87.2 | 100.0 | 51.5 | 64.0 | 36.7 | 39.5 | 39.7 | 22.2 |
| Ether extract | 2.3 | 5.8 | 0.8 | 8.3 | 19.9 | 3.1 | 1.6 | 12.8 | 0.2 |
| NDF | 54.5 | 79.2 | 81.2 | 51.7 | 70.2 | 48.2 | 63.5 | 29.7 | 56.2 |
| ADF | 38.1 | 69.8 | 64.8 | 46.4 | 62.8 | 16.4 | 32.6 | 30.6 | 37.4 |
| Lignin | 11.4 | 18.3 | 20.4 | 17.9 | 13.6 | 5.6 | 13.0 | 24.8 | 22.3 |
| Ash | 7.1 | 5.8 | 5.9 | 10.3 | 4.1 | 6.9 | 5.0 | 5.6 | 10.2 |
| Calcium | 1.22 | 0.14 | 1.43 | 0.19 | 0.12 | 0.45 | 1.16 | 0.66 | 2.00 |
| Magnesium | 0.32 | 0.51 | 0.08 | 0.30 | 0.20 | 0.17 | 0.12 | 0.27 | 0.45 |
| Phosphorus | 0.21 | 0.74 | 0.06 | 0.21 | 0.18 | 0.18 | 0.06 | 0.20 | 0.71 |
| Silica | 0.27 | 0.35 | — | 0.27 | — | 1.16 | 0.06 | — | 1.31 |

[a] From Wohlt et al. (1981).
[b] Plant residue remaining after essential oils and oleoresins were extracted.
[c] Dry matter basis. Soluble nitrogen and nitrogen bound to ADF are expressed as a percentage of the total nitrogen content.
[d] A grass-legume hay given to a fistulated Angus heifer from which rumen digesta were used in fermentation studies *in vitro* and *in vivo*.

TABLE 9.2.

Digestibility of Residues from the Essential Oil Industry[a]

| Plant residue | Dry matter digestibility (%) | |
|---|---|---|
| | In vitro[b] | In vivo[c] |
| Nutmeg hulls | 25.6 ± 0.2 | 21.5 ± 1.6 |
| Cinnamon bark | 29.6 ± 0.8 | 13.1 ± 0.3 |
| Black pepper | 55.1 ± 0.4 | 42.7 ± 1.3 |
| Vanilla bean | 42.7 ± 0.4 | 36.4 ± 0.8 |
| Ginger root | 65.0 ± 0.8 | 60.5 ± 2.7 |
| Coffee bean | 52.1 ± 1.7 | 68.8 ± 0.8 |
| Cocoa bean | 56.8 ± 3.8 | 92.7 ± 0.8[d] |
| Celery seed | 58.4 ± 1.0 | 43.8 ± 0.9 |

[a]From Wohlt et al. (1981).
[b]Rumen fermentation in vitro and NDF procedure of Goering and van Soest (1970).
[c]Dacron bags in rumen over a 24-hr period.
[d]High value due to loss of powdered cocoa bean residue through Dacron cloth (particle size less than mesh of dacron bags).

protein bound to ADF, respectively. Residues from nutmeg hulls, cinnamon bark, vanilla bean, and ginger root contained less than 9% crude protein, with 37–100% of the protein bound to the ADF fraction. The protein bound to ADF would be relatively resistant to microbial degradation. Supplementation of nitrogen in diets is indicated when the residues are fed to ruminants. Feedlot trials with ruminants are needed to evaluate further the value of these residues.

## XX. LIGNOCELLULOSE MATERIALS TREATED WITH ALKALI

Ready availability of NPN sources such as urea increases the importance of obtaining as much energy from plant materials as possible. Much waste wood, straw, bagasse, and other by-products are available as sources of energy for ruminants if economical means are utilized to make their carbohydrates available to ruminal microbes. Alkaline treatments increase the availability of carbohydrates in lignocellulosic materials without removal of lignin.

### Wood

Pew and Weyna (1962) improved the digestibility of wood by treating it with aqueous sodium hydroxide (NaOH) and observed more pronounced effects with

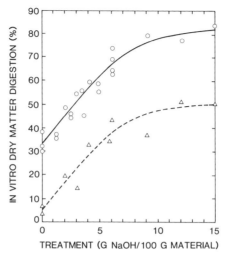

**Fig. 9.5.** Effect of NaOH pretreatment on the percentage of dry matter digestion *in vitro* with rumen microorganisms. ○, wheat straw; △, poplar wood (From Wilson and Pigden, 1964).

hardwoods than with softwoods. The *in vitro* rate of digestion of hardwoods by rumen microbes measured by succinic acid production was markedly increased by pretreatment with 1–5% NaOH (Stranks, 1959). Wilson and Pigden (1964) evaluated the effect of varying amounts of NaOH on wheat straw and poplar wood by *in vitro* rumen microbial fermentation (Fig. 9.5). The digestibility of both wheat straw and poplar wood improved linearly with increasing NaOH up to about 7 g NaOH per 100 g straw, but gains at higher levels of NaOH were small.

Saponification of uronic acid polymeric units and acetate esters of polymeric xylans occurs with alkaline treatment. The liberation of the free carboxyl groups of uronic acids can be assayed by calcium ion exchange. As shown in Fig. 9.6, the carboxyl content of sugar maple wood upon treatment with NaOH increases to a maximum calcium ion exchange value of 24 mEq/100 g within approximately 3 hr of saponification with 1 or 2% NaOH (Tarkow and Feist, 1969). Polymeric xylan also contains acetyl groups that saponify as rapidly as glucuronic acid groups (Timell, 1967).

Tarkow and Feist (1969) reported that the reaction between aqueous ammonia ($NH_3$) and maple wood was very slow. When they treated sugar maple with liquid ammonia at 25°C for 8 hr, the nitrogen content was 0.36 g/g, which was equivalent to about 25 mEq of nitrogen per 100 g treated wood or 25 mEq of total carboxyl. This value is in good agreement with carboxyl values found by calcium exchange for NaOH-treated wood.

Feist *et al.* (1970) utilized rumen microbes to determine the *in vitro* digestibility of black ash, American basswood, yellow birch, eastern cottonwood, American Elm, silver maple, sugar maple, red oak, white oak, and quaking

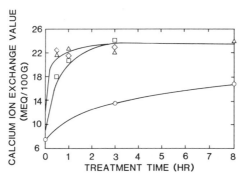

**Fig. 9.6.** Effect of treating sugar maple with aqueous NaOH at room temperature on values for calcium ion exchange (free carboxyl content). Percentages of NaOH: ○ = 1.5; □ = 0.5; △ = 1.0; ◇ = 2.0

aspen after room temperature treatments with 1% NaOH for 4 hr. Digestibility of all samples increased due to the alkali treatment but varied inversely with lignin content and was independent of wood species. Maximum digestibility of aspen was about 50% compared to 15% for red oak.

Gharib et al. (1975) studied the effect of sodium hydroxide (NaOH), NaOH + sodium sulfide ($Na_2S$), sodium sulfite ($Na_2SO_3$), $Na_2SO_3$ + sodium carbonate ($Na_2CO_3$), calcium oxide (CaO), CaO + NaOH, sodium hypochlorite (NaClO), and sodium chlorite ($NaClO_2$) at levels of 0–30 g/100 g on poplar bark in regard to *in vitro* digestibility of the bark by rumen microbes. One-day treatments with the higher concentrations of NaOH, CaO, and CaO + NaOH decreased lignin from 21.1 to 16.9, 19.7, and 20.7% respectively, while NaClO and $NaClO_2$ decreased lignin from 21.1 to 12.4 and 12.4%, respectively. *In vitro* digestibility of the bark was increased (from 24 to 33%) to the greatest extent by NaOH, CaO, CaO + NaOH, NaClO, and $NaClO_2$ treatments. Sodium sulfide, $Na_2SO_3$, and $Na_2SO_3$ + $Na_2CO_3$ had little or no effect on digestibility of the poplar bark.

Keith and Daniels (1976) treated hardwood sawdust with 1.0–2.5% solutions of NaOH or sulfuric acid ($H_2SO_4$) for 24 hr and observed that *in vitro* dry matter digestibility was increased by both alkali and acid treatments. However, Holstein steers fed diets containing 25% untreated, 1.0% NaOH-treated or 1.0% $H_2SO_4$-treated sawdust for 85 days had no differences in weight gains, feed intake, or feed efficiency. Gharib et al. (1975) determined the effect of five concentrations of NaOH (0, 3, 6, 9, and 12 g NaOH per 100 g dry matter), three temperatures (25, 50, and 75°C), and two reaction times (1 and 20 days) on *in vitro* digestibility of poplar bark. Hemicellulose and lignin contents of the bark were decreased with 9–12 g NaOH per 100 g bark in 1 day of treatment and were influenced by temperature. *In vitro* digestibility was significantly increased by the 9- or 12-g NaOH levels.

## XXI. ENHANCING THE UTILIZATION OF CONCENTRATE FEEDS

When ruminants are fed high-concentrate finishing diets (90% concentrate or more), special precautions need to be taken (NRC, 1976a). Such diets should be self-fed and available at all times, since rumen dysfunction may occur after excessive intake of readily fermentable carbohydrates. A sudden increase in readily digestible carbohydrates may result in high lactic acid production and a drop in rumen fluid pH, producing symptoms of acute indigestion. Some animals may die, others may have poor performance due to depressed feed intake, and recovered animals may have laminitis. A high performance level of ruminants on finishing diets can best be maintained by including 10–15% or more roughage. Generally, limited levels of roughage in the diets of finishing cattle do not require any more concentrate consumption to reach a given degree of finish; however, longer periods of feeding are required.

It is the fiber in the roughage that benefits rumination. The definition of dietary fiber depends on the methodology of its chemical determination (van Soest, 1975). *Crude fiber (CF)* refers to the residue of roughages that is resistant to acid and alkali used in proximate analysis (American Association of Analytical Chemists [AOAC], 1975). Crude fiber is considered by many to be unsatisfactory as a measure of total fiber because of the proportions of hemicellulose and lignin in various plant species and because the recovery of these substances in the analysis is highly variable. *Acid detergent fiber (ADF)* separates cellulose and lignin, but hemicellulose is dissolved in the acid utilized in its determination (Harris *et al.*, 1968). The hemicellulose in legumes, cool- and warm-season grasses, and by-product feeds varies; for this reason, ADF does not satisfactorily represent the total fiber of feeds. *Neutral detergent fiber (NDF)*, frequently called *cell walls,* includes cellulose, hemicullulose, lignin, and insoluble ash (van Soest, 1975). The NDF is related to most factors associated with intake-limiting characteristics of the diet such as digestibility, rumination, bulk density, and rate of passage through the alimentary tract. The physical form of roughages also affects chewing time and rumination. Many investigators believe that NDF may be the best single measure of fiber in the formulation of diets for ruminants.

Variations in proportions of hemicellulose, cellulose, and lignin components of NDF and physicochemical factors associated with processing produce different responses in ruminants fed equivalent levels of fiber from different sources. The effective fiber of feedstuffs was related to cottonseed hulls as the standard in regard to its ability to maintain milk fat by Harris (1975). He recommended an effective fiber requirement of 17% (air-dried basis) for complete diets. The term *roughage factor* was used by McCullough (1973) to compare different feeds to long hay as a source of true fiber in complete diets. The total chewing time and

**TABLE 9.3.**

**Expressions of Fiber Adequacy**[a]

| Ingredients | Dry matter basis (%) | | | Roughage factor[d] | Minutes of chewing time per kg dry matter[e] |
| --- | --- | --- | --- | --- | --- |
| | NDF[b] | CF[b] | Effective fiber[c] | | |
| Alfalfa | | | | | |
|   Long | 48 | 31 | 31 | 1.0 | 61.5 |
|   Chopped | 48 | 31 | — | — | 44.3 |
|   Pelleted | 48 | 31 | — | 0.2 | 36.9 |
|   Dehydrated meal | 45 | 28 | 28 | 0.2 | 36.9 |
| Corn silage | 50 | — | — | — | 59.6 |
| Sugarcane bagasse | 87 | 48 | 54 | — | 18.0 |
| Soybean hulls | 67 | 39 | 27 | 0.0 | 8.4 |
| Corn cobs | 89 | 35 | 44 | 0.6 | — |
| Citrus pulp | 23 | 14 | 0 | 0.2 | 30.9 |
| Cottonseed hulls | 90 | 50 | 50 | 1.0 | 30.1 |
| Barley | 19 | 6 | 6 | 0.1 | 15.0 |
| Oats | 31 | 12 | 12 | 0.1 | 12.0 |
| Corn and cob meal | 30 | 9 | 9 | 0.2 | — |
| Corn, ground | 9 | 3 | — | 0.0 | 8.3 |
| Brewers grains | 42 | 16 | 16 | 0.2 | — |
| Distillers dried grains | 43 | 13 | 13 | 0.1 | — |

[a]Jorgensen et al. (1981).
[b]NRC (1978).
[c]Harris (1975).
[d]McCullough (1973).
[e]Sudweeks et al. (1979).

milk fat percentage of dairy cows were observed to be decreased as grain replaced long hay in diets (Balch, 1971). Sudweeks et al. (1979) determined the roughage index values of various feedstuffs, expressed as minutes of chewing time per kilogram of dry matter feed intake. Table 9.3 presents fiber values in various feedstuffs expressed as NDF, CF, effective fiber, roughage factor, and minutes of chewing time per kilogram of dry matter, according to Jorgensen et al. (1981).

Jorgensen et al. (1981) pointed out that chewing time index values change not only with forage quality and total dry matter intake but with particle length as well. They conducted studies utilizing the mean particle length of forages to adjust forage or intake in order to predict roughage index values. As the mean particle length of forage increased, total chewing time per kilogram of forage increased, due largely to greater rumination time. As forage intake increased, total chewing time decreased.

Cole *et al.* (1976b) fed steers fitted with ruminal and abomasal cannulae whole shelled corn diets containing 0, 7, 14, or 21% cottonseed hulls and found total starch digestibilities to be 96.4, 94.7, 92.2, and 95.4%, respectively. Total digestion percentages for starch, dry matter, and cellulose tended to be lowest with the diets that contained 14% hulls, apparently due to a decrease in the digestion of the concentrate portion of the diet caused by an increased rate of passage through the alimentary tract.

## XXII. INFRARED REFLECTANCE SPECTROSCOPY

Infrared reflectance spectroscopy has been investigated as a means to evaluate forage quality. Norris *et al.* (1976) recorded the near-infrared-reflectance spectra of 87 samples of ground dry alfalfa, tall fescue, alfalfa-bromegrass mixtures preserved as hay, silage, and fresh-frozen forages, as well as Bermuda grass and Pangola digitgrass. Reflectance (RP) spectra were recorded as log (1/R) versus wavelength and transformed to the second derivative of log (1/R) versus wavelength for correlation with nutritional and compositional data. Multiple linear-regression techniques were used to determine the optimum wavelengths for predicting each *in vitro* and *in vivo* analysis. Correlation coefficients were 0.99 for crude protein, 0.98 for NDF, 0.96 for ADF, 0.96 for lignin, 0.95 for *in vitro* dry matter digestibility, 0.88 for *in vivo* dry matter digestibility, 0.80 for dry matter intake, and 0.85 for DE intakes by ruminants. The investigators concluded that infrared reflectance has a potential for rapid evaluation of forage quality.

### Electron Microscopy Scanning

Brazle and Harbers (1977) used a scanning electron microscope to observe digestion of mature alfalfa leaves and stems suspended in nylon bags in the rumen of steers. Digestion began with random hydrolysis on the adaxial leaf surface. After 24 hr, the leaf cuticle and epidermis were sloughed, allowing massive digestion of the mesophyll. The external stem surface was attached by sloughing of the cuticle and epidermis, exposing a dense matrix (cortex) that was partially hydrolyzed during the first 24 hr in the rumen. After 48 hr, fiber cells were exposed, and by 72 hr only vascular tissue remained.

# 10

# Effects of Processing Feedstuffs on Nitrogen and Energy

## I. INTRODUCTION

Physical preparation of cereal grains and other feedstuffs by grinding, crushing, flaking, rolling, popping, micronizing, soaking, and ensiling is practiced by livestock feeders. Riggs (1958), upon reviewing 50 years of beef cattle nutrition, concluded that rolling or crushing had little advantage over grinding. Flaking of corn reduced the ratio of acetate to propionate in the rumen and increased feed efficiency with lambs (Phillipson, 1952). Feeding of pelleted alfalfa hay plus steam flaked corn and linseed meal to Holstein calves gave increased gains and feed efficiency compared to a diet of chopped alfalfa hay, ground corn, and linseed meal (Shaw *et al.*, 1960). Keating *et al.* (1965) reported that cooking milo for 9 hr at 82°C significantly decreased the digestibility of protein by cattle compared to dry-rolled milo. Moist heat treatment of corn starch resulted in more rapid fermentation by rumen microbes (Salsbury *et al.*, 1961). Hale and Theurer (1972) listed 18 different processing methods for grain and classified them according to wet or dry methods. Processing appears to benefit cattle more than sheep, which have a greater ability to digest whole grain. Most of the studies on processing involve corn, sorghum, barley, and alfalfa.

## II. CORN-PROCESSING STUDIES

Corn was demonstrated by Matsushima and Montgomery (1967) to be improved in digestibility by steam processing and flaking. They observed an improvement in feed efficiency with finishing steers when fed a thin-flaked corn grain (0.8 mm) compared to a thick flake (2.0 mm). Johnson *et al.* (1968) found higher dry matter digestibility for steam-flaked than cracked corn. If corn was steamed, dried and then cracked, the dry matter and protein digestibilities were lower than those of corn that was only flaked.

Utley and McCormick (1975) compared steers fed a cracked corn diet containing 20% peanut hulls, an all-concentrate diet containing unground dry shelled corn (DC), and an all-concentrate diet containing ensiled high-moisture shelled corn (HMC). Total digestible nutrients (TDN) were similar for the DC and HMC diets and approximately 19% greater than in the diet containing 20% peanut hulls. Steers fed DC and HMC diets had fewer protozoa and more bacteria in the rumen fluid than those fed the peanut hull-containing diet. Cole et al. (1976c) compared steam-flaked and dry-rolled corn-containing diets with either 0 or 21% cottonseed hulls (CSH)-fed steers. Microbial protein synthesized per 100 g dry matter digested was 7.0 and 6.3 g for the flaked and rolled corn without CSH and 10.5 and 11.2 g with CSH, respectively.

Ware et al. (1977) compared dried corn with high-moisture corn treated with propionic acid and ammonium isobutyrate in diets fed yearling steers. On a poststorage dry matter basis, steers fed propionic acid-treated grain had daily dry matter consumption and average daily gains equivalent to those of the steers fed dried grain. Steers fed the ammonium isobutyrate-treated grain had daily gains equivalent to those of steers fed dried grain and were more efficient in converting grain dry matter to gain in weight.

Lee et al. (1981) fed feedlot steers 85% concentrate diets containing whole shelled (W) and steam-flaked (SF) corn in the following proportions: 100W : 0SF, 75W : 25SF, 50W : 50SF, 25W : 75SF, and 0W : 100SF, and concluded from weight gains and feed efficiency data that whole shelled corn can substitute for up to 25% of steam-flaked corn in conventional feedlot diets containing 15% roughage with no detrimental effect on steer performance. Turgeon et al. (1981) tested the effect of particle size of corn on starch digestion utilizing cannulated steers offered a diet ad libitum. Whole (W), cracked (C), and a 50 : 50 mixture of W : C corn were compared in a diet containing 85% corn, 10% alfalfa-brome hay, and 5% pelleted supplement fed steers. Total starch digestion was equivalent for the three diets, but ruminal starch digestion was greatest in the diets containing cracked corn.

Hale and Prouty (1980) reviewed many feeding trials conducted at 11 state experiment stations in which 80% or greater concentrate diets that contained corn processed either by being dry rolled, flaked, whole shelled, early harvested, or reconstituted were fed to feedlot cattle (Table 10.1). Since all trials were not conducted concurrently on grain processed by the various methods, it was necessary to correct or adjust all values to a common processing system, and dry-rolled grain was selected as a reference. Steers ranging from 400 to 750 lb were generally utilized in the trials. Daily gain was not affected by flaking compared to dry rolling, but feed efficiency was improved by 8.4%. Whole shelled corn diets gave gains and feed efficiency equivalent to those that contained cracked corn. Weight gains were slightly higher with early-harvested grain than with the

**TABLE 10.1.**

**Comparison of Corn-Processing Systems**[a-c]

| Processing method | Dry rolled | Flaked | Whole shelled | Early harvested and reconstituted |
|---|---|---|---|---|
| Avg. daily gain, lb | 2.75 | 2.75 | 2.75 | 2.84 |
| Avg. daily feed, lb[b] | 19.0 | 17.4 | 18.9 | 18.3 |
| Reduction, % | | 8.4 | | 3.7 |
| Feed per 100 lb gain, lb[b] | 691 | 633 | 687 | 644 |
| Grain level, % | 74 | 74 | 78 | 80 |
| Improvement in diet efficiency, % | | 8.4 | | 6.8 |
| Improvement in gain efficiency, % | | 11.4 | | 8.5 |

[a]From Hale and Prouty (1980).
[b]Average, 150 days on feed; initial weight of cattle, 550 lb.
[c]90% dry matter basis.

other three processing methods, and feed intake was 3.7% less than with dry-rolled grain diets.

## III. PROCESSING OF SORGHUM GRAINS

Hale *et al.* (1966a) reported that steam flaking of sorghum grain increased performance and feed efficiency in feedlot cattle. This was confirmed by Newson *et al.* (1968) when they obtained a 15% improvement in utilization of sorghum grain by steam processing and flaking. Husted *et al.* (1968) conducted two trials with cattle fed 77% sorghum grain diets in which the grain was processed by several techniques. In the first trial, digestion was not improved by fine grinding compared to rolling.

Digestibility of nitrogen-free extract was improved by 14% when sorghum grain was steam processed and flaked or pressure cooked and flaked compared to dry-rolled sorghum grain. Protein digestibility was somewhat better with dry-rolled grain than with sorghum grain exposed to heat. In a second trial, steam processing and flaking the sorghum grain improved nitrogen-free extract digestibility by 18% over that obtained by dry rolling, but no comparative benefit was obtained when the grain was steam processed and cut or water soaked and cut.

Steam processing and flaking of sorghum grain significantly improved digestibility by cattle of starch and reducing sugars compared to fine grinding or dry rolling, with no difference in digestibility between coarse and finely ground sorghum grain (Buchanan-Smith *et al.*, 1968). Reconstitution markedly improved the digestibility of nonprotein organic matter of grain by cattle to about the same extent observed in steam-processed and flaked sorghum grain (Buchanan-Smith *et al.*, 1968; Riggs and McGinty, 1970).

McNeill *et al.* (1971) reported greater total digestion of starch by steers fed reconstituted and steam-flaked sorghum grain diets compared to diets containing micronized and dry ground grain. Hinman and Johnson (1974) conducted trials with steers fitted with rumen and abomasal cannulae and fed high-concentrate diets containing sorghum grain either dry rolled or micronized to densities of 412, 322, and 232 g/liter. No significant differences in starch digestion occurred, but 95% of the micronized sorghum was digested prior to entering the intestines compared to 51% with the dry-rolled sorghum. The degree of gelatinization increased from 9 mg maltose/g for dry-rolled sorghum to 112 mg/g for the high micronized sorghum. A smaller ratio of acetate to propionate with greater total VFA were observed in rumen fluid with the micronized sorghum.

Holmes *et al.* (1970) determined the digestion of starch in diets containing 80% milo steamed either at atmospheric pressure or at 3.5 kg/cm$^2$ prior to rolling utilizing abomasal fistulated cattle and slaughter studies with sheep. Starch fermented in the rumen ranged from 90 to 95% with both types of pressure steaming. Kiesling *et al.* (1973), in a comparative slaughter trial with yearling steers, found that net energy of maintenance plus gain ($NE_{m+g}$) was greater with reconstituted (38% moisture) rolled than with dry-rolled sorghum grain, but dry matter intake was lower due to reconstitution. In a commercial feedlot study, Schake *et al.* (1972) reported that feedlot steers offered steam-flaked, whole reconstituted, or rolled reconstituted sorghum grain diets gained similar weights and produced similar carcasses. Steers fed whole reconstituted and rolled grain diets had better feed efficiency than those fed steam-flaked sorghum diets. Similar steers fed either steam-flaked or micronized sorghum grain diets in another trial had equivalent gains and carcasses. The steers fed micronized grain diets consumed more feed and had lower feed efficiency than those fed steam-flaked grain.

Riggs *et al.* (1970) dry heat-treated sorghum grain containing about 15% moisture at approximately 178°C and obtained 45% maximum popping, with a bulk density of 98 g/liter compared to 779 g/liter for the unpopped grain. Feeding a popped grain mixture, 100% popped grain or partially popped grain, all crimped, in all-concentrate diets to finishing steers resulted in decreased feed intake but increased feed efficiency compared to steers fed diets containing nonheated, dry-rolled grain. Steers fed diets with the highest feed efficiency had higher levels of rumen propionic acid relative to acetic and isovaleric acids. Micronization improved the nitrogen-free extract digestibility of sorghum grain by cattle but had no effect on protein digestibility.

Hale *et al.* (1980) reviewed many trials conducted at various state experiment stations with cattle fed approximately 80% concentrate sorghum grain diets, the

**TABLE 10.2.**

**Comparison of Milo-Processing Systems**[a–c]

| Processing method | Dry rolled | Flaked | Reconstituted | Popped, exploded, and micronized |
|---|---|---|---|---|
| Avg. daily gain, lb | 2.56 | 2.76 | 2.76 | 2.76 |
| Avg. daily feed, lb[c] | 18.7 | 17.8 | 17.4 | 17.8 |
| Reduction, % | | 4.8 | 7.0 | 4.8 |
| Feed per 100 lb gain, lb[c] | 730 | 645 | 630 | 645 |
| Grain level, % | 74 | 74 | 78 | 74 |
| Improvement in diet efficiency, % | | 11.6 | 13.6 | 11.6 |
| Improvement in gain efficiency, % | | 15.7 | 17.6 | 15.7 |

[a]From Hale et al. (1980).
[b]Average, 140 days on feed; initial weight of cattle, 550 lb.
[c]90% dry matter basis.

sorghum grain being either dry rolled, flaked, reconstituted, popped, exploded, or micronized. A summary of these effects is presented in Table 10.2. All feed intake and feed requirement values were corrected to a 90% air-dried basis and, to make comparisons valid since trials of different processed grains were not carried out simultaneously, the control was dry-rolled grain. Values are based on actual weights without shrink of animals.

## IV. PROCESSING OF BARLEY

Steam processing and flaking improved the utilization of barley when digestibility was low (Parrott *et al.*, 1969), and increased the daily gains but not the feed efficiency of feedlot cattle (Hale *et al.*, 1966b). Improvement in feed requirements with feedlot steers was obtained when barley was pressure cooked at 1.4 kg/cm$^2$ and incorporated in diets (Garrett *et al.*, 1966). Osman *et al.* (1970b) conducted *in vitro* digestibility studies on barley and sorghum grain with porcine pancreatin and observed that if grains were steamed and not flaked, starch digestion was reduced. Greater thinness of flakes resulted in improved starch digestion.

Broadbent (1976) fed steers diets that contained wet distillers grains or grass silage supplemented with 0 or 2.5 kg barley dry matter daily, offered either whole or rolled. The barley was prepared for storage by drying to 84.6% dry matter and by treatment with propionic acid at 79.9% dry matter. Dried whole barley was also fed at 50% dry matter after soaking for 24 hr in water. The barley supplement increased liveweight gains, carcass weight, and carcass fat, and these effects were enhanced if the barley was rolled compared to whole barley grain. Nordin and Campling (1976) studied the susceptibility of whole and rolled cereal grains to rumen microbial attack in nylon bags in fistulated cows. Whole intact seeds were less susceptible to attack by microbes than naked grains, and breaking the grains considerably increased the loss of dry matter.

## V. PROCESSING OF WHEAT

Aimone and Wagner (1977) conducted two feedlot trials with steers to evaluate micronized wheat versus dry-rolled wheat in 95% wheat diets. In the second trial, micronized hard red winter wheat gave greater dry matter intake (7.16 versus 7.81 kg) and gains (1.45 versus 1.59 kg) than the dry-rolled wheat, but no differences were obtained in the first trial. Feed efficiency was not influenced in either trial. *In vivo* digestion coefficients for protein (79.3 versus 79.8%) and starch (99.2 versus 99.3%) were similar for the two wheat treatments. Prasad *et al.* (1975) observed that flaking and extruding red winter wheat reduced *in vitro* protein hydrolysis and increased *in vitro* starch availability but had no significant effect on nitrogen-free extract digestion and nitrogen retention with Holstein steers. Flaking or extruding of wheat gave similar values for feed intake, weight gains, and feed efficiency. In a second trial, pressure cooking decreased *in vitro* protein hydrolysis, but compacting and pressure cooking increased starch availability. *In vivo* nutrient digestibilities with Holstein bulls were not affected by treatments.

## VI. COTTONSEED MEAL PROCESSING

The effect of autoclaving cottonseed meal on ruminal protein degradation and rumen bypassing was studied by Broderick and Craig (1980). One protein fraction of cottonseed meal was degraded *in vitro* at rates 2.0–3.5 times greater than that of casein, while a second fraction was degraded at a slower rate. Ruminal escape increased with each increment of heat, and intestinal protein digestibility increased to a maximum at 60 min of autoclaving and then declined. Intestinal protein digestibility averaged 30.6 and 50.3% for solvent-extracted and screw-pressed meals, respectively. It was suggested that heat treatment decreased degradation in the rumen partly by blocking reactive sites for bacterial proteolytic enzymes and partly by decreasing the solubility of cottonseed protein.

## VII. SOYBEAN MEAL PROCESSING

Netemeyer *et al.* (1980) studied the effect of particle size of soybean meal (coarse or finely ground) on protein utilization in diets fed steers and lactating cows. The diets consisted of prairie hay and ground corn, supplemented with either 3.2% urea or 22.3% soybean meal, either finely ground or of a coarser particle size. Bypass of nitrogen in steers was 61.4 and 44.7% for finely ground and coarse soybean meal, respectively. Particle size had no effect on milk yield, rumen ammonia, and blood urea.

## VIII. PROCESSING OF FORAGES

Balch (1950) reported that processing decreased cellulolytic activity by rumen microbes. Blaxter et al. (1956) observed that increasing fineness of grinding led to decreased time spent in the rumen and decreased digestion of dry matter and organic matter, including crude fiber, nitrogen-free extract, and crude protein. Variation in the affects of grinding and pelleting on digestion have been attributed to changes in chemical composition as well as inadequate definition of the degree of fineness of grinding (Moore, 1964; Osbourn et al., 1976).

Goering (1976) analyzed 111 samples of commercial dehydrated alfalfa from 48 U.S. states for overheating by both acid detergent-insoluble nitrogen (ADIN) and pepsin-insoluble nitrogen (PIN) determination. ADIN exceeded 0.29% in 89% of samples and PIN exceeded 0.79% in 32% of samples. Such values indicated that these samples were overheated and had a mean relative loss of 22–24% of digestible nitrogen using ADIN or a loss of 15–24% using PIN. This is a large loss of nitrogen to livestock in commercially dehydrated alfalfa.

Osbourn et al. (1981) fed lambs dried early- and late-cut alfalfa, red clover, Italian and perennial ryegrasses, and timothy in the long form or three processed forms. Digestion of the cell content fraction of diets was not influenced by species, fineness of grinding, or level of feeding, but digestion of organic matter and cell walls was depressed by increased levels of feeding and increased coarseness of grinding. Nocek et al. (1980) fed Holstein bull calves chopped hay, ground hay, and all-concentrate diets, each with three levels of ruminal degradable nitrogen. Rumen tissue samples were obtained by biopsy after 8 weeks of feeding and at slaughter (20 weeks). It was found that acetate transport increased across the rumen epithelium between 8 and 20 weeks in calves fed the ground and chopped hays but not in those fed the concentrate. Propionic acid transport was highest in those fed chopped hay and lowest in those fed concentrate at both 8 and 20 weeks. Transport of both acetate and propionate was increased by increasing ruminal degradable nitrogen.

## IX. STEAM TREATMENT OF CROP RESIDUES

Hart et al. (1981) steam treated rice straw, sugarcane bagasse, and sugarcane field trash at pressures ranging from 7.0 to 42.2 kg/cm$^2$ with aqueous sodium hydroxide, ammonia, and urea as additives up to 4.5% of the dry weight. Effects of treatments on digestibility were determined by an *in vitro* enzymatic assay. The highest digestibilities were reached at 21.1 kg/cm$^2$ pressure, both with and without the alkali and urea treatments. After 1.5 min at this pressure, rice straw digestibility improved from 26% initially to about 47 and 64%, without and with additives, respectively.

# 11

# Production Practices Affecting Nitrogen and Energy Nutrition

## I. INTRODUCTION

Many production practices have been investigated to determine the influence on beef production of such factors as backgrounding with forages prior to finishing with concentrates, lowering of protein in diets toward the end of the finishing period, natural protein compared to nonprotein nitrogen (NPN) sources, and compensatory gains. When forage quantity is adequate for maximum consumption and animals have the potential for production, feeding of limited amounts of concentrate will have two effects: additive and substitutive. The additive effect is expressed in greater daily gains, and the substitutive effect is demonstrated by decreased forage comsumption. The extent of either additive or substitutive effect depends on the quality of the forage (Peacock *et al.*, 1964; Coleman *et al.*, 1976). The additive and substitutive effects occur simultaneously when concentrates are fed with most forages (Mott *et al.*, 1968; Vohnout and Jimenez, 1975; Golding *et al.*, 1976b).

## II. SPRING VERSUS FALL CALVING

Some of the benefits of fall calving compared to spring calving include reduced incidence of disease in young calves, better control of the cow herd at breeding, and the production of calves that are large enough to utilize pasture efficiently at its peak quality in the spring (Mueller and Harris, 1967; Raleigh *et al.*, 1970). Kartchner *et al.* (1979) conducted a study with fall- and spring-calving cows that grazed crested wheatgrass pasture over a 125-day period from April through August. The forage dry matter intake for the fall- and spring-born calves averaged 4.25 and 0.94 kg, respectively, with corresponding gains of 1.00 and 0.75 kg daily. There was no difference in forage intake by the cows in the two groups. The spring-calving cows gained an average of 0.65 kg daily and produced 5.24 kg of milk daily compared to the fall-calving cows, which gained 0.98 kg daily and produced 2.47 kg milk per day. On the basis of forage consumption, the fall-calving pair was approximately 25% more efficient than the spring-calving pair.

## III. GROWTH RATE AND REPRODUCTION OF BEEF HEIFERS

Extremes in management, ranging from provision of full feeding from weaning of creep-fed heifers to provision of a slightly better than maintenance diet after weaning, have been practiced with undesirable results. High levels of feeding have been shown to result in reduced life span and impaired milking ability (Swanson, 1960; Pinney et al., 1962). Low levels of feeding have resulted in poor reproductive performance (Turman et al., 1964; Earley et al., 1977), reduced milk production, and poor weaning weights (Neville et al., 1960; Pinney et al., 1962).

Fleck et al. (1980) reported that heifers with high weight gains during the first winter as weanlings had greater breeding efficiency when bred as yearlings, had a larger pelvic area as 2-year-olds, gave birth to larger calves, had less calving difficulty at first parturition, and had higher breeding efficiency the following year. Heifers with excellent second-winter gains had less difficulty in calving during first parturition and slightly greater milk production and breeding efficiency when rebred, but the second-winter gain of the dam had a high negative relationship with calf growth to weaning. Mangus and Brinks (1971) reported that a high preweaning nutritional level and heavy weaning weight of heifers were correlated negatively with subsequent calf productivity.

Holloway and Totusek (1973a) utilized Angus and Hereford heifers to determine the effects of three preweaning planes of nutrition imposed by (1) weaning at 140 days, (2) weaning at 240 days, and (3) creep feeding and weaning at 240 days. The creep feed consisted of 0.45 kg of cottonseed meal and 0.90 kg ground milo per head daily from 140 to 240 days of age. At 240 days of age, the calves weighed 176, 207, and 226 kg for the respective treatments. After weaning, all females were treated alike under range conditions. The creep-fed heifers gained the least and the 140-day weaned heifers the most in structural size between 240 days and 2 years of age. The creep-fed females lost more weight and condition during their first pregnancy than those on the other treatments and had a lower skeletal growth rate. The 140-day weaned heifers tended to remain smaller for all variables studied, but at 1.5 years of age the differences were small.

## IV. FORAGE GRAZING AND CONCENTRATE SUPPLEMENTS

The effect of the concentrate supplement on the gains of grazing ruminants is dependent on the quality of forage and supplement, the extent to which intake meets the requirements of rumen microbes, and the animal's overall requirements for protein and energy. The extent to which forage intake decreases with concentrate supplementation is a factor of digestible energy (DE) intake (Raymond, 1969; Waldo and Jorgensen, 1981) and reduction of apparent digestibility of cellulose due to grain feeding (El-Shazly *et al.*, 1961; Milne *et al.*, 1981) Any decrease in the degradation rate of cellulose in the rumen would decrease roughage intake (Campling, 1966). Depression of forage intake due to given amounts of supplemental energy has been found to be greatest with high-quality forage (Blaxter and Wilson, 1963; Leaver, 1973).

Golding *et al.* (1976b) fed sheep ad libitum four Bermuda grass hays of 4-, 6-, 8-, and 10-week regrowth, with and without limited grain supplementation. When hays were fed alone, gross energy (GE) intake decreased with advancing maturity. Depression in DE of hays due to grain in diets was greatest with the highest-quality hay and least with the lowest-quality hay (10-week regrowth). Animals grazing very-low-quality pasture consume a minimum of forage and do not significantly reduce their consumption of forage as higher levels of concentrate are fed. Peacock *et al.* 1964) fed four levels of concentrates to heifer calves grazing very-low-quality Pensacola bahiagrass during the fall and winter, and obtained a gain daily of 0.04 kg with 0.4 kg concentrate supplementation per day compared to a daily gain of 0.55 kg with 3.1 kg concentrate per day. The 0.51-kg difference in average daily gain during the pasture period resulted in a conversion rate of 6.5 kg of concentrate for each additional kilogram of gain, suggesting that the effect of feeding concentrate was additive and that the heifers were not substituting concentrate for pasture.

Coleman *et al.* (1976) supplemented steers grazing moderately high-quality St. Augustine grass pasture and obtained data suggesting that the concentrate had both additive and substitutive affects. They observed that 4.5 kg concentrate per steer daily gave an increase in daily gain of 0.25 kg, with a conversion of 18 kg concentrate per kilogram of gain. While forage consumption was not measured, the fact that the gain per animal was increased indicates an additive effect. The poor conversion of concentrate to gain indicates that the steers must have reduced their consumption of forage.

Feeding of supplemental concentrate to grazing cattle would be expected to reduce the amount of forage utilized per animal to reach a given weight. If the forage plus grain diet is of greater quality than the forage alone, daily gain should be increased so that it takes less time to reach a given weight. Mott *et al.* (1967)

supplemented grazing steers with molasses and observed that the control steers consumed approximately 4.76 kg total digestible nutrients (TDN) daily from pasture compared to molasses-supplemented steers that consumed 4.08 kg TDN from pasture and 1.08 kg TDN from molasses daily (5.08 kg total). It was estimated that 1.19 steers consumed as much pasture when supplemented as 1.00 steer did in the control group; that is, 19% more steers could be carried on equivalent pastures if they were supplemented at this level.

Weaning of calves earlier than the usual age of 7–8 months has been of interest due to research results that indicate that transfer of energy in the form of milk as part of the calves' diet is about 34% (Neville, 1974), that calves weaned early will gain as fast as similar calves nursed by dams on pasture, and that the stocking rate per unit of land can be increased by stocking pastures with dry cows (Harvey et al., 1975). Also, dams of early-weaned calves have generally a shorter interval between calving and subsequent estrus and conception than cows nursing calves (Laster et al., 1973).

Neville et al. (1977) weaned Angus and Polled Hereford calves at about 48 days of age and group-fed them ad libitum in drylot for 17 weeks with four diets that varied in crude protein level (14.5, 18.9, 23.7, and 28.5%) and had 14% crude fiber on a dry matter basis. In terms of gain and feed efficiency, the diets that contained 18.9 and 23.7% protein were best during the first 6 weeks on the diets; the 18.9% protein level was best during the second 6 weeks; and the 14.5% protein level had a slight advantage during the final 5 weeks of the trial. Angus bull calves had better gains than Hereford bulls during the first 12 weeks, but daily gains were the same during the last 5 weeks. Hereford bulls gained faster than Hereford females.

## V. CALORIC EFFICIENCY OF COW–CALF PRODUCTION

The total energy requirement for the production of carcass beef from conception to slaughter generally is not available. Various aspects of the production process have been reported by Gibbs (1966), Kress et al. (1969), Garrett (1971), and Klosterman et al. (1972). Research in this area usually has involved grazing programs in which feed intake was unknown. However, Schake and Riggs (1975) conducted a study in which data were obtained using confinement cow–calf production and comparative slaughter and carcass weights to determine total caloric efficiency from conception to slaughter. Slightly over 6% of the total GE intake by a cow–calf pair was converted to empty body energy (EBE) of the calf upon being slaughtered at 351 days of age. Gross efficiency of the EBE gain by the calf from birth to slaughter was 16.7%, and net efficiency of nonlactating cows for EBE gain was estimated to be 15.7%. Gross and net efficiencies of milk production were 4.9 and 23.5%, respectively.

## VI. CREEP FEEDING

High weaning weights are an important factor in the gross income of cow–calf operations. Supplementing calves by creep feeding is a means of obtaining greater rates of growth during the preweaning period. Marlowe et al. (1955) observed that bull calves grew about 7% faster than steer calves, which, in turn, grew about 6% faster than heifer calves; these differences were somewhat greater when all calves were creep fed.

Holloway and Totusek (1973b) compared Angus and Hereford heifer calves that were exposed to three planes of nutrition prior to 240 days of age (weaned at 140 days, weaned at 240 days, or creep feed followed by weaning at 240 days) on subsequent performance through three calf crops. The creep-fed females weaned slightly lighter calves in each of the three calf crops, and their calves had the lowest condition scores in calf crops 1 and 3. The creep-fed females tended to have the lowest milk yields. Creep feeding of the females had no significant effect on the percentage of their calves that were weaned over a 3-year period.

Martin et al. (1981) reported on a 21 year study to determine the effects of creep feeding on the growth of Angus male and female calves and on subsequent cow productivity. The creep diet consisted of two parts ground corn and one part crushed oats, and was fed for 90–120 days before weaning. Creep-fed male calves were heavier at weaning and at 1 year of age. Heifers were heavier at weaning if creep-fed but lost their advantage by 1 year of age. Creep feeding of replacement heifers had a detrimental effect on their performance as cows, based on the number of calves weaned, the birth weight of calves, calf 210-day weight, and lifetime productivity. Progeny weaning weights were less at all ages of dam when the cow had been creep-fed as a calf.

Stricker et al. (1979) conducted a 4-year study to determine if calf production could be economically increased by the use of nitrogen fertilizer on tall fescue-ladino clover pasture and/or creep feeding. Nitrogen fertilizer was applied at rates of 0, 112, or 224 kg/ha, and calves received or did not receive creep feed. The creep feed consisted of 55% whole oats, 30% cracked corn, 5% dry molasses, and 10% protein supplement (32% protein). Pastures with creep feed had 0.90 metabolic animal unit months/hectare carrying capacity during the summer phase, but on a year-long basis, the increase in carrying capacity with creep feed was only 0.3 metabolic animal unit months/hectare. Calves that were creep fed weighed 32.2 kg more at weaning than control. The increased carrying capacity obtained by applying nitrogen fertilizer was not sufficient to offset the additional cost. Creep feeding appeared to be more practical than fertilization, but the economic advantage would depend on calf prices and creep feed costs.

## VII. FEEDING FREQUENCY

Gibson (1981) analyzed 15 published reports on trials in which feeding frequency was tested for effects on growth and feed efficiency on cattle. He concluded that, on the average, increasing the feeding frequency from one or two to four or more times daily enhanced average daily gains by 16.2 ± 4.8% and feed efficiency by 18.7 ± 6.0%. The response appeared to be greater when gains were low and levels of concentrates were high in the diets. Young animals responded more readily to greater feeding frequency. Different sexes responded in a similar manner. The data were too limited for accurate assessment of possible breed differences. More limited data with sheep indicate that they respond similarly to cattle with increased frequency of feeding.

Bond et al. (1976), utilizing three sets of monozygotic twin beef steers, studied the effect of deprivation and reintroduction of feed and water on the feed and water intake of beef cattle. The steers were fed ad libitum 0, 30, or 88% forage diets and were deprived of feed, water, or both for 12, 24, 36, or 48-hr periods. Feed intake was decreased about 50% regardless of the type of diet when water was withheld. When feed, water, or both were withheld for up to 2 days and then the same feed was reintroduced to the steers, there were very small changes in their intake patterns.

## VIII. COMPENSATORY GROWTH

Since most cattle are fed at some stage on pasture, attempts to obtain maximum benefit from compensatory growth are a very important objective in the search for a solution to the problem caused by low seasonal production of forages. Knowledge of the different factors that affect compensatory growth is an important aid toward efficient understanding and handling of the problem. Osborne and Mendel (1916) reported accelerated growth of rats upon full feeding after a long period of food restriction. Later, young cattle were observed to have their highest summer gains on range pastures after being wintered on low planes of nutrition (Nelson and Cambell, 1954; Heinemann and van Keuren, 1956; Bohman and Torell, 1956). The physiological reason for compensatory growth has not been explained, but Sheehy and Senior (1942) postulated that it was due to deposition of more protein and less fat during recovery. Increased feed intake relative to metabolic size during recovery may be partially responsible for compensatory growth (Tayler, 1959; Ashworth, 1969); however, increased energy

## VIII. Compensatory Growth

utilization independent of feed intake was found during compensatory growth (Meyer and Clawson, 1964; R. M. Meyer et al., 1965).

Fox et al. (1972) conducted comparative slaughter trials with steers and determined their growth, body composition, feed intake, efficiency of energy, and protein utilization as affected by plane of nutrition (5–6 months maintenance, then full fed or continuously full fed) and by high-energy corn–based versus medium-energy soybran flake–based diets. The steers that made compensatory gains had more protein and lower fat content at 364 kg body weight but were equivalent in body composition to continuously full-fed steers at 454 kg body weight. The data indicate that steers that have compensatory gains deposit relatively more protein and less fat than controls during the first part of full feeding but deposit relatively more fat than controls during the last part of full feeding. Compensatory growth was due to increased efficiency of protein and energy utilization during the full-feeding period. There was a trend for higher net energies for maintenance ($NE_m$) and gain ($NE_g$) in diets fed the compensatory steers. A longer time was required by the compensatory steers to reach 364 or 454 kg body weight than the controls, but the total metabolizable energy (ME) required to reach 454 kg was only slightly increased. Total protein intake required by steers making compensatory gains to reach 454 kg was less than that required by controls.

Sanhidet and Verde (1976) conducted a study with Angus steers that were 8 months of age and weighed 185 kg. They subjected the steers to varying degrees of energy intake by feeding them 1.45–2.85 Mcal of ME per kilogram of dry matter (DM) for 16 weeks. Then all steers were fed ad libitum the diet containing 2.85 Mcal ME per kilogram of DM, and feed consumption rapidly became quite similar with all steers. When compared at equal liveweight, the restricted animals had a greater feed intake than the controls, but at an equal age feed intake was practically the same. Age accounted for 65% of the variability in feed intake, liveweight accounted for 43%, and both parameters jointly explained 73% of the variation in feed intake.

Winchester et al. (1967) studied the effects of undernutrition and malnutrition on 17 pairs of monozygotic twin beef-type cattle from about 6 to 12 months of age. The least liberal intakes of energy and protein were about 16 and 40% of the requirements for rapidly growing cattle. Then the cattle were fed ad libitum until they reached choice grades and slaughtered. Overall efficiency of feed utilization, carcass grade, and meat quality were similar for the twins. In a similar study with twins, Winchester and Howe (1955) concluded that beef cattle under conditions of feed scarcity between the ages of 6 and 12 months of age could be fed at a maintenance level without a later loss in feed efficiency or meat quality. However, cattle fed the restricted diets generally required 10–20 weeks more time on full feed to obtain the same slaughter weight and carcass grade as those that were continuously on full feed. Lake et al. (1974) observed that the time required to finish steers in the feedlot decreased with prior increased energy supplementation on irrigated pasture, but carcass characteristics were not affected by energy supplementation on pasture.

Horton and Holmes (1978) conducted a study to determine the effects of realimentation on grass and on a pelleted diet, making use of pasture and of dehydrated alfalfa in place of high grain concentrates. Gains were held at 0.22 or 0.58 kg/day for 20 weeks by restricting the intakes of yearling beef cattle. During the subsequent 8 weeks on pasture, gains and intakes were greater for the limited-fed group. Similar results were observed in a second trial.

Folman et al. (1974) fed bull calves a maintenance diet for 90 days from either 180 or 270 days of age and found compensatory growth upon ad libitum feeding of the previously restricted animals. They concluded that feeding a maintenance diet for 90 days may prolong the finishing period but may give feed efficiency equal to or better than that of continuously fed controls. Drori et al. (1974) restricted Israeli-Friesian bull calves to maintenance or 1.25 times maintenance feed intakes for 80 or 90 days, starting at 180 days of age. They observed considerable compensatory growth on realimentation, but bulls fed the 1.25 maintenance level during restriction of feed had better overall performance in terms of daily gains and feed efficiency. They concluded that a uniform feeding of 85–90% of the ad libitum level was the most efficient method of raising dairy bull calves to slaughter at 500 kg liveweight.

Shirley et al. (1963) fed Brahman-British crossbred heifers for 148 days on winter pastures with four different levels of protein and energy, and then for 140 days on a growing-finishing diet in the feedlot. The highest level of feed supplementation on pasture gave gains of 1.4 lb compared to 0.3 lb per day for the lowest level of feed supplementation (Fig. 11.1). Average daily gain in the feedlot was greatest for those heifers that had the lowest daily gain on pasture.

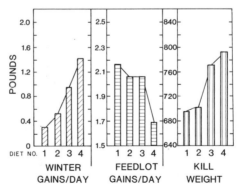

**Fig. 11.1.** Pounds of gain per day by heifers. *Left,* under varying levels of protein and energy during 148 days on winter pastures. *Center,* subsequently during 140 days in the feedlots. *Right,* slaughter weights of the heifers (From Shirley et al., 1963).

## VIII. Compensatory Growth

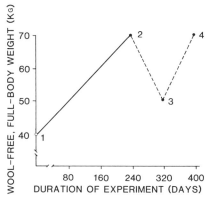

**Fig. 11.2.** Growth patterns of continuously fed and deprived-refed sheep (From Burton et al., 1974).

However, heifers fed the highest level of supplemental feed on pasture had the heaviest weight when slaughtered.

Growing sheep maintained in continuously positive energy balance are different in body composition from similar animals that have had a period of compensatory growth to reach equivalent body weights. The growth patterns of sheep that were fed continuously and those that were deprived of feed to the extent that they lost about 30% of their body weight and were then realimented for compensatory growth are shown in Fig. 11.2 (Burton et al., 1974). Points 1, 2, 3, and 4 in the figure designate the weights of the sheep at the time of slaughter. The corresponding weights of the chemical components of the tissues are summarized in Table 11.1. The sheep in group 1 (50 kg) that had been grown continuously from 40 kg had the same composition as those that had been grown continuously

**TABLE 11.1.**

**Chemical Composition and Energy Value of Ewes Grown by Different Patterns**[a]

| Growth group[b] | Components of wool-free, empty body | | | | | Energy (Mcal) |
|---|---|---|---|---|---|---|
| | EBW (kg)[c] | Water (kg) | Protein (kg) | Fat (kg) | Ash (kg) | |
| 1 | 45.6 | 21.4 | 6.2 | 16.3 | 1.5 | 190.9 |
| 3 | 46.8 | 22.1 | 6.3 | 16.4 | 1.8 | 190.9 |
| 2 | 65.6 | 26.0[d] | 7.5[d] | 29.7[d] | 2.0 | 327.3[d] |
| 4 | 63.5 | 29.6[d] | 8.3[d] | 22.9[d] | 2.0 | 265.9[d] |

[a] From Burton et al. (1974).
[b] 1, 2 = grown continuously from 40 kg; 3 = grown continuously from 40 to 70 kg, starved to 50 kg; 4 = grown continuously from 40 to 70 kg, starved to 50 kg, and then grown to 70 kg.
[c] EBW = empty body weight, i.e., body weight minus ingesta.
[d] Differences within components between groups 2 and 4 ($P < .01$).

to 70 kg and then deprived of energy to lose body weight and return to 50 kg (group 3). Some sheep in group 3 were refed to reach 70 kg (group 4), and were found to contain more water and protein but less fat and energy than their group 1 mates that were fed continuously to reach 70 kg (group 2).

## IX. FINISHING OF CATTLE IN DRYLOT

Many factors that affect the feed intake, feed efficiency, and carcass characteristics of feedlot cattle, such as different sources of dietary energy, nitrogen, and roughage, processing of feedstuffs, feeding frequency, days on concentrate feed, breed, and sex of animals, have been investigated. Danner *et al.* (1980) conducted studies on three types of cattle (Angus × Hereford yearling steers, Charolais × British steer calves, and Hereford heifer calves) fed in drylot on three different feeding systems utilizing high-moisture corn, corn silage, and soybean meal as major feed ingredients. Steers were compared in the following feeding systems: A = 85% concentrate; B = 40% concentrate; C = two phases, with a switch from all-corn silage to 85% concentrate in the middle of the feeding period; D = same as C, except that the switch was made late in the feeding period; E = all-corn silage. Diets fed heifers consisted of the following ratios of corn silage to concentrate: low energy = 89:11, medium energy = 67:33, and high energy = 0:100. Increasing the percentage of corn in the diets resulted in most cases in increased daily gains. Increasing dietary energy increased fat thickness and reduced the retail product percentage, with no effect on quality grade in yearling steers and heifer calves. The two-phase system (C) slightly improved energy efficiency in yearling steers, while steer calves had maximum energy efficiency with the 85% concentrate diet (A). Hereford heifers were most efficient on medium-energy diets.

### A. Effect of Protein Withdrawal

The National Research Council (NRC, 1976a) recommends varying levels of dietary protein ranging from 18% protein for steers weighing 100 kg to 9% for steers weighing 500 kg. Corresponding levels recommended for heifers vary from 18 to 7%. Preston and Cahill (1972, 1973, 1974) observed that when the diets of feedlot steers weighing 345 kg were decreased from recommended levels of crude protein to approximately 8.4%, the steers performed as well as those that were fed recommended levels of protein for the entire finishing period. However, the following year (1975), when supplemental protein was withdrawn after the steers had been in the feedlot for 84 days, weight gains were depressed. This was probably due to a lower level of protein in the corn than in previous years (Preston *et al.,* 1975).

Heifers fed no supplemental protein during the last 91 days of a 161-day finishing period gained as well as those fed at recommended levels throughout the trial (Farlin and Guyer, 1974). Thomas *et al.*, (1976b) reported that steers fed no supplemental protein (7.48% dry feed basis) during the first 70 days in the feedlot gained less (0.34 kg/day) than steers fed supplemental protein (10.82% dry feed basis). When protein was withdrawn from steers after 70 days on feed at a weight of 368 kg, the daily gain (0.17 kg) was less than that of steers fed supplemental protein, which gained 1.03 kg/day for the next 84 days. Feed conversions changed from 6.67 to 28.69 kg dry feed per kilogram of gain when steers were deprived of protein supplementation. In a second trial, steers weighing 334 kg and fed no supplemental protein initially (8.66%) gained less during the first three consecutive 28-day weigh periods and had lower feed efficiency (7.26 versus 8.78) overall than those fed a 12.3% crude protein diet.

Dartt *et al.* (1978) studied the effect of supplemental protein withdrawal and monensin in diets fed finishing steers for 168 days. They utilized four treatments: (1) control diet; (2) control diet for the first 84 days and soybean meal removed from the diet for the last 84 days; (3) control diet plus 200 mg of monensin per steer daily; and (4) 200 mg monensin per steer daily with soybean meal fed for the first 84 days and removed from the diet for the last 84 days. Overall gains for the respective diets were 0.98, 0.78, 1.07, and 0.92 kg daily. Corresponding efficiency of gain values were 7.02, 8.28, 5.83, and 6.14 kg feed per kilogram of gain. All treatment groups had similar dressing percentages, marbling scores, and carcass grades.

Young (1978) conducted finishing trials with steers to determine the effect of protein withdrawal at different weights on performance when the steers were fed high-grain (80% corn, 20% corn silage, dry matter) or high-silage (40% corn, 60% corn silage, dry matter) diets. Withdrawal of protein supplement had no effect on the rate of gain, feed efficiency, or carcass merit of steers fed either diet compared to those of steers fed protein continuously.

## B. Preconditioning of Cattle for Feedlots

The feedlot care and feeding of newly arrived cattle at feedlots suffering from stresses of marketing and shipping require special managerial skills. Most new cattle going into feedlots for finishing have not previously been fed grain or corn silage and generally do not readily consume concentrate diets. Adequate feed consumption upon arrival at the feedlot is necessary to obtain satisfactory performance. Hay is usually consumed readily.

Lofgreen *et al.* (1975b) utilized calves subjected to marketing and shipping stress to test various energy levels in receiving diets. The diets tested contained 20, 55, 72, and 90% concentrates (0.84, 1.01, 1.10, and 1.19 Mcal $NE_g$/kg). Performance of stressed calves improved with increasing concentrate or energy levels up to the 72% level. At the 90% concentrate level, calves started more slowly than those fed at the 72% level. Stressed calves fed the 72% barley-containing concentrate diet showed no advantage when provided with free access to long-stem alfalfa hay. However, Lofgreen *et al.* (1980) reported that stressed calves fed receiving diets containing flaked milo (50 and 75% concentrates) plus free-choice alfalfa hay had less death loss and fewer required medications, and the number of treatments per medicated calf was reduced. Free-choice feeding of alfalfa may result in reduced intake of high-energy diets, with a slower recovery from shipping shrink, as well as losses from bloat (Lofgreen *et al.*, 1981).

Calves subjected to stresses of weaning, marketing, shipping, and processing ate more feed, gained more weight, and gained more efficiently when fed a 75% concentrate receiving diet, alone or with free-choice hay, than when fed hay alone (Lofgreen *et al.*, 1981). Greater bulk or lower energy densities in diets did not promote greater feed intake in the stressed calves, as occurs in unstressed animals. There were no differences in the number of calves treated, days sick per calf, death loss, or relapses due to the receiving diets. Feeding a 55% concentrate diet before shipment reduced mortality of transported calves, and 1.3–2.2% potassium additions to the diet after arrival improved their performance in the feedlot (Hutcheson *et al.*, 1984).

Pate and Crockett (1978) determined the cost and returns of preconditioning to the ranch and the effect of preconditioning on subsequent animal performance and disease in the feedlot. During a period of 2–3 days, calves were weaned, sold through a public auction (often sorted and resold), trucked as far as 3000 miles, and placed in feedlots. It was concluded that (1) when calves are fed 3–4 weeks postweaning at the ranch, weight gains alone may not offset feed and overhead costs; (2) the higher shipping shrink of calves preconditioned with feed would be a disadvantage to the feeder calf buyer if the purchase price were determined on a ranch weight and standard pencil shrink basis; and (3) preconditioning of calves with feed prior to shipment improves the rate of gain and reduces sickness and death loss in the feedlot, but savings in the feedlot will not offset the premium price required to recover the preconditioning cost.

A primary asset to a stocker operation is good-quality forages. Generally, due to poor forages during the winter, stocking rates of pastures need to be reduced or supplementation practiced. Haines *et al.* (1965) grew calves on pasture for 1 year to study the value of limited supplementation during the period prior to finishing in the feedlot. Steers fed 1.3–2.3 kg concentrate during the winter and spring after fall weaning gave the highest net return from supplementation over the stocker and finishing periods. Stocker steers that were supplemented remained on inventory for a shorter time (Chapman *et al.*, 1967) and reached a higher slaughter grade when finished (Peacock *et al.*, 1964). Heifer calves wintered for 120 days on pasture at different levels of supplementation prior to a 140-day finishing period showed a positive relationship between supplementation, winter gain, and carcass quality when slaughtered (Peacock et al., 1964). However, greater winter gains resulted in lowered daily weight gain and reduced feed efficiency in the feedlot period.

Pate et al. (1972) determined the value of feeding stocker steers various levels of supplement on pasture. Steers were fed concentrate (snapped corn, citrus pulp, cottonseed meal) at a rate of 0, 5, and 10 lb per head daily or on a free-choice basis during the stocker period and then placed in a feedlot. Another trial involved feeding a molasses-Vegefat mixture free choice on pasture until steers were slaughtered. Rate of gain increased and stocking time decreased as the quantity of supplement fed on pasture increased. The concentrate supplement was utilized more effectively when limited to 10 lb per head daily compared to free-choice feeding, but supplementation had no consistent effect on feedlot performance. An economic analysis showed that it was unprofitable to feed steers concentrate at any level during the stocker period.

Preston et al. (1972), by decreasing corn silage over a 10-day period, brought steers to a full feed of corn. The steers performed satisfactorily in two trials, but in a third trial they performed poorly and had more liver abscesses. In the two successful trials, corn silage was fed at a level of 11–14 kg per head daily previous to adapting the steers to high-concentrate diets, but in the less satisfactory trial, only 7 kg of corn silage was fed per head daily prior to increasing the level of concentrate in diets. It was postulated that the higher level of corn silage, which contained lactic acid, may have helped the cattle adapt to the high-concentrate diets.

Kunkle et al. (1976) studied the effect of hay, hay plus lactic acid, corn silage, and concentrate on performance and disease incidence among new feeder calves during the initial 30- to 60-day period in the feedlot. Gains were 20–160% faster on hay than on corn silage diets during the first 5 days after the steers were received, but steers fed corn silage gained faster after the fifth day than those fed hay. The incidence and severity of disease were equivalent in steers fed hay or corn silage receiving diets. Feeder steers changed to corn grain diets within a 19-day period had greater gains but an increased incidence and severity of sickness than steers fed hay, hay plus lactic acid, and corn silage (0.96, 0.53, 0.58, and 0.86 kg average daily gains over 61 days, respectively). Overall gains during the finishing period, however, were similar regardless of the initial dietary treatments.

A 10-day period was considered to be adequate for bringing steers to a full feed of barley without digestive disorders (Hironaka, 1969). Tremere et al. (1968) reported that at least 21 days were required for heifers to adapt to diets containing wheat without going off feed.

# 12

# Nitrogen and Energy in Animal Excreta

## I. INTRODUCTION

It has been estimated there are about 2.2 million metric tons of nitrogen produced yearly in livestock wastes in the United States and that about 50% is collectible for recycling in animal diets (Yeck et al., 1975). Smith and Wheeler (1979) concluded in a review that animal excreta contains 48–73% total digestible nutrients (TDN), 20–31% crude protein, and 13–20% crude fiber on a dry matter basis. They estimated the value of poultry litter, dehydrated poultry excreta, and swine and cattle excreta as sources of protein (relative to the cost of soybean meal) and energy (relative to the cost of soybean meal, corn grain, and corn silage). Poultry litter is a slightly better source of digestible nitrogen and energy for most classes of cattle than dehydrated poultry excreta. Swine excreta is similar to dehydrated poultry excreta as a source of energy but is lower in available nitrogen. Cattle excreta is similar to swine and dehydrated poultry excreta as a source of energy but is a less available source of nitrogen (Smith and Wheeler, 1979). McCaskey and Anthony (1979) reviewed the hazards of pathogenic microbes and chemical residues in animal excreta. While no serious health problems have resulted from feeding poultry litter, there is apprehension concerning the possible dangers of pathogenic organisms.

## II. PROCESSING AND HANDLING OF ANIMAL EXCRETA

Animal excreta must be collected regularly to preserve the nutrients. Adriano et al. (1971) reported that 35% of total nitrogen in animal excrement was lost by the second week and 50% by the 10th week in laboratory storage at ambient temperature. Smith (1973) presented data shown in Table 12.1 for the distribution of nitrogen in the excreted feces and urine of several classes of livestock and poultry. Beef cattle and sheep have equivalent distribution in feces and urine,

**TABLE 12.1.**

**Distribution of Nitrogen in the Feces and Urine of Livestock**[a]

| Species | Percent of total Nitrogen | |
|---|---|---|
| | Feces | Urine |
| Beef cattle | 50 | 50 |
| Dairy cattle | 60 | 40 |
| Sheep | 50 | 50 |
| Swine | 33 | 67 |
| Poultry | 25 | 75 |

[a]From L. W. Smith (1973).

and dairy cattle have somewhat more in feces than in urine. Swine have about two-thirds of their excreted nitrogen in the urine. Poultry have about three-fourths of the excreted nitrogen in the urine, but it is excreted mixed with the feces. Arndt *et al.* (1979) discussed the advantages and disadvantages of natural and heated air drying, ensiling, liquid–solid separation, chemical treatment, oxidation ditch aerobic liquid treatment, and mechanical composing of animal manure.

## III. UTILIZING THE NUTRIENTS IN ANIMAL EXCRETA

As suggested by Arndt *et al.* (1979), the chemical composition and digestibilities of the nutrients present should be determined before diet formulation. These should include especially dry matter, protein equivalents, gross energy, and minerals. The high mineral content of animal excreta due to the addition of soil during the collection of bedding material is probably a performance-limiting factor. Otherwise, diets should be formulated on the basis of the nutritive content and digestibility of the excreta as well as the other ingredients.

## IV. POULTRY LITTER

Noland *et al.* (1955) successfully utilized ground chicken litter as a source of nitrogen for gestating-lactating ewes and finishing steers. Ammerman *et al.* (1966) collected broiler manure on citrus pulp and fed the dried mixture in diets to sheep in a metabolism trial. The manure had digestibilities of organic matter, nitrogen, fat, and crude fiber of 80.7, 82.0, 85.6, and 72.8%, respectively, when calculated by difference.

## IV. Poultry Litter

Harmon et al. (1975b) successfully fed poultry litter ensiled with corn forage in diets to sheep, and Caswell et al. (1977) reported on the fermentation, nitrogen utilization, and palatability of poultry litter ensiled with high-moisture corn. The high nitrogen content of the poultry litter supplemented the low nitrogen content of the corn. Vijchulata et al. (1980b) conducted a study with steers to evaluate the effects of 0–30 g monensin per metric ton in diets that contained either dried citrus pulp or corn as a principal source of energy, and either cage layer manure or soybean meal as a source of supplemental nitrogen. Steers fed soybean meal gained faster than those fed cage layer manure (1.21 versus 1.03 kg daily), but monensin and the source of energy had no effect on daily gain. Monensin improved the feed efficiency of diets (7.22 versus 8.42 kg feed per kilogram of gain).

In a metabolism trial, Oltjen and Dinius (1976) fed steers diets that contained 50% of their nitrogen from uric acid, sodium urate, or a processed poultry waste product that contained 28% uric acid. Dry matter, fiber, and nitrogen were digested equally well in the three diets. In a growth trial, steers gained weight more rapidly when fed diets that contained poultry waste than with diets containing uric acid or sodium urate. Evans et al. (1978) compared the composition of excreta from laying hens fed at a low or high plane of nutrition over a 252-day period. Concentrations of nitrogen and ash were higher and gross energy, acid detergent fiber (ADF), and neutral detergent fiber (NDF) were lower in fresh manure originating from hens on the high plane of nutrition.

O. B. Smith et al. (1978) reported higher crude protein, ash, and digestible dry matter but lower ADF and NDF in excreta from caged layer hens fed on a high plane of nutrition. Feed efficiency was lower with a diet containing excreta from hens fed on a low plane of nutrition.

### A. Dehydrated Poultry Manure

Both dehydrated poultry manure without bedding and broiler manure have been fed successfully to ruminants. The layer manure generally has greater ash and calcium levels, but otherwise, the two poultry excrement sources are quite similar in composition.

Cullison et al. (1976) evaluated diets fed finishing steers that contained either pure dried broiler manure, broiler manure with a wood shavings base, broiler manure with a peanut hull base, or dried layer manure. The four manures had crude protein levels of 34.5, 22.5, 24.9, and 40.4%, respectively, on a dry basis. Diets that contained the dried broiler excreta at levels that provided either all or one-half of the supplemental protein in the steer diets gave weight gains similar to those of the control diet. When dried caged layer hen manure was used, a lower rate of gain was observed compared to that produced by diets containing broiler manure.

Lowman and Knight (1970) fed diets to wethers that ranged from 100% dried poultry excreta to 100% barley in steps of 25% inclusion. The poultry excreta supplied 20.2% of apparently digestible protein and 1.57 or 1.74 Mcal of metabolizable energy (ME) per kilogram of dry matter. Oliphant (1974) compared dried poultry manure with soybean meal and fish meal in beef cattle diets at 14.5% crude protein levels. When the crude protein of the poultry manure was low (24%), it was necessary to include 27.6% in the diet, and this level resulted in poor performance. Manures that contained 30% crude protein resulted in animal performance similar to that of controls. Tagari *et al.* (1976) reported on the nutritional value of pelleted finishing concentrate diets fed bulls and rams in which heat-sterilized poultry litter was included at levels of 0, 15, 25, and 35%. There were no significant differences in liveweight or carcass gains between the dietary groups except for decreases with the 35% level of poultry litter.

Smith and Calvert (1976) compared dehydrated broiler excreta with soybean meal as a source of protein in diets fed sheep. There were no significant differences in digestibility or nitrogen retention with the two sources of dietary protein. Smith and Lindahl (1977) compared dehydrated caged layer excreta with alfalfa in diets fed lambs. The protein supplements contributed 60–65% of the total dietary nitrogen. Gains were the same with the two types of diets. Smith *et al.* (1979) reported that pelleted diets of corn meal supplemented with dehydrated caged layer excreta were consumed by wethers and Holstein steers as readily as diets supplemented with cottonseed meal. Steers fed the cottonseed meal diets gained 5% faster on 8% less feed than steers fed the excreta-containing diet. Dressing percentage, tenderness, juiciness, quantity of connective tissue, and flavor of steaks did not differ with the two dietary groups.

## B. Formaldehyde-Treated Poultry Waste

Koenig *et al.* (1978) fed formaldehyde-treated caged layer manure to lambs and crossbred steers. The poultry waste was treated with 1% formaldehyde on a wet basis and then incorporated at either 0 or 10% levels in diets on a dry basis. The treated manure-containing diets had no effect on the weight gains and feed efficiency of lambs. The cattle fed the treated manure-containing diet had less feed intake but gained the same as controls.

## C. Ensiling of Poultry Litter

Harmon *et al.* (1975a) ensiled corn forage harvested at two stages of maturity with broiler litter making up 15, 30, or 45% of the dry matter. All mixtures appeared to have typical fermentation characteristics. The crude protein content was 7.8% for the corn silage compared to 10.5, 12.3, and 16.9% for silages containing 15, 30, and 40% broiler litter, respectively. Some of the silages had higher bacterial counts than others, but coliform counts were generally lower in silages containing litter than in the control silages.

Cross et al. (1978) ensiled broiler litter in an oxygen-limiting structure at 36% moisture. The silage was evaluated in diets fed steers that contained either 70% corn silage, 60% corn silage plus 10% litter silage, 40% corn silage plus 30% litter silage, or 20% corn silage plus 50% litter silage. All diets had 30% concentrate mixed in. Average daily gains were greatest for steers fed diets containing 30% litter silage. Carcass and organoleptic evaluations were not affected by litter silage. Cross and Jenny (1976) observed that Holstein heifers fed a diet that contained 15% turkey litter silage plus 75% corn silage had better daily gains (0.58 versus 0.42 kg/day) than heifers fed a 90% corn silage diet.

Labosky et al. (1977) studied the potential use of softwood and hardwood bark broiler litters as a feedstuff for ruminants. In vitro dry matter digestibility was higher for hardwood bark litters than for softwood bark litters or softwood planer shavings. Ensiling of the bark litters had no effect on in vitro digestibility. There were no differences in the digestibilities of ash, phosphorus, ether extract, nitrogen, total energy, and ADF among the treatments.

## V. SWINE EXCRETA

Hennig et al. (1972) fed pelleted diets that contained 40% dried swine excreta to bulls and obtained a 1.1-kg average daily gain per animal over the 48 days of the trial. Flachowsky (1975) fed cattle in finishing trials diets that contained 30 or 50% solid material from semiliquid swine excreta and observed gains of 1.2 and 1.0 kg/day, respectively.

## VI. CATTLE EXCRETA

Anthony (1970) reported that diets containing cattle manure were readily consumed by steers and supported gains essentially equal to those of cattle fed control diets. Cooking or washing the cattle manure before mixing it with concentrate for feeding did not enhance its feeding value. Carcass data were similar for cattle fed manure and those fed control diets.

Newton et al. (1977) found that heifers fed a diet that contained 40% wet cattle manure gained 1.27 kg daily compared to 1.34 kg per day for those on a control diet. Lucas et al. (1975) fed steers cattle manure that contained 13.2% crude protein, 31.4% crude fiber, 2.8% ether extract, 5.4% ash, 47.2% nitrogen-free extract, and 44.8% ADF on a dry basis at a 20% level in diets. Apparent digestibility of dry matter in the manure-containing diet was 57.4% compared to 68.2% in the basal diet. Apparent digestibilities of fecal crude protein and dry matter as calculated by difference were 24.4 and 16.6%, respectively. The manure contained 763 kcal digestible energy (DE) and 485 kcal of ME per kilogram of dry matter.

## A. Screened Cattle Manure Solids

Richter et al. (1980) fed beef steers diets that contained manure roughage processed on a commercial vibrating screen and screw press to separate it from the liquid fraction. The manure roughage at levels of 0, 20, 40, and 60% (wet basis) in the diets stimulated feed intake and body weight gains even though it depressed the ME and TDN of the diets. The roughage manure contained 26.6% dry matter. Steers fed these levels over 124 days gained 1.10, 1.53, 1.58, and 1.51 kg/day, respectively. Net energy (NE) for gain was not depressed in diets with increasing levels of manure roughage. Dressing percentage, carcass grade, marbling, and organoleptic evaluations were not affected by the manure roughage. Armentano and Rakes (1982) fed nonlactating, nonpregnant Holstein cows diets that contained 22% screened cattle manure solids that were either untreated, treated to contain 7% sodium hydroxide (NaOH) and fed immediately, or treated to contain 7% NaOH and stored 7 days prior to feeding. The alkali treatment reduced the hemicellulose of the manure solids and increased the digestibilities of crude fiber and cellulose in the complete diets. Digestibilities of crude protein and soluble carbohydrates were decreased in the stored material. It was concluded that untreated screened manure solids can be used successfully in complete diets for dairy cattle, but treatment with NaOH does not appear to be justified.

## B. Ensiling of Cattle Manure

Harpster et al. (1978) compared diets that contained ensiled cattle waste to corn silage in sheep metabolism and steer growth trials. The ensiled cattle waste contained approximately 60% cattle excreta and 40% chopped hay on a fresh weight basis. Nitrogen retention and digestibilities of dry matter, organic matter, ether extract, nitrogen-free extract, and energy were lower for the ensiled cattle waste than for the corn silage. Daily weight gains were higher for steers fed the corn silage diets. Cornman et al. (1981) ensiled cattle waste with ground rye straw in the following proportions, wet basis: 30:70, 40:60, 50:50, 60:40, and 70:30. There were no reductions in soluble carbohydrate content or changes in pH and lactic acid after the first week of ensiling. All total and fecal coliforms, *Salmonella, Shigella,* and *Proteus* organisms were destroyed in the mixtures after ensiling for 1 week.

Aines et al. (1982) collected waste from cattle fed high-roughage or high-concentrate diets and mixed it with rye straw in a 60:40 ratio, wet basis. The mixtures were adjusted with water to 30, 40, or 50% dry matter and treated with 0 or 4% NaOH on a dry basis before ensiling. Ensiling eliminated coliforms and *Proteus* organisms in the untreated silages but not in the NaOH-treated silage. Digestibility of dry matter by lambs was greater in silages containing waste from cattle fed high-concentrate diets than in silages containing waste from cattle fed a high roughage diet. The alkali treatment increased dry matter and organic matter digestibilities but decreased crude protein digestibility.

### C. Loss of Volatile Compounds on Storage

Kellems *et al.* (1979) reported that dry matter and crude protein of cattle feces were not correlated with initial ammonia volatilization. The urea content and specific gravity of urine were positively correlated with an initial loss of ammonia. Peppermint oil (0.25%) in the diet reduced the relative offensiveness associated with the fecal waste. Addition of sagebrush (1 and 1.5%) had no effect on the olfactory evaluation of the waste.

## VII. FEEDING VALUE OF METHANE FERMENTATION RESIDUE

Harris *et al.* (1982) determined the nutritive value of the methane fermentation residue (MFR) from the effluent of a large-scale methane generator in diets fed steers in metabolism and feedlot trials. The substrate of the methane generator was manure from feedlot cattle. The MFR contained 22.2% dry matter and 21.9% crude protein (dry basis). Two diets containing 10.6% (dry basis) MFR were formulated using the urea fermentation potential (UFP) system such that in one diet, nitrogen was in excess (−1.6 UFP), while in the other diet, energy was in excess (+2.6 UFP). These two diets were compared in a California Net Energy Trial with a feedlot diet (−0.3 UFP) containing the same ingredients except MFR. Digestibilities of dry matter, organic matter, crude protein, ADF, and ash, as well as TDN and ME, were depressed (all $P < .05$) in the MFR-containing diets. Rates of gain and feed efficiency were depressed in steers fed the MFR-containing diets. NE values for maintenance and gain were slightly lower for the MFR-containing diets than for the control diet. Crude protein digestibility for the MFR calculated by difference for the −UFP and +MFP diets was 37.8 and 50.7%, while corresponding values for TDN were 28.8 and 12.8%, respectively.

## VIII. SEWAGE SLUDGE IN LIVESTOCK DIETS

The chemical composition of sewage sludge indicates that it has potential for supplying nutrients for livestock. Beaudouin *et al.* (1980) reported data on 22 samples that ranged in crude protein from 22 to 37%; crude fiber, 0.7–15%; ether extract, 0.01–1.1%; nitrogen-free extract, 9–47%; and ash, 27–57%.

However, when female swine were fed corn-soybean grower diets that contained 0, 10, or 20% sewage sludge in a metabolism trial, the mean values for TDN were 79.4, 73.7, and 55.0%, respectively. Corresponding ME values were 3.36, 2.25, and 1.15 Mcal per kilogram of diet, and those for nitrogen retained were 42.8, 44.0, and 25.3%, respectively.

Kienholz et al. (1976) fed steers 15% dried sewage sludge in feedlot diets for 94 days and observed decreased body weight gains, a 10-fold increase in lead and mercury, and a 2-fold increase in cadmium and copper in their kidneys and livers. Smith et al. (1976) fed sewage sludge treated with thermoradiation to sheep and reported that 24% of the nitrogen in the sludge was digested.

Ammerman and Block (1964) fed a corn meal-cottonseed diet containing 20% of either a precomposted or composted sewage-oak sawdust mixture or Bermuda grass hay to lambs for 70 days. Feed intake, body weight gains, and digestibility of crude protein, fat, crude fiber, and nitrogen-free extract were lower with the diets containing the sludge-sawdust mixture than with the diet containing hay. Improved feed intake occurred by composting the sludge-sawdust mixture. It was concluded that these waste materials may find a place in animal feeding, especially under conditions where sources of nitrogen and roughage are limiting.

Bertrand et al. (1980) fed steers one of three diets for 141 days to determine the effects of digested municipal sludges on performance, carcass quality, and levels of selected potentially toxic metals in muscle, liver, and kidneys. They consisted of (1) a corn diet, (2) a corn diet plus 500 g per head per day of digested sludge from Chicago, Il, and (3) a corn diet containing corn produced from soil treated with surface applications of liquid digested sludge from Pensacola, Fl (7.6 cm/ha) prior to planting. The digested sludge from Chicago or corn grown on soil treated with liquid digested sludge from Pensacola in the diets had no effect on the performance and carcass quality of beef steers. Higher levels of cadmium and nickel were found in tissues of steers fed the Chicago sludge-containing diets. Bertrand et al. (1981) grazed steers on pastures with the following treatments: (1) control-normal fertilization, (2) liquid digested sludge (16 metric tons of solids per hectare), and (3) liquid digested sludge (30 metric tons per hectare). There were no differences due to treatments in animal performance and carcass characteristics.

# 13

# Minerals and Water in Nitrogen and Energy Nutrition

## I. INTRODUCTION

Maximum utilization of cellulose in ruminant diets is important in view of the large quantities of fibrous feedstuffs available. The rumen environment must have a balance of all of the nutrients necessary for optimal microbial activity. Interest in the mineral nutrition of rumen microbes was stimulated by observations that either alfalfa ash or a mixture of trace elements improved ruminant digestion of fibrous diets. Martinez and Church (1970) determined the extent of stimulation or inhibition of numerous ions on cellulose digestion by washed rumen microorganisms. Elements and concentrations (ppm) observed to be stimulatory to cellulose digestion were: cobalt, 3; iodine, 20; iron 3–5; manganese, 5–30; molybdenum, 10–100; zinc, 5–7; rubidium, 20; cadmium, 5; chromium, 2; and strontium, 10–15. Depression of cellulose digestion occurred at the following concentrations (ppm); barium, 30; boron, 300; cadmium, 10; chromium, 7; cobalt, 7; copper, 1; fluorine, 0.5; iron, 100; manganese, 100; nickel, 0.5; selenium, 7; strontium, 200; vanadium, 5; and zinc, 20. Bromine levels of 1–1000 ppm, iodine up to 1000 ppm, molybdenum at 500 ppm, and rubidium at 1000 ppm had no detrimental effect on cellulose digestion.

Evans and Davis (1966) conducted *in vivo* and *in vitro* trials to determine the effects of mineral interrelationships on cellulose digestion by rumen microbes. Sulfur (65 ppm) increased microbial digestion of cellulose. Phosphorus stimulated digestion except at 0.54% in the diet. Sulfur and phosphorus gave additive effects *in vitro*. Cobalt in vitamin $B_{12}$ helps to synthesize propionate through its role in methylmalonyl-coenzyme A isomerase (Underwood, 1977). Since ruminants largely depend on volatile fatty acid (VFA) synthesis by rumen microbes from cellulose and other carbohydrates, inadequate dietary cobalt leads to energy starvation.

Cellulose digestion and urea utilization were stimulated *in vitro* when sulfate or methionine sulfur was added to rumen fluid media (Hunt et al., 1954). Martin et al. (1964) fed purified diets with and without magnesium as magnesium oxide (MgO) and sulfur as sodium sulfate ($Na_2SO_4$) to fistulated steers and diets with and without magnesium to intact lambs. Cellulose digestion was decreased in steers fed diets without magnesium and sulfur. In vitro cellulose digestion was reduced when rumen fluid from either the steers or the lambs was used in the fermentation. Voluntary intake of the magnesium-deficient diet was significantly reduced by the third day of feeding. Oral administration of magnesium oxide to the lambs by capsule restored appetite, but intravenous injection of magnesium sulfate had a lesser effect.

In vitro digestibility of ground milo stalks with various levels of calcium (10, 100, and 190 ppm), phosphorus (225, 350, and 475 ppm), and magnesium (50, 100, and 150 ppm) have been made to determine the effects on dry matter digestibility (Bales et al., 1979). Calcium at 190 ppm decreased the disappearance of dry matter, but phosphorus had no effect at the levels tested. Magnesium lowered digestibility in linear fashion. Iron, strontium, and zinc had no effect on in vitro dry matter digestibility. The addition of manganese beyond 4 ppm tended to reduce digestibility and 24 ppm had a definite depression. Cadmium at 3 ppm stimulated digestibility over the 0-, 1-, or 2-ppm levels, whereas 4 and 5 ppm gave intermediate values. When sodium chloride (NaCl) replaced sodium bicarbonate ($NaHCO_3$) as part of artificial rumen fluid, the in vitro dry matter digestibility (IVDMD) of milo stalks declined from 48 to 46% when the chloride ion concentration reached 5.8 g/liter.

Trace minerals may be essential in all concentrate barley and milo diets for cattle (Brethour and Duitsman, 1966; Raun et al., 1965). However, Oltjen et al. (1959) observed no response with steers from trace minerals when the finishing diet contained milo, but did when corn was utilized.

## II. NITROGEN AND SULFUR

Pendlum et al. (1976) evaluated the performance of Holstein steers fed diets having different sources of nitrogen and sulfur. Daily gains averaged 1.61 kg for steers fed the soybean meal diets compared to 1.51 kg for those consuming urea diets. Sulfur had no effect on daily gains or feed efficiency. However, plasma concentrations of valine, lysine, and threonine were higher in the soybean meal dietary group, and glycine was higher in the plasma of steers fed urea diets. Spears et al. (1976) determined the effect of adding elemental sulfur or L-methionine on in vitro fermentation of wood cellulose (Solka-floc), Kentucky-31 tall fescue, Kenhy tall fescue, and orchardgrass. Wood cellulose digestion was increased by both sources of sulfur, but methionine was more effective than elemental sulfur. Both sources of sulfur increased the digestion of the fescue forages but had no effect on the digestion of orchardgrass. The incorporation of inorganic sulfur into ruminal microbial protein has been demonstrated (Block et al., 1981; Thomas et al., 1951).

Supplemental sulfur was shown to increase cellulose digestion and nitrogen retention in purified diets (Bray and Hemsley, 1969; Hume and Bird, 1970). Bull and Vandersall (1973) found increased nitrogen retention and acid detergent fiber (ADF) digestion in steers by increasing the sulfur content of a high-corn silage diet from 0.20 to 0.32%.

Willis et al. (1981) fed diets to steers that contained soybean meal or urea as their primary sources of nitrogen, with sulfur added at 0 and 0.3%. Daily weight gains of soybean meal-supplemented steers averaged 1.28 kg compared to 1.17 kg for those fed urea over 111 days. Feed : gain ratios were slightly improved with urea. Steers supplemented with sulfur had slightly improved gains and feed efficiency (7.12 versus 7.59 kg feed per kilogram of gain). Blood and ruminal ammonia concentrations were not influenced by sulfur supplementation.

A number of studies have demonstrated that inorganic sulfur can substitute for organic sulfur sources such as methionine (Barton et al., 1971; Gil et al., 1973a; Bull and Vandersall, 1973). Johnson et al. (1971) found elemental sulfur to be less digestible than sulfate or methionine and determined that greater dietary amounts of elemental sulfur were needed to meet ruminants' requirement.

## III. EFFECT OF CALCIUM

Wheeler et al. (1975) attributed the depression of high-energy corn-based diets to the passage of undigested starch in ruminants. In digestion trials, Fernandez et al. (1982) fed Holstein cows diets that contained 0.75, 1.00, 1.25, and 1.50% calcium in complete diets having a 35 : 65 ratio of forage sorghum-Sudan hay or sorghum silage to concentrate. There were no differences among the treatments in digestibility of dry matter, starch, energy, protein, and cell walls, or in fecal pH.

Two sources of limestone that varied in their requirement for acid during neutralization in feedlot diets were fed to steers and wethers (Harmon et al., 1981). Each source was fed at three levels (0.35, 0.70, and 1.05%). Weight gains and feed efficiency were not altered by the level or source of the calcium when fed in diets to the steers and wethers.

Ward et al. (1979) followed the fate of calcium-containing crystals in alfalfa through the bovine digestive tract by scanning electron microscopy. Most crystals were released from the alfalfa postruminally. Some crystals appeared partially eroded and some intact in the feces. Most crystals were calcium oxalate

($CaC_2O_4$), a few were potassium oxalate ($K_2C_2O_4$) and some contained both oxalates. These observations account for the poorer utilization by cattle of calcium from alfalfa than from inorganic sources.

## IV. NITROGEN AND TRACE MINERALS

Feedlot trials with steers and heifers were made to compare urea and soybean meal with a combination of zinc, magnesium, iron, and copper (Clark *et al.*, 1970). The addition of the trace elements improved performance only when a soybean meal supplement was used. No improvement was observed with the trace minerals when the basal or urea supplements were fed. Byers and Moxon (1980) conducted a growing trial with Hereford steers as they gained from 230- to 340 kg body weight. The diets contained 55% corn silage (dry basis) with shelled corn plus supplementation with soybean meal and sodium selenite ($Na_2SeO_3$) to provide three levels of protein, each with basal (0.05 ppm) or added (0.08 ppm) selenium. The steers fed each diet responded positively to selenium in growth rate and feed efficiency. The greatest response was on the low-protein diet. Plasma selenium levels increased about twofold when supplemented with the element. When the steers were fed finishing diets containing shelled corn and 7% corn silage with basal or added protein as soybean meal or linseed meal with or without supplemental selenium, the element had no effect on steer performance. However, liver selenium levels increased with either sodium selenite or linseed meal sources of selenium in the diets.

## V. WATER IN RUMINANT NUTRITION

The water content of ruminants as well as other animals is relatively constant at 68–72% of the total body weight on a fat-free basis. The body water level usually cannot change appreciably without severe consequences to the animal. Body water is excreted in urine, feces, milk, and sweat, as well as by evaporation from the lungs and skin. Minimal water requirements of the animal are influenced by anything that affects these modes of water loss. Excess salt or protein consumption will increase urine excretion in order to eliminate the salt or end products of protein metabolism. Fecal loss of water varies with the diet and species. Cattle excrete a higher percentage of water in feces than sheep. Loss of water from skin and lungs is increased by elevated temperature and exercise.

In addition to environmental and physiological factors, water in feed and metabolic water are factors in determining the drinking water requirements of animals. Silages and other feedstuffs may supply considerable water in the diet. Metabolic water liberated from the oxidation of 100 g of protein, fat, and carbohydrates amounts to 41, 107, and 60 g, respectively (NRC, 1974). Tissue protein is associated with approximately 3 g of water; this amount is liberated when 1 g of protein is metabolized.

Water balance trials have shown that the requirement of various species is a function of their body surface area rather than their weight. Adolph (1933) concluded that a convenient liberal method of expressing the total water requirement for nonruminants is 1 ml per kilocalorie of heat produced, which would automatically include the increased requirement associated with activity. Cattle excrete large amounts of water in their feces, and their requirement would be somewhat higher (1.29–2.05 ml/kcal) than that of other species (NRC, 1974).

Winchester and Morris (1956) worked out tables of water consumption for cattle of various ages utilizing data in the literature. When water intake per unit of dry matter consumed was plotted against ambient temperature, it was found to change very little from $-12$ to $4°C$, but it did accelerate from 4 to $38°C$. A 450-kg steer would drink 28, 49, and 66 liters daily at 4, 21, and $32°C$, respectively. Huffman and Self (1972) reported that yearling feedlot cattle consumed more water in summer than in winter (31.2 versus 19.0 liters/day) and 8% less under shelter in summer. Shelter apparently had no effect during the winter. Processing water containing 417 ppm hardness through an ion-exchange unit had no effect on water intake or rate of gain.

## A. Nutrients and Toxic Elements in Water

All of the minerals essential as dietary nutrients are present to some extent in water (Shirley, 1970). Generally, it is believed that the availability of the elements in water solution is at least equal to that in solid feeds or dry salt mixes. Radioactive salts of $^{32}P$ and $^{45}Ca$ dissolved in water and administered to steers in a drench were found to be absorbed at levels equivalent to those of isotopes incorporated into forages from fertilizer (Shirley et al., 1951, 1957).

The concentrations of 14 nutrient elements found in U.S. surface waters have been summarized (NRC, 1974). Average concentrations of NaCl in surface waters could supply beef cattle with approximately 34% of their daily requirement, the lactating dairy cow with 19%, growing heifers and cows on maintenance diets with 40%, and lactating ewes with 6–7% (NRC, 1974). Generally, salt needs to be provided in the mixed diet or given in free-choice salt mixes or salt blocks. In arid areas and in some wells, water may contain excessive levels of salinity, and animals restricted to such waters may suffer physiological upset and death. The ions most commonly involved in highly saline waters are sodium, chloride, calcium, magnesium, bicarbonate, and sulfate. Other ions such as nitrate may at times occur at toxic levels. Hardness of water is dependent on the concentrations of calcium and magnesium present.

According to the NRC (1974), water that contains less than 1000 mg per liter of total soluble salts should present no serious problem to livestock; levels of 1000–2999 ppm may cause some temporary and mild diarrhea in livestock but should not affect their health or performance; levels from 3000–4999 ppm may cause temporary diarrhea or may be refused at first by animals not accustomed to them, but should be satisfactory for ruminants; levels of 5000–6999 ppm probably should be avoided by pregnant and lactating ruminants, but can be used with reasonable safety with other types of cattle and sheep; levels of 7000–10,000 ppm involve considerable risk when water is consumed by pregnant and lactating ruminants and, in general, should be avoided; and levels above 10,000 ppm cannot be recommended for use under any conditions.

Phosphorus, iron, zinc, copper, iodine, and selenium have generally been found to be present in water at such low levels that less than 1% of the dietary requirements of ruminants would be met from that source. Somewhat more calcium, magnesium, potassium, sulfur, cobalt, and manganese has been reported, but apparently not in sufficient amounts to be considered significant in meeting the requirements of commercial ruminants (NRC, 1974). Addition of magnesium sulfate to the drinking water of sheep increased the feed intake (13–15%) of timothy hay grown on either unfertilized or magnesium sulfate fertilized land (Reid et al., 1984).

## B. Effect of Water Intake on Utilization of Feed

Feed intake is closely related to water consumption in ruminants (Leitch and Thompson, 1944), and Kay and Hobson (1963) concluded that 2–4 kg of water is normally consumed for each kilogram of dry matter eaten. Wilson (1970) reported that restricting the intake of feed to sheep will decrease their water intake. Meyer et al. (1955) deprived sheep of water and feed, and then reintroduced water after 36 hr and feed after 45 hr. Water consumption did not return to normal after deprivation until feed was reintroduced. Bond et al. (1976) deprived monozygotic twin beef steers of water, feed, or both for periods of 12, 24, 36, or 48 hr when fed diets containing 0, 30, or 88% forage diets. Water deprivation caused a 47% depression in feed intake regardless of diet.

Nutrient digestibility has generally been found to increase as water intake is restricted (Balch et al., 1953; Brown, 1966; Utley et al., 1970). The increase in digestibility may be due to decreased feed intake during water restriction (Brown, 1966; Thornton and Yates, 1968). Balch et al. (1953) reduced water intake to 60% of that drunk in a control period and reported that five of six cows refused part of their hay diet and lost body weight. Restriction of water to 60% of free choice with Angus steers resulted in a decrease in voluntary feed intake and slight increases in the digestibility of dry matter, protein, and nitrogen-free extract, as well as nitrogen retention (Utley et al., 1970). Steers were used to determine the influence of prefast feed intake on recovery from feed and water deprivation by Cole and Hutchison (1985). The data indicated that increased prefast feed intake will provide a greater reserve of energy, water, and electrolytes during deprivation and will give a shorter postfast recovery period.

Cole and Hutcheson (1981) observed fistulated beef steers during sequential short periods of feed and water deprivation. The steers were adapted to the diets, deprived of feed and water for 24 hr, fed and given water for 24 hr, deprived for another 24 hr, and again fed and watered. Rumen fermentation capacity was significantly reduced by deprivation. After 5 days of refeeding, fermentation capacity was still below prefast levels. Upon refeeding, rumen pH returned to prefast levels within 24 hr, whereas 3-5 days were required for rumen VFA to return to prefast levels. In a trial with nonfistulated steers adapted to all-alfalfa pellet or 40% concentrate diets and deprived of water and feed at intervals, the steers fed the alfalfa pellets recovered from deprivation more rapidly. The investigators concluded that pre- and posttransit diets of market-transit stressed calves may be important in reducing stress and improving performance.

## C. Effect of the Water Content of Feedstuffs

In the formulation of diets, ingredients must be evaluated on a dry matter basis in addition to the as-fed basis. Only limited data are available on the effect of high moisture feeds on intake and their utilization. Renton and Forbes (1973) compared the performance of Friesian steers fed a fortified barley supplement that was air-dry or suspended in water. There were no significant differences in liveweight gain, feed intake, or conversion ratios, although there were slightly lower digestibility coefficients with the liquid supplement. Hinks and Whittemore (1976) fed Friesian steer calves diets that contained 67% of either cooked potato or cooked corn in the form of a dry, coarse meal or finely ground in liquid suspension. Daily gains were equivalent with the two moisture levels.

Holzer *et al.* (1976) fed Israeli-Friesian bulls diets that contained two levels of roughage (25 and 45%), two particle sizes (6- and 12-mm sieves) of the roughage component, and three levels of moisture (10, 50, and 75%). Diets that contained 10, 50, and 75% moisture resulted in liveweight gains of 956, 1080, and 1025 g/day and carcass gains of 516, 584, and 563 g/day, respectively. The particle size and roughage content of diets had no effect on gains. The total VFA concentration was greater, with more propionic acid relative to acetate, in rumen fluid of bulls fed high-moisture diets.

Shirley (1982) concluded that there is very little change in dry matter intake by cattle when the diet contains 20% or less moisture. He pointed out that dry matter intake would be approximately 95, 90, and 85% as much at 30, 40, and 50% moisture levels as that obtained with 20% or less moisture. This reduction in dry matter intake should be taken into account when feeding high-moisture corn, silages, and other high-moisture feedstuffs. Shirley noted that, as a rule of thumb, a 5% decrease in dry matter intake reduces gain by approximately 10%.

## D. Effect of Water Temperature

Temperature is a marked variable with pure clean water and may sometimes affect intake. Sheep were reported to have reduced water intake at 0°C or with snow water (Bailey et al., 1962; Butcher, 1966). Cunningham et al. (1964) reported a decreased intake in 0°C water by nonlactating dairy cows. Brod et al. (1982) conducted studies with wethers provided water at 0, 10, 20, and 30°C. Rumen temperature was depressed most by 0°C water. When wethers drank 0, 10, 20 and 30°C water, 108, 96, 96, and 72 min were needed for the rumen to return to its initial temperature. There were no significant effects on the nitrogen retention or digestibility of dry matter, crude protein, or crude fiber, although the lowest coefficients were found for the 0°C water treatment.

Lofgreen et al. (1975a) reported that British cattle in pipe and cable corrals in a hot environment consumed more feed, gained more weight, and in 3 out of 4 years had improved energy utilization when provided water cooled to 18°C compared to 32°C. The level of roughage in diets did not affect the response to cold water. Brahman × British crossbred cattle did not have the depression in feed intake and energy utilization with the warm water that occurred with the British cattle.

## VI. WATER INTAKE IN DAIRY COWS

Winchester and Morris (1956) reported that dairy cows have water intakes that vary with the weight of the cow, kilograms of milk produced daily, fat in the milk, and ambient temperature. They concluded that a lactating cow weighing about 450 kg at 21°C would consume approximately 30 liters of water daily plus 2.7 liters per kilogram of milk produced with 4% fat. Cows producing 40 kg of milk per day may drink as much as 110 liters of water when fed dry feeds. Heifers fed fresh forage and silage may obtain about 20% of their water needs from their feed. Dairy cattle will suffer more quickly and severely from a lack of water than from a deficiency of any other nutrient (NRC, 1978).

## VII. WATER INTAKE IN SHEEP

Asplund and Pfander (1972) fed sheep water : feed ratios of 1.75 : 1 and 1 : 1 at two levels of feed intake during a 7-day metabolism trial. Sheep that received the high feed : water diet had impaction of ingesta in the rumen, and only one of four replicates completed the trial. Sheep generally consume three times as much water as dry feed, but many factors, such as ambient temperature, activity, age, stage of production, plane of nutrition, composition of feed, and type of pasture, may alter the intake of drinking water (NRC, 1974). During winter, ewes eating dry feed require 4 liters of water per head daily before lambing and 6 or more liters per day when nursing (Morrison, 1959).

# 14

# Body Composition versus Nutritional and Other Factors

## I. INTRODUCTION

The chemical composition of the whole empty body of an animal is due principally to species, heredity, environment, and nutrition. Changes in the concentration of fat, protein, and carbohydrates in the body reflect energy gain or loss. Change in body energy is utilized as a criterion of the energy value of feeds for maintenance, growth, and/or finishing livestock.

Haecker (1920) concluded that the amount of water decreases and the amounts of ash and protein increase as the animals approach maturity, especially if finishing does not intervene. Fat and water are variable, depending largely on the amount and kind of feed consumed, and together generally constitute 75–79% of the whole empty body. It appears that the finishing process consists principally of replacing water by fat, or that the percentage body composition is markedly influenced by the deposition of large amounts of fat that contains very little water.

Moulton *et al.* (1922) studied the fat-free composition of cattle ranging in development from early embryonic stages to maturity. The water content of the fat-free body decreased rapidly from the time of conception to birth, and then decreased less rapidly until a relatively constant concentration of water was reached at 5–10 months of age. Moulton (1923) introduced the concept of *chemical maturity,* which he defined as the age at which the concentrations of water, protein, and minerals in the fat-free tissues of the animal become practically constant. Different animals vary in their age of chemical maturity. That is, there is some decrease in the proportion of water and an increase in the proportions of protein and ash as animals become older.

Body weight changes are commonly utilized to indicate responses in many feeding experiments with livestock. In experiments of short duration, changes in rumen fill can result in appreciable errors. The use of body weight changes assumes that weight gained or lost has the same composition irrespective or treatment. Such an assumption was early pointed out to be in error. By comparing casein and urea as sources of dietary nitrogen for cattle, Watson *et al.* (1949) showed that although the animals fed urea diets gained only 70% as much weight as those fed casein, they gained 81% as much energy.

The total digestible nutrient (TDN) content or net energy (NE) values are commonly used criteria of the energy value of feed ingredients in the United States. It is recognized that the TDN of one feed does not necessarily produce a response of the same magnitude as an equal amount of TDN from another feed. This is especially true when comparing TDN values of concentrates and roughages. Evaluating mixed feeds utilizing roughages for ruminants on a TDN basis is more complex than it is for nonruminants fed relatively roughage-free diets. Evaluation of feed on the basis of NE is less complex than it is for TDN in that a unit of NE in one feedstuff produces the same amount of a given response in a given animal as that produced by a unit of energy from entirely different types of feeds. NE values of feedstuffs and diets for maintenance and gain may be determined by metabolism and comparative slaughter feedlot trials. In such studies, the energy of whole empty animal bodies at the beginning and end of the feeding period is determined directly on tissues by calorimetry or indirectly by specific gravity (Lofgreen and Garrett, 1968; Garrett and Hinman, 1969). The specific gravity values are related to the fat and protein in tissues by chemical analysis. Empty bodies of cattle generally contain less than 0.5% carbohydrates (Trowbridge and Francis, 1910). Tissue carbohydrates are disregarded in dietary energy studies on feedlot cattle that involve sufficient time for considerable weight gains.

## II. EFFECT OF DIET, BREED, AND SEX

Blaxter (1962) concluded, from a biochemical appraisal of metabolic pathways involved in the deposition of protein and fat in cattle, that it seemed unlikely that one breed would be more energetically efficient than another. However, Garrett (1971) conducted comparative slaughter experiments with Holstein and Hereford cattle and compared the net efficiency of energy utilization by the two breeds. The results indicated that Herefords were 12 and 20% more efficient than Holsteins in converting feed energy consumed above maintenance to energy stored as protein and fat, respectively. The beef steers had gains that contained more fat, and consequently more energy, than corresponding gains by the dairy steers. The two breeds had equivalent gains of protein per unit of feed above maintenance.

Nichols *et al.* (1964) compared Holstein bulls and steers for meat production from birth until slaughter at 800–1000 pounds. Bulls had slightly lower dressing

## II. Effect of Diet, Breed, and Sex

percentages due to heavier hides and less fatty carcasses. Steers had higher marbling and more outside fat covering than bulls. Martin et al. (1978a) fed Holstein steers diets containing three levels of energy. The higher levels of dietary energy produced faster gains, more efficient conversion of dry matter to gain, greater deposition of fat, and higher carcass grade.

Fortin et al. (1980a) studied the effects of energy intake, breed, and sex on muscle growth and distribution in Holstein and Angus bulls, steers, and heifers. The animals were fed energy at two levels, ad libitum and 60–70% of ad libitum. In both breeds, irrespective of energy intake, sex had no influence on the growth rate of muscle in the thoracic and pelvic limbs relative to carcass side or total muscle. Neither breed nor energy intake altered the growth rate of muscle in the three joints relative to carcass side or total muscle. Fortin et al. (1980b) determined the chemical composition of Angus and Holstein bulls, steers, and heifers fed at two energy levels and slaughtered at weights ranging from 121 to 706 kg. The accretion rate of fat was not affected by sex, but accretion rates of protein and ash were more rapid in Holsteins than in Angus cattle in the high-energy dietary groups. Angus bulls had a faster accretion of fat than Holstein bulls.

Fortin et al. (1981) concluded from composition studies with Angus steers, bulls, and heifers that the growth rate of fat relative to carcass side was greater in a high-energy than in a low-energy dietary group, whereas with corresponding sex groups of Holsteins, the growth rate of fat was not altered by the level of energy intake. The breed (Angus versus Holstein) generally did not influence the growth rate of muscle and bone plus tendon.

A number of studies have demonstrated that bulls gain more rapidly and convert feed to lean meat more efficiently with less fat than steers (Cahill, 1964; Prescott and Lamming, 1964; Field, 1971). Bulls generally produce meat with less marbling, coarser texture, darker color, and less tenderness, depending on their age. However, organoleptic studies with young bull meat indicate a high level of palatability.

Arthaud et al. (1977) compared carcasses of bulls and steers fed at two levels of energy and slaughtered at 12, 15, 18, and 24 months of age for composition and taste panel evaluation. Bulls made faster gains on less feed per unit of gain and had less fat in the carcass than steers at all ages. Bulls exhibited a slight but consistent tendency toward more maturity than steers of the same age. Quality grades and taste panel scores of meat were generally higher for steers than for bulls, but the differences were small. Taste panel scores for bulls were acceptable.

Jesse et al. (1976a–c) determined the empty body and carcass protein, fat, water, and ash of Hereford steers slaughtered at 227, 341, 454, and 545 kg body weight. The diets consisted of the following ratios of corn to corn silage (dry basis): 30 : 70, 50 : 50, 70 : 30, and 80 : 20. The composition of the empty body and the carcass gain by comparative slaughter were not affected by the diet. The greatest increase in fat occurred after the cattle reached 341 kg, and the percentages of empty body and carcass gains from 454 and 545 kg were 63.4 and 68.3%, respectively.

Aberle et al. (1981) studied the effect of the following dietary treatments of steer calves: (1) high-energy diet for 210 days, (2) low-energy diet for 77 days, followed by high-energy diet for 140 days, (3) low-energy diet for 153 days, followed by high-energy diet for 70 days, and (4) low-energy diet for 230 days. The low-energy diet was intended to give gains of 0.68 kg/day. The steers were slaughtered at the same age. Those fed the low-energy diet for longer periods had less carcass fat, smaller ribeye areas, and lower-quality grades than steers fed the high-energy diet for longer periods. Longissimus muscle from steers fed the low-energy diet had lower collagen solubility than muscle from those fed the high-energy diet for 70 days or more. It was concluded that the growth rate of cattle before slaughter affects meat palatability, particularly tenderness, and that the growth rate may be a more important determinant of tenderness than the length of time cattle are fed high-energy diets.

Rouquette and Carpenter (1981) observed the carcass characteristics of weanling calves grazed at three levels of forage availability to an average age of 259 days. The grazing areas supported 6.27, 4.52, and 2.72 cow-calf units/hectare and calf weaning weights of 236, 251, and 289 kg, respectively. Carcasses of calves that grazed the high-forage availability area were fatter than those of calves that grazed the lowest available pasture area (5.08 versus 1.27 mm over the longissimus muscle), and their longissimus muscle areas were more than 18% larger. Steer carcasses showed greater average daily gains and larger longissimus muscle areas than heifer carcasses, but heifer carcasses had more fat thickness over the longissimus muscle and more kidney, pelvic, and heart fat.

Martin et al. (1978b) fed varying levels of crude protein to bulls in the weight range of 220–410 kg and found that near-maximal weight gains were obtained by feeding a continuous level of 11.1% crude protein. However, bulls fed diets containing 13.3 or 15.5% crude protein gained significantly faster during the first 56 days of the 168-day feedlot trial. During the last 84 days, the bulls fed the 11.1% crude protein diet appeared to compensate, so that the total gain did not differ among the three protein dietary groups. Lemenager et al. (1981) studied the effects of four dietary protein levels (9.6, 10.7, 12.5, and 13.9%) on Angus bulls provided early and late in 140-day feedlot trials. Bulls fed the 12.5 and 13.9% crude protein diets did not differ in gains, but both groups gained more rapidly than those fed the lower protein levels.

Broadbent et al. (1976) reported that Angus steers had carcasses with more fat, less bone, and a higher lean : bone ratio than Ayrshire or British Friesian steers. Carcass data on the crosses of the three purebreds fell midway between their parent values and did not differ significantly. The investigators concluded that their data refuted the hypothesis that crossbred cattle produced carcasses with higher muscle : bone ratios. Lindsay and Davies (1981) fed Friesian steers barley-based diets with different levels of crude protein (formaldehyde-treated soybean meal) and slaughtered them at 365 or 465 kg liveweight. Steers fed 23.7 g nitrogen per kilogram of dry matter and slaughtered at 365 kg liveweight had more fat depth at the 12th rib than those fed 27.9–38.5 g nitrogen per kilogram of dry matter in their diets. Animals slaughtered at 465 kg liveweight were not affected by dietary nitrogen in regard to fat depth at the 12th rib.

Chigaru and Topps (1981) observed the effects of reducing feed intake for winter-calving cows to the maintenance level for 6 weeks from week 10 of lactation. Changes in body water estimated by the dilution of tritiated water and deuterium oxide at the end of each feeding period were used to calculate changes in body composition. The weight losses during "underfeeding" consisted mainly of fat, but some cows apparently mobilized relatively large amounts of protein. More fat was mobilized by heifers than by cows, and cows appeared to mobilize more protein than heifers. Only a few cows were able to achieve complete tissue repletions upon "refeeding."

Loveday and Dikeman (1980) slaughtered steers of Angus × Hereford (AHX) reciprocal crossbreeding and Brown Swiss sired-steers out of AHX reciprocal crossbred dams at an energy efficiency and endpoint of either 9.0 Mcal $NE_p$ per kilogram of gain (EEP) or 11.5 Mcal $NE_p$ per kilogram of gain ($EEP_2$). There were no differences in the chemical composition of adipose fat between EEP and $EEP_2$ steers or in lipogenic enzyme activities of 6-phosphogluconate dehydrogenase, glucose-6-phosphate dehydrogenase, or nicotinamide adenine dinucleotide phosphate (NADP)-isocitrate dehydrogenase.

## III. REALIMENTATION OF CULL COWS

Beef cows grazing range pastures normally undergo cyclic loss and gain in weight due to seasonal variation in availability and quality of forage. Cows culled during poor grazing periods are generally quite thin, and the usefulness of their carcasses is limited. Such cows that have undergone malnutrition generally have higher than normal rates of gain when realimented, and compensatory gains are usually associated with better feed efficiency (Wilson and Osbourn, 1960). Short-term feeding of cull cows to obtain compensatory gain should increase their salvage weight and improve carcass quality.

Swingle et al. (1979) conducted several trials with cull range cows to determine their level of performance during realimentation. Composition of carcass gains was estimated by comparative slaughter values before and after realimentation. Concentrates at various levels (80, 40, and 22%) were fed during realimentation and the cows were slaughtered at different final body conditions. Across all trials, average carcass gain contained 51% fat, 14% protein, and 35% moisture. The percentage of protein in most wholesale cuts decreased and that of fat increased in all wholesale cuts when culled cows were realimented (Wooten *et al.*, 1979). Marchello *et al.* (1979) utilized data obtained on carcasses of realimented range cows to develop the following equations for predicting carcass protein or lipid from plate composition:

Carcass protein, % = 18.59 − (0.008 × cold carcass weight) + (.029 × ribeye area) − (0.31 × % of kidney, pelvic, and heart fat) − (0.1 × % of plate bone) − (0.54 × % of plate lipid) + (0.189 × % of plate protein). $R^2 = 0.79$.

Carcass lipid, % = 2.2 + (0.022 × cold carcass weight) − (0.07 × ribeye area) + (0.492 × 12th rib fat thickness) + (0.639 × % of plate lipid). $R^2 = 0.97$.

## IV. DIETARY FATS

It has been established that the major fatty acids of dietary fats will appear in the body fat stores of nonruminant animals (Channon *et al.*, 1937; Bhalerao *et al.*, 1961; Alsmeyer *et al.*, 1955). Only small changes have been found in the body fat of ruminants fed normal additions of either a saturated or a highly unsaturated fat at a level of 4–8% of the diet (Willey *et al.*, 1952; Edwards *et al.*, 1961; Erwin *et al.*, 1963; Roberts and McKirdy, 1964). It has been suggested that dietary fat is deposited in the fat depot of ruminants as in nonruminants, provided the fatty acids are not subjected to the hydrogenation of microbes in the rumen (Ogilvie *et al.*, 1961).

Dryden and Marchello (1973) made a fatty acid analysis of subcutaneous and intramuscular fat at 10 locations in the carcass and four muscles of Charolais-Hereford steers fed diets that contained various levels of safflower oil (0, 5, 10, and 15%) or 6% animal fat. In longissimus, triceps brachii, semimembranous, chuck intramuscular fat, pericardial, mesentary, kidney, brisket, and caudal tissues, safflower oil in the diet caused the depot fat to become more unsaturated than that of controls. Brisket fat was not affected by safflower oil. Intramuscular fat from steers fed diets supplemented with animal fat became more saturated, while the safflower oil caused the fatty acids of these tissues to become more unsaturated. Skelley *et al.* (1973) evaluated the effects of vitamin A, corn silage, and raw soybeans on the fatty acid composition of carcass depot fat in steers. Raw soybeans in the diets decreased the total amount of saturated fatty acids, but vitamin A and corn silage had no effect on them. Stearic acid accounted for a great deal of the saturated fatty acids, while oleic acid accounted for much of the unsaturated fatty acids. Fat thickness and yield grade had little relationship to the percentages of individual fatty acids.

## V. IMPLANTS

Harris *et al.* (1979) implanted Hereford steers with 36 mg zeranol or 200 mg progesterone and 20 mg estradiol benzoate at the initiation of a feedlot trial and 87 days later. The steers were slaughtered when they reached choice live grade. The zeranol-implanted steers gained less than the progesterone-estradiol–implanted steers during the first 87 days and yielded carcasses with less external fat, more kidney fat, and higher marbling scores. Average daily gains were the same for the two implanted groups overall to the time of slaughter. Cross and Dinius (1978) compared steers fed 94% ground alfalfa hay- or dehydrated alfalfa meal–containing diets, with and without Synovex S (progesterone plus estradiol) implants, for 120 days. The implant decreased the amount of marbling of longissimus muscle and the U.S. Department of Agriculture (USDA) quality grade of carcasses but increased the ribeye area. The implant had no effect on fat thickness. The type of alfalfa did not affect fat thickness or the percentage of kidney, pelvic, and heart fat or ribeye area. Pelleting of the diet reduced the amount of marbling.

## VI. pH OF MUSCLE

Sodium pyrophosphate injected antemortem into cattle increased the glycolysis and lowered the initial pH of tissues (Howard and Lawrie, 1957). Huffman *et al.* (1969) injected four lots of mature sheep with pyrophosphate or hexametaphosphate antemortem and observed the muscle pH to be lower initially for the treated sheep than for the controls. No correlation was observed between tenderness of loin chops and muscle pH, but it was noted that the meat from animals fed a low-calcium : phosphorus diet was more tender than that from control sheep and that the initial pH of low-calcium : phosphorus-fed animals was lower than that of controls.

## VII. PREDICTING THE COMPOSITION OF BEEF CARCASSES

The most accurate method of determining carcass composition is by chemical analysis of the whole carcass. Obviously, this is time-consuming and expensive, and renders the carcass unsalable. Hankins and Howe (1946) reported that the chemical composition and separable physical components of the 9–10–11th rib cut of slaughter steers were highly related to the composition of the entire carcass, and developed equations for predicting carcass composition. Reid *et al.*

(1955) summarized composition data on beef versus dairy cattle, males versus females, and age of bovines. Regression equations were given relating (1) the age of animal to the protein, ash, and water of the fat-free body and (2) the water content of beef, dairy, male, and female animals to the fat content of the whole empty body when the body's water content was determined. The estimation of the chemical composition and energy content of whole animal bodies by Reid et al. (1955) utilized the system of Kraybill et al. (1952) in which the carcass fat was estimated from both the 9–10–11th rib cut using the prediction equation of Hopper (1944) and the body water content using the antipyrine method (Soberman et al., 1949). These relationships are generally regarded as crude estimates.

Garrett and Hinman (1969) related the specific gravity of Hereford steer carcasses to the water, fat, nitrogen, ash, and energy contents determined by chemical analyses of whole empty bodies. The investigators concluded that the highly significant coefficients of correlation and the low standard errors indicate that carcass density is a good index of the carcass and empty body composition of steers. Carcass fat and offal fat had identical caloric values of 9.385 ± 0.08 kcal/g, and the dry fat-free organic matter (protein) of the carcass and empty body had caloric values of 5.532 and 5.539 kcal/g, respectively (Garrett and Hinman, 1969).

Vance et al. (1971) observed that the chemical composition of meat sawdust, obtained by sawing wholesale cuts, was highly correlated with the composition of the carcass. Preston et al. (1974) evaluated steers of varying weights and degrees of fatness to determine the relationship between carcass specific gravity and carcass composition, and the effect of bone proportionality on this relationship. Bone proportionality (11.7–18.6% separable bone) did not alter the relationship between carcass specific gravity and carcass composition. Specific gravity of the carcass was highly correlated with carcass fat (−0.96), protein (0.89), and water (0.94). Ledger et al. (1973) subdivided sides of beef carcasses into 11 cuts, which were then dissected into fat, muscle, and bone. They developed a table from which the carcass composition of the animal was read once the specific gravity of the 10th rib was determined.

Mata-Hernandez et al. (1981) utilized feedlot steers varying in genetic background, feeding regimen, and management to evaluate specific gravity measurements and USDA quality and yield grade factors as predictors of carcass chemical composition. Steers were slaughtered when they reached 1 cm of fat thickness at the 12th rib or had been on feed for 220 days, in the case of those fed as calves, or 120 days, in the case of those fed as yearlings. Specific gravities of the left side quarters and standard wholesale cuts were determined after a 72-hr chill. The soft tissue was separated from bone and analyzed for fat, protein, and water. Prediction equations for composition from specific gravity and carcass traits as independent variables gave higher coefficients of determination than equations for specific gravity alone. The specific gravity of various wholesale cuts possessed predictive characteristics equivalent to those of the specific gravity of the entire side.

Jones et al. (1978) reviewed the literature on measurement of bovine carcass density and its relationship to fatness. They suggested that density measurement of carcasses would be more repeatable if they were determined in water on the killing chain prior to final washing and shrouding. The estimate of carcass fatness could then be included in the grading scheme.

## VIII. BODY COMPOSITION OF LIVE ANIMALS

There is interest in the use of objective procedures to determine the body composition of live animals. Among such methods are the sonoray utilized by Hedrick et al. (1962), electronic meat measuring equipment (Domermuth et al., 1973), and $^{40}K$ (Lohman et al., 1966; Frahm et al., 1971).

Clark et al. (1976) evaluated the body composition of steers ranging in weight from 183 to 574 kg from $^{40}K$ measured in a whole body counter. After the animals were monitored for $^{40}K$, they were slaughtered and tissues were analyzed. Independent variables, liveweight and carcass weight, live and carcass $^{40}K$, and carcass specific gravity were used to predict the dependent variables, carcass or empty body fat, nitrogen, gross energy, and water. The $R^2$ values between $^{40}K$ and nitrogen, fat, gross energy, and water were lower than when specific gravity was compared to these dependent variables. When $^{40}K$ and liveweight were used to predict the nitrogen, fat, gross energy, and water of the carcass, $R^2$ values of 0.87, 0.87, 0.84, and 0.84, respectively, were obtained.

# 15

# Endocrines and Nitrogen and Energy Nutrition

## I. INTRODUCTION

The partitioning of dietary nitrogen and energy between tissues of the body varies with physiological state, genetic makeup, age, and amount of feed intake. A function of the endocrine system is partitioning of nutrients to different tissues for deposition and to the mammary gland for milk. A number of reviews have dealt with the role of hormones in regulating the metabolism of ruminants (Bassett, 1978; Bines and Hart, 1978; Trenkle, 1978, 1981).

## II. CROSSBREEDING AND HORMONES

Eversole *et al.* (1981) observed the effects of crossbreeding, cattle type, and dietary energy level on feedlot performance and serum anabolic hormone concentrations. Over 3 years, feedlot cattle representing four cattle types, i.e., unselected Hereford (UH), selected Hereford (SH), Angus × Hereford × Charolais (ACH), and Angus × Hereford × Holstein (AHH), were fed either an all-corn silage or a high-grain diet. Steers fed the high-grain diets had higher insulin levels in serum than steers fed the high-silage diets, but cattle type had no consistent effect on insulin level. Growth hormone levels in serum were not affected by diet or type of cattle. Concentrations of insulin in serum combined across diet and cattle types were highly correlated with daily weight gain. However, growth hormone in serum assessed on the same basis was not related to average daily gain. Average daily protein and fat gain were positively related to insulin and negatively related to growth hormone in serum.

## III. REPRODUCTION AND NITROGEN AND ENERGY INTAKE

Infertility in beef heifers is commonly associated with undernutrition (Warnick et al., 1960; McClure, 1961; Wiltbank et al., 1964). The balance between the hypophysis and the gonad can be altered by inadequate nutrition and may upset the estrogen–progesterone relationship (Lutwak-Mann, 1958; Leathem, 1961). Synergistic action of progesterone and estrogen upon the uterus is essential for placentation and embryo survival (Wynn, 1967).

Wordinger et al. (1972) reported that fertilization and pregnancy rates were lower in undernourished Hereford and Angus heifers. Glycogen granules and moderate amounts of acidic mucosubstances were observed in the glandular epithelial cells and the endometrial surface of 8- and 13-day pregnant and nonpregnant undernourished heifers.

Bedrak et al. (1969) fed diets of Pangola grass hay, cottonseed meal, crude cane sugar, and mineral-vitamin mix to Angus, Hereford, and Angus-Hereford crossbred heifers to determine the effect of varying dictary protein levels on reproduction. Yearling heifers initially weighing 226 kg with daily consumption of 0.48, 0.29, 0.13, and 0.04 kg crude protein gained 0.42, 0.32, 0.01, and −0.33 kg daily, respectively. Two-year old heifers, initially weighing 306 kg, consumed 0.61, 0.48, 0.32, and 0.28 kg crude protein daily and gained 0.34, 0.24, −0.03, and −0.10 kg/day, respectively. At 44 days postbreeding, there were 10 normal embryos from 10 yearling heifers and 7 normal embryos from 10 2-year-old heifers fed the two higher intakes of protein. Eight of the 10 yearling heifers did not ovulate, and there were only 4 normal embryos from 7 2-year-old heifers that consumed the lower-protein diets.

## IV. MILK PRODUCTION

Increased milk production in beef cows can be accomplished most rapidly by crossbreeding with dairy breeds. Deutscher and Whiteman (1971) reported that Angus × Holstein cows produced more milk than Angus cows under range conditions. Hereford × Holstein cows produced more milk than Hereford cows (Holloway et al., 1975). The higher costs of the greater energy and protein requirements of Holstein crossbred cows may offset the advantage in calf response. When energy intake was limited, beef × dairy crossbreds had poor reproductive performance (Deutscher and Whiteman, 1971; Holloway et al., 1975).

Wyatt et al. (1977a) compared the performance of winter-calving 4- and 5-year-old Hereford, Hereford × Holstein crosses, and Holstein cows under tall grass native range conditions. Digestible energy and digestible protein intake increased as the percentage of Holstein breeding increased. Greater forage intake was accompanied by greater milk production. All cows except the Holstein regained their winter weight losses during the summer months. Wyatt *et al.* (1977b) reported that a high level of milk consumption increased weaning weights by 20% on range and by 19% in drylot when Hereford × Angus cross calves were fostered by Friesian cows instead of their own dams.

## Hormones

Hamermik *et al.* (1985) found no differences in feed efficiency or carcass characteristics of (1) intact (2) intact plus melengestrol acetate (3) hysterectomized, and (4) ovariectomized treated heifers in feedlots. High-yielding dairy cows during early and mid-lactation divert more dietary energy and mobilize body tissue to meet the high demands of the lactating udder than low-yielding cows fed the same diet. The low-yielding cows will partition varying proportions of nutrients into body weight gain. Hart *et al.* (1979) measured hormones and metabolites in the blood plasma of high- and low-yielding cows matched for diet and stage of lactation. Interrelationships between milk yield, body weight, and concentrations of prolactin, growth hormone, insulin, and thyroxine were determined. The total milk production of high-yielding cows was correlated positively with ratios of growth hormone to thyroxine and glucose to thyroxine and was negatively correlated with thyroxine. Changes in growth hormone and its ratio to insulin and changes in milk yield were positively correlated for all cows. Changes in thyroxine were negatively correlated with changes in milk yield. Liveweight changes were positively related to insulin and lactic acid and negatively to changes in growth hormone and nonesterified fatty acids.

Feeding lactating cows diets containing thyroprotein (15 g protamine daily) caused cows fed the thyroprotein for 5 weeks to give 2.2–3.3 kg more milk daily than controls. Those fed the thyroprotein for 13 weeks gave 0.95–2.5 kg more milk per day than controls for 60 days, but between 60 and 91 days, milk production was 0.9 kg/day less than that of controls (Shaw *et al.*, 1975). Blaxter *et al.* (1949) made an extensive review of the early work on the role of thyroidal materials and synthetic goitrogens in animal production and their practical use.

# 16

# Effects of Ambient Temperature on Utilization of Nutrients

## I. INTRODUCTION

Body weight loss occurs due to loss of feces and urine and to insensible losses due to perspiration of water and differences in the weight of carbon dioxide ($CO_2$) expired and oxygen ($O_2$) consumed. Since a fasting animal gives off approximately 0.7 mole of $CO_2$ for each mole of $O_2$ absorbed, the weight is about equal, since $0.7 \times 44$ (molecular weight of $CO_2$) = 30.8 g of $CO_2$ for each $1 \times 32$ (molecular weight of $O_2$) = 32 g of $O_2$. The average of 77 calorimetric measurements on steers showed that the insensible loss of gases and water ranged from 2 to 18 kg/day (Kriss, 1930). Cole and Kleiber (1945) found the insensible weight loss of cows on pasture to range from 0.5 to 1.5 kg/hr; they determined this by preventing their water and food intake with muzzles and collecting and weighing their feces and urine. Insensible weight loss of cattle should be estimated in trials in which feed intake is measured at the start and end of a grazing period.

## II. FASTING BODY WEIGHT LOSS

According to Chossat (Kleiber, 1975), fasting animals burn up their body and die when about one-half of their body mass is consumed. The body substance is used as a fuel, as inferred from the following observations: (1) consumption of body substance during fasting increases with increasing thermostatic heat requirement; (2) the life-prolonging property of body substances can be expressed in terms of calories or as heat of combustion; and (3) heating the environment saves body substance and prolongs life in starvation (Kleiber, 1975). The last is true only up to a certain environmental temperature, designated as the *lower critical temperature;* above this critical temperature level, there is no further saving of body substance for fuel. The rate of heat production becomes independent of further increases in environmental temperature above the lower critical temperature until the *upper critical* temperature is reached, at which point the animal begins to lose its homeothermic power. At this upper critical temperature the body temperature increases and, along with it, the metabolic rate. The level of feed intake, productivity of an animal, and thermal insulation of hair coat and superficial tissue have a marked effect on the lower critical temperature. In the range between the lower and upper critical environmental temperatures, the fasting and resting animal has a minimal rate of heat production called the *basal metabolic rate (BMR)* (Kleiber, 1975).

## III. METABOLIC BODY RATE

The BMR can be estimated from the body weight by the equation (Kleiber, 1975): BMR, kcal/day = 70 $W^{0.75}$. The $W^{0.75}$ is expressed as metabolic body weight in kilograms and is approximately the weight of body tissues without water and minerals. The BMR of a 465-kg steer may be calculated by the use of logarithisms since log $a^n$ = $n$ log $a$, and BMR = 70 × $n$ × log $a$ = 70 × 0.75 log 465 = 70 × 0.75 × 2.6675 or 70 × 2.000 or 70 × 100 = 7000 kcal = 7.0 Mcal. The antilog of 2.000 = 100. At times it may be simpler to take the three-fourths root of the whole body weight to obtain the metabolic body weight (Kleiber, 1975).

## IV. CALORIGENIC EFFECT OR HEAT INCREMENT

Rubner (Kleiber, 1975) measured heat production of a dog during a day of hunger, and then again after feeding 2 kg of meat (which contained 1926 kcal energy), and observed the following:

Heat produced during 1 day of hunger = 742 kcal
Heat produced during 1 day of feeding = 1046 kcal
Difference  304 kcal

This increased heat production due to food is called the *calorigenic effect,* or *specific dynamic effect,* or, more commonly, the *heat increment of food.* It amounts to 304/1926 × 100 = 15.8% of energy in meat.

## V. INTERACTIONS WITH ENVIRONMENTAL TEMPERATURE

Most animals have evolved in such a way that they are capable of survival within a wide range of temperatures, with responses that are dietary, metabolic, physiological, and behavioral. As the temperature decreases, the animal enters a zone in which the response is largely neurophysiologic; hair may become erect with vasoconstriction of the skin, and animals frequently move closer together. This requires little effort and energy expenditure. Further lowering of the temperature makes it necessary to generate additional heat to maintain body temperature; this is accomplished by an increase in metabolic rate and shivering. If the temperature decreases more and the maximum metabolic rate is reached, the body temperature decreases and the animal can survive for only short periods (Newton, 1980).

As the temperature increases above the comfort zone, physiologic responses such as vasodilation within the skin, increased respiration rate, sweating, panting, dispersion of animals, and seeking shade begin. Water intake may double. Further increases in temperature cause the metabolic rate to drop, especially with cattle. If the heat input is greater than the rejection of heat, body temperature rises with increases in metabolic rate and survival time is limited in this zone. Food intake is markedly affected by environmental temperature, i.e., increases in intake occur with cooling below the comfort zone and decreases with heating above the comfort zone.

Dietary changes should improve the feed efficiency of cattle once it is determined that they are below their critical temperature. To do this, the energy concentration in the diet should be increased by replacing roughage with grain or by decreasing the protein. Newton (1980) suggested a dietary change for a 225-kg steer that involved the addition of 0.45 kg of corn and the deletion of an equal amount of roughage dry matter for each 45-g decrease in daily gain or for each 5.5°C effective temperature decrease below the critical temperature. The protein adjustment for 225-kg cattle would be 6.6 g of protein removed for each 45-g decrease in daily gain or for each 5.5°C effective temperature decrease below the critical temperature.

## VI. HEAT STRESS

Shorthorn heifers maintained on a high-energy diet had decreased feed intake of 3–5% during the first 2 weeks of feeding at 32°C, while gains were reduced 50–75% (Newton, 1980). Maximum gains at 32°C were obtained after 75–170 days at the elevated temperature and were maximized at 0.7 kg/day on a feed intake per $W^{0.75}$ that had produced 0.81 kg/day at 17–21°C. According to Newton (1980), it should be pointed out that decreased feed intake of cattle at high daytime temperatures in commercial feedlots is often not reduced to the same extent as occurs in constant high-temperature trials. Immature animals are not affected as much as old cattle, and well-fed animals survive longer during cold stress, while fasted ones survive longer under acute heat stress. Apparently, the production problem during cold stress is heat production, while that of heat stress is feed intake. Air movement or wind increases cold stress but generally helps to relieve heat stress.

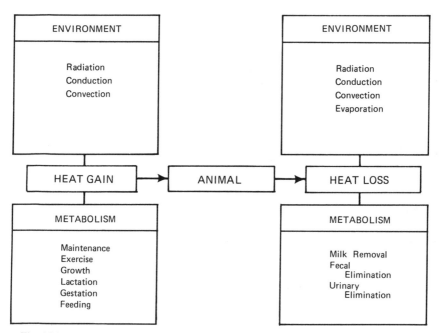

**Fig. 16.1.** Increments of heat gain and heat loss, which are balanced to constitute thermoregulation in an animal (From Fuquay, 1981).

Thermoregulation involves a balance between heat gain and heat loss and is illustrated in a general manner in Fig. 16.1 (Fuquay, 1981). Metabolic heat includes that required for maintenance plus increments for exercise, feeding, lactation, gestation, and growth. In addition to the heat of metabolism, during daylight hours heat is gained from solar radiation and from convection and conduction if the air temperature is higher than the skin temperature of the animal. Loss of heat occurs through loss of feces, urine, and milk, as well as radiation, when the environment is cooler than the animal. Conductive and convective heat losses are greater at cooler air temperatures but are not very effective at high temperatures. Evaporative loss of heat occurs when the dew-point temperature of the air is lower than the temperature of the evaporative surfaces of the skin and respiratory passages. At high ambient temperature, evaporation becomes the major means of heat loss.

## A. Production Problems at High Temperatures

The primary problem is how to maintain high nutrient intake at high temperatures by reducing metabolic heat production and increasing metabolic heat loss. Ruminants with a high plane of nutrition are more sensitive to hot conditions than those with a low plane (Yeates, 1956). A high plane of nutrition is necessary for good performance, and at elevated temperatures, declining feed intake is the major cause of reduced milk production in dairy cows (Davis and Merilan, 1960; Wayman et al., 1962; Johnson and Yeck, 1964). McDowell et al. (1969) observed that milk energy decreased about twice as much as digestible energy (DE) intake at elevated temperatures.

Wayman et al. (1962) force-fed dairy cows through a rumen fistula when stressed at 32°C and observed milk production to be 10% below that of controls at 18°C. The reduction in efficiency may have been due to the energy expended in removing the excess heat load by increased respiration, heartbeat, and other activities. Digestibility of feed has been observed to increase at higher temperatures, probably due to decreased intake resulting in a slower rate of passage through the alimentary tract (Davis and Merilan, 1960). Cows fed a low-fiber diet during the summer produced more milk and had a lower rectal temperature and respiration rate than those fed a high-fiber diet. Protein and energy levels of the diets were about equal and produced no differences when fed in cooler temperatures.

High-fiber diets in hot climates are generally unavoidable for ruminants dependent on local forages. Deinum et al. (1968) observed a negative relationship between ambient temperature and digestibility of forages native to northern Europe. The forage in summer had higher cell walls and crude fiber but lower soluble carbohydrates than that in early spring. Tropical forages tend to be lower in quality than temperate forages due to more rapid maturation (van Soest et al., 1978). Moore and Mott (1973) concluded that tropical plants have higher lignin, silica, and cell walls, all contributing to less digestibility, slower passage through the alimentary tract, and decreased intake. The higher ratio of structural to soluble components in forages contributes to the heat stress problems of ruminants. Van Soest et al. (1978) suggested that frequent clipping to allow ruminants to feed on new vegetative growth should help alleviate the problem. Rapidly growing sorghum-Sudan hybrids or millet should provide a summer forage high in soluble carbohydrates. Other alternatives are corn silage, alfalfa hay, or other stored forages.

Reid et al. (1958) estimated that energy expended by grazing cattle resulted in maintenance requirements 40–50% greater than those of cattle confined and fed in a barn, indicating that the total heat load is increased by grazing. Graf and Peterson (1953) reported that increases in rectal temperature and respiration rate from forced activity are generally short-lived, and Hutson et al. (1971) found that they had no effect on appetite. Selective grazing of low-fiber portions of low-quality forage by cattle may require work energy equivalent to the decreased dietary energy present in greenchop forage containing both vegetative and stem parts.

Bond and McDowell (1972) studied the reproductive efficiency of beef cows that had been adapted to temperatures of either 1.2 or 24.4°C and then exposed to 32°C and 60% relative humidity. The cows previously adapted to the cold became anestrous but generally reestablished their cycles by the 16th week, then conceived and delivered normal calves. Those previously accustomed to a temperature of 24.4°C had uninterrupted cycles at 32°C but later became anestrous at 38°C.

## B. Effect of Water in Heat Stress

Increasing water intake should aid metabolic heat loss, and Macfarlane (1965) observed that feed intake was closely associated with water intake. Cattle encountering heat stress have increased water consumption and evaporative water losses but decreased fecal water (McDowell *et al.*, 1969). In lactating cows, the increased water consumption during heat stress may be modulated by declining milk production (Johnson and Yeck, 1964). At high temperatures, provision of cooled water improved the rate of gain in beef cattle (Kelly *et al.*, 1955). Bhattacharya and Warner (1968) found that cold water in the rumen of dairy cows increased feed intake by 24% and lowered rectal and tympanic membrane temperatures. On the other hand, Gengler *et al.* (1970) observed that heating the rumen with a heating coil depressed feed intake by 15%. It is likely that this heat effect on appetite is due to heating of the blood, which warms the hypothalamus and supports the concept of a thermostatic mechanism for regulation of feed intake (Brobeck, 1960).

## C. Feedlot Nutrition of Cattle in Warm Weather

During the 1950s, a rapid expansion occurred in large beef cattle feedlots in the southern United States. It had long been thought that temperatures in this area were too high. The basis for this belief is the high heat production of steers on finishing diets. In cattle, the primary absorbed products for energy use are the VFA from rumen fermentation, while in swine glucose is the principal energy source. The volatile fatty acids (VFA) have a greater heat increment than glucose when utilized for production. Feedlot cattle have no great problem in eliminating heat under moderate conditions. Cattle on full feed can withstand extremely cold temperatures, due largely to their high heat production. At high ambient temperatures, cattle must balance heat loss with heat production. This may be done by physical processes such as providing shade and air movement and by nutrition. Physical means regulating heat loss may involve radiation, conduction, convection, and evaporation.

At 16°C the majority of heat loss is by convection and radiation. At 38°C essentially all of the heat loss is by evaporation (Bond et al., 1952). In hot, arid areas with low humidity, evaporation is an important means of cooling cattle. In warm climates, means must be utilized to take advantage of heat loss by evaporation; the most common method is well-designed shade. Without shade in hot climates, cattle are subject to heavy solar heat loads, as well as their own heat production. Properly designed shade for cattle may reduce the radiant heat load by more than 50% (Ittner et al., 1955). Such protection usually results in additional weight gain and improved feed efficiency. Hale (1978) cited studies done in the Imperial Valley in California in summer months in which shade increased gains by 40% and decreased feed requirements by 14% with cattle maintained in wire corrals. Daily gains were better with wire corrals than with wood corrals in shade; presumably this was due to differences in air movement.

Lippke (1975) utilized four fistulated steers at 21 and 32°C and 70% relative humidity to determine the effect of temperature on dry matter and fiber digestibilities, rectal temperature, respiration rate, and heart rate; the results are presented in Table 16.1. The steers experienced considerable heat stress, as indicated by the increased rectal temperature and respiration rate and the decreased heart rate. However, dry matter and neutral detergent solubles digestibilities were greater at the high ambient temperatures even though feed intake was held constant at 2.3% of body weight. Drastic depression in feed intake by cattle under serious heat stress was established by Ragsdale et al. (1948). Reduced intake, in turn, results in increased digestibility of diets (Brown, 1961).

TABLE 16.1.

Digestion Coefficients, Rectal Temperature, Respiration Rate, and Heart Rate for Steers Maintained at 21 or 32°C Ambient temperature[a]

| | Steers | |
|---|---|---|
| Factor | 21°C | 32°C |
| Digestibility, % | | |
| Dry matter | 62.2[b] | 67.5[c] |
| Neutral detergent fiber | 47.9 | 54.3 |
| Acid detergent fiber | 44.2 | 51.1 |
| Neutral detergent solubles | 71.8[d] | 75.8[e] |
| Rectal temperature (°C) | 38.7[d] | 40.3[e] |
| Respiration rate (min) | 26[d] | 115[e] |
| Heart rate (min) | 82[d] | 69[e] |

[a]From Lippke (1975).
[b-e]Different letter superscripts indicate differences ($P < .1$).

## D. Effect of Shade

Shade is not used in many feeding areas in which there are moderately high daytime temperatures, since nighttime temperatures are low and will dissipate the heat load. Roman-Ponce et al. (1977), working in Florida, found that the use of sheet metal shade that was painted on top and insulated on the underside with insulation board gave a 10.7% increase in the milk production of dairy cows compared to that of cows with no shade. Efficiency of metal roofs as radiation shields for cattle was improved by painting the topside white and the underside black (Bond et al., 1958), by adding insulation (Daniel et al., 1973), and by sprinkling the roofs with water (Brown et al., 1973).

## E. Effect of Sprinkling and Refrigeration

Morrison et al. (1973) sprinkled cattle under shades at temperatures above 27°C for 1 min every 30 min and found a significantly higher feed consumption and rate of gain but only slightly improved feed efficiency compared to corresponding cattle not sprinkled. Sprinkling was as effective as a refrigerated barn at 24°C in one trial and more effective in a second trial. Sprinkling and refrigeration provided more comfort to the cattle, as shown by the lower respiratory rates and body temperatures observed in the afternoon. Space per animal is important at high temperatures; both noncooled and cooled cattle consumed more feed and gained more with 3.7 m$^2$ per head than with 1.86 m$^2$ per head of space. Sprinkling would be expected to be more effective on slotted floors than on ground feedlots. Fox et al. (1977) estimated that deep mud in the feedlot increased net energy (NE) requirement for maintenance (NE$_m$) of cattle by 30%.

## F. Season and Eating Habits of Cattle

Ray and Roubicek (1971) observed that cattle ate more in late afternoon or early morning during the summer, but during the winter ate more during daylight hours. Hale (1978) reported that cattle in Arizona consumed about two-thirds of their daily feed intake from 5 p.m. to 8 a.m. during hot weather. He concluded that at such periods cattle should not receive the feed in the morning, since it would be subject to heat during the day and would not be as acceptable as fresh feed in the late afternoon. Hale (1978) compared winter and summer feeding of diets to steers at different roughage levels. Feed requirements were increased by 5% during the summer, but were 15% higher with 20% roughage diets than with 10% roughage diets.

## G. Effect of Roughages on Heat Increment

The heat increment is a greater percentage of the metabolizable energy (ME) for roughages than for concentrates, but on a kilogram-to-kilogram basis of comparison, concentrates produce more total calories of heat increment than roughages. For this reason, it is unwise to add roughages at the expense of concentrates during cold stress.

Difference trials with concentrates such as corn or barley show that NE is the sum of the NE contents of the digestible protein, fat, and carbohydrates contained in them. However, when a roughage such as barley straw is added to the basal diet, the effect is much less than the sum of the NE contents of the digestible nutrients in the roughage. For a given amount of DE, roughages have a greater heat increment than concentrates have. The heat increment of roughages is directly proportional to the crude fiber content of the diet, and each gram of crude fiber decreases NE by 1.36 kcal (Kleiber, 1975). Kellner (Kleiber, 1975) suggested that half of the effect of crude fiber resulted from action by fermentation in the rumen and that the other half was due to the work of chewing.

## H. Feeding Dairy Cattle in Warm Weather

High temperatures are responsible for heat stress in dairy cows and affect milk yield, especially in mid-lactation (Maust *et al.*, 1972). The authors observed that early-stage milking cows consumed less feed energy than mid-state ones during summer heat stress (15.0 versus 24.0 Mcal of estimated NE energy per day) but had higher milk production. This indicates that during early lactation, cows utilize body reserves rapidly to offset thermal stress and maintain peak milk production. In a review on heat stress, Bianca (1965) concluded that most studies indicated that the decrease in feed consumption with high temperatures accounted for decreased yields of milk.

Kelley *et al.* (1967) reported that total VFA decreased from 153 to 66 mEq per liter of rumen fluid as the temperature increased from 4 to 38°C with nonlactating Holstein cows. Lactic acid increased significantly in the rumen fluid of lactating Holstein cows fed a controlled feed intake at 40°C compared to that at 19°C (Mishra *et al.*, 1969).

## VII. EFFECT OF COLD

Records of 1970 steers fed in Saskatchewan, Canada, over a 7-year period were reviewed by Milligan and Christison (1974). As shown in Table 15.2, it was found that during December, January, and February, with a mean temperature of −17°C, daily gain was 30% lower, feed per unit of gain was 49% higher, and ME per unit of gain was 40% higher than the mean for March to November. Daily gain was correlated with ambient temperature, days below −23°C, wind chill, and dew point. Feed intake was lower in midwinter than in the fall. Lower production and greater energy needs suggest that dietary adjustments should be helpful during very cold temperatures.

TABLE 16.2.

Effect of Season of the Year on Steer Performance in a Cold Climate[a]

| Factor | Mar.–May | June–Aug. | Sept.–Nov. | Dec.–Feb. | Mean |
|---|---|---|---|---|---|
| No. of steers | 1320 | 1226 | 1976 | 2692 | — |
| Average wt., kg | 384 | 374 | 430 | 420 | 403 |
| Daily gain, kg | 1.33 | 1.50 | 1.59 | 1.03 | 1.36 |
| Predicted gain, kg | 1.14 | .99 | 1.22 | 1.04 | 1.10 |
| Feed/day, kg | 9.17 | 8.03 | 10.68 | 8.91 | 9.20 |
| Feed/gain | 7.17 | 5.68 | 7.21 | 9.75 | 7.46 |
| ME/kg gain, Mcal | 18.0 | 14.4 | 17.5 | 23.2 | 18.32 |

[a]From Milligan and Christison (1974).

Climatic cold stress has been the subject of several reviews (Thompson, 1973; Monteith and Mount, 1974; Young, 1981). In order to ascertain the effect of adverse climate on finishing cattle kept outside, Young (1981) summarized data accumulated over several years from various research centers located in the Great Plains area of North America. Included are data on steers weighing 350–500 kg, fed 0–90% grain diets, and not involved in experiments in which performance was different from that of control groups. The summary data emphasize the marked seasonal spreads in the rate of gain and feed efficiency of steers fed in cooler regions.

In Canada, ME requirements for wintering beef cows outside have been found to increase by 30–70% due to adverse climatic conditions (Jordan et al., 1968; Hironoka and Peters, 1969; Young and Berg, 1970; Lister et al., 1972), and cows are usually fed over winter to maintain weight or to allow some loss in body weight. The effect of cold stress on beef cows appears to be due most to increases in ME requirements, since conceptus development and calf weight are not affected (Wiltbank et al., 1962; Hironoka and Peters, 1969). If energy intake is grossly inadequate or protein is deficient, there may be a greater incidence of weak calf syndrome at calving in the spring (Bull et al., 1978). Wiltbank et al. (1962) reported that cows in thin body condition have reduced lactation potential and delays in rebreeding. Thompson et al. (1981) determined the body composition of cows at the initiation and termination of a winter feeding trial in Minnesota. They observed that fatter cows had a lower ME requirement than thinner Angus-Hereford crosses and Angus cows, but the opposite occurred with Angus-Holstein cross cows.

Shelter during inclement weather generally improves the rate of gain of feedlot cattle and reduces the ME requirements of cows. However, Leu et al. (1977) compared the performance of finishing steers in Iowa over a 4-year period and found that those in confinement, in open lots with shelter, or in open lots without shelter consumed 6.29, 6.67, and 6.67 kg per day of grain, respectively. Steers in open lots with access to shelter gained 1.22 kg, those without shelter gained 1.14 kg, and those in confinement gained 1.12 kg/day, with no difference in feed efficiency.

McDonald and Bell (1958) observed that the milk yields of dairy cows fed ad libitum began to decline at $-4°C$ and that a marked decrease in yield occurred at $-23°C$. Since appetite is increased by cold, increased feed intake may delay decreases in milk yield, but increased feed intake with no corresponding increase in milk yield, as usually occurs in very cold weather, will reduce feed efficiency.

In addition to air temperature, other components of the environment such as air movement, precipitation, and radiation are factors in cold stress. Wind and moisture destroy the insulative value of the hair coat and increase surface heat loss. Ames and Insley (1975) studied the wind-chill factor for cattle and sheep at temperatures ranging from $-23$ to $2°C$ and at wind velocities ranging from 0 to 56 km/hr. They concluded that the wind-chill factor for winds above 40 km/hr had a cubic function for animals with hair and wool compared to a quadratic function for humans. High winds destroy the external insulation of animals.

## VIII. HOUSING AND MANAGEMENT

High-producing animals whose nutrient needs have been met still have weather constraints. Selection of housing and management requires consideration of alternatives and evaluation of the effects of (1) no change, (2) a change in the management of animals, (3) provision or modification of housing to alter the effect of weather, or (4) combined alterations in management and housing to limit weather effects (Hahn, 1981).

Williams *et al.* (1981) worked with bull calves reared on a once-daily milk replacer system from 6 to 72 days of age in four naturally ventilated, unheated calf houses. The houses had differences in design and degree of insulation, and produced significant differences in house temperature, humidity, and air movement, but had no measurable effect on daily gains. They concluded that naturally ventilated, unheated calf houses were suitable for rearing calves in southwest Scotland.

Adverse environments can be imposed by livestock structures designed for purposes related to laborsaving and waste management. Expending resources to improve the climatic environments for livestock may be ineffective if such factors as the genetic potential of animals, nutrition, and disease control are not adequately considered.

# 17

# Composition of Feeds for Ruminants

## I. INTRODUCTION

Studies have defined the nutrient requirements of ruminant livestock, and from this information, diets can be formulated from feedstuffs and supplemental ingredients that should give healthy and productive cattle and sheep. Preston (1980) discussed variations in the composition of feeds for cattle and sheep, and emphasized that feeds are not of constant composition and that one can expect crude protein, crude fiber, ether extract, acid detergent fiber (ADF), and neutral detergent fiber (NDF) to vary by as much as ± 15%, mineral constituents by as much as + 30%, and energy values by at least 10% from those listed in composition tables. Obviously, chemical analysis of ingredients should be made prior to formulating diets, if practical. Composition tables give values that serve as guidelines in meeting the dietary requirements of ruminants (NRC, 1975, 1976a).

## II. DRY MATTER

Dry matter in feeds can vary greatly due to varying moisture levels and accounts for much of the variation in the composition of diets on an as-fed basis. Since the dry matter content of the diet is a principle factor in regulating total feed intake, dietary formulation on the dry matter basis is more sound than on the as-fed basis.

## III. PROTEIN

Crude protein values in composition tables are calculated from Kjeldahl nitrogen determinations of feedstuffs, followed by multiplication of the nitrogen by 6.25, based on the average nitrogen content of protein (16%). Considerable nonprotein nitrogen (NPN) is present in some feedstuffs, and this is included in a crude protein value. Digestible protein has been included in many composition tables, but Preston (1980) concluded that because of the large contribution of body protein to the apparent protein in the feces, a digestible protein value is more misleading than that of crude protein. Metabolizable protein and net protein systems offer promising guidelines for the utilization of dietary nitrogen.

## IV. CRUDE FIBER AND NDF

Crude fiber is declining in popularity as a measure of low-digestible material after many years of use. A problem with crude fiber is that variable amounts of indigestible lignin are removed from feeds by the crude fiber procedure. In the crude fiber analysis, the material extracted is called *nitrogen-free extract*. This has been considered more digestible than crude fiber even though many feeds have been shown to have a greater digestibility for crude fiber than for nitrogen-free extract. In determining total digestible nutrients (TDN) by the proximate analysis method, crude fiber needs to be evaluated as well as crude protein, ether extract, and nitrogen-free extract. Newer techniques for the determination of cell walls (NDF) and cell contents promise better correlations with high-roughage diets and animal performance. NDF is related to voluntary feed intake and net energy (NE) of diets. If TDN is replaced by NE values for maintenance ($NE_m$) and gain ($NE_g$) and NDF evaluations, there will be little use for crude fiber determinations of feedstuffs.

## V. ENERGY

Four measures of energy values, i.e., TDN, digestible energy (DE), $NE_m$, and $NE_g$, are commonly presented in feed composition tables. Preston (1980) stated that TDN is presented simply because there are more TDN values for feeds, and this is standard practice in expressing the energy values of feeds for cattle and sheep. The principal problem with TDN is that there is a poor relationship between crude fiber and nitrogen-free extract digestibility in certain feeds, and it overestimates the value of roughages compared to concentrates. Many studies demonstrate a constant relationship between TDN and DE, and there are approximately 4.4 Mcal of DE per kilogram of TDN. The ability of TDN and DE to predict animal performance is exactly the same. Emphasis on NDF as a replacement for crude fiber and the use of the bomb calorimeter to measure DE directly should gradually decrease the use of TDN in diet formulation.

Development of the California NE system has stimulated the use of NE in feed formulation due to improved predictability of results, depending on whether the feed energy is being utilized for maintenance ($NE_m$) or the growth ($NE_g$). A problem in using these NE values is predicting feed intake and therefore the proportion of diet used for maintenance or production. Some diet formulators use only $NE_g$ values. It should be apparent that this practice suffers from the equal but opposite criticism mentioned for TDN, i.e., $NE_g$ alone will overestimate the feeding value of concentrates relative to roughages.

Other diet formulators use the average of $NE_m$ and $NE_g$ values, which would be satisfactory only when cattle or sheep are eating twice the maintenance requirement. Preston (1980) recommended that a more accurate result would be achieved by using the $NE_m$ value and a multiplier times the $NE_g$ value, all divided by 1 plus the multiplier. The multiplier value is the level of feed intake above maintenance relative to maintenance. For example, if 700-lb cattle are expected to eat 18 lb of feed, 8 lb of which will be required for maintenance, then the NE value of the diet would be:

$$NE = \frac{NE_m + 10/8\ NE_g}{1 + 10/8}$$

This value for NE can be introduced into the computerized programs designed to formulate diets and predict performance. Shirley (1980) applied an $NE_g : NE_m$ ratio to facilitate calculations for an animal's expected rate of gain on a diet using the NE system.

## VI. MINERALS

As the level of concentrate in the diet is increased, potassium generally decreases, since grains are lower in potassium than forages. Also, potassium is not supplied when NPN as urea is substituted for natural plant or animal protein. Sulfur also becomes increasingly deficient as NPN is increased or replaces natural protein in diets. Many trace elements, such as iodine and selenium, are limited in feedstuffs by their level in the soil on which the plants are grown. Where there are deficiencies, trace mineralized salt and premixes are generally utilized to supplement diets.

## VII. VITAMIN A

Feeding vitamin A or carotene to cattle and sheep in supplements is generally dependent on the maturity and length of storage of feedstuffs. For economic reasons, it is probably not wise to depend entirely on harvested feeds for the vitamin, and green forages will probably meet the animals' requirements. However, in Florida, Burns *et al.* (1973) reported that the pregnancy rate of 2-year-old heifers was 80.9% compared to 65.4% due to injection of 6 million IU of vitamin A at 6-month intervals from weaning when grazing green pastures. Older cows did not respond to the vitamin. Meacham *et al.* (1970) gave a single injection of 6 million IU of vitamin A to cows during late gestation and observed a significant increase in calf survival to weaning.

Chapman *et al.* (1971), during a finishing trial on summer pasture, injected steers with 25,000 IU of vitamin A at 28-day intervals and obtained increased gains of 0.19 kg/day over controls. However, a similar treatment with steers on winter pasture had no benefit, presumably due to high storage in the liver during the previous summer while grazing. Perry *et al.* (1967) compared oral and injected vitamin A for finishing calves in a 210-day trial and observed significant increases in gains and feed efficiency from both methods of administering the vitamin.

## VIII. TABLES ON THE COMPOSITION OF FEEDSTUFFS

Composition data are presented for the protein, energy, minerals, and vitamins of common feedstuffs consumed by cattle (NRC, 1976a). More complete tabulation of data on the composition of feedstuffs is presented in the "Atlas of Nutritional Data on United States and Canadian Feeds" (NRC, 1971b) and in the book entitled "Feeds and Nutrition" by Ensminger and Olentine (1978). Separate tables are presented by Ensminger and Olentine (1978) on energy feeds (high in energy, under 18% fiber, and generally less than 20% protein), protein feeds (more than 20% protein or protein equivalent), dry forages (low in weight per unit volume, relatively low in energy, and more than 18% crude fiber in the dry state), silages and haylages (ensilage is fermentable, high-moisture forage stored under anaerobic conditions in a silo, and chopped; when stored, it contains 65–70% moisture; haylages are low-moisture silages made from grasses and/or legumes wilted to 40–55% moisture content before ensiling), pasture and range plants (including grass, browse, and other plants harvested by grazing ruminants), mineral supplements (including natural and synthetic compounds needed for certain essential body functions), vitamin supplements (including rich synthetic and natural feed sources of vitamins needed by animals for health and good performance), and amino acids (gives known amino acid composition values of certain feeds).

# Glossary

**Absorbed energy**  The heat of combustion of compounds absorbed from foods that pass through the walls of the intestinal tract.

**Apparent digestibility**  The amount of a nutrient ingested minus the amount excreted in the feces expressed as a percentage of the intake (regardless of whether the fecal material is from the feed or of endogenous origin):

$$\text{Apparent digestibility} = \frac{\text{amount in feed} - \text{amount in feces}}{\text{amount in feed}}$$

This value is often termed the *digestibility coefficient*.

**Apparent digestible energy**  The gross energy (heat of combustion) of a feedstuff minus the gross energy of the feces, expressed as megacalories or kilocalories per kilogram of total feed.

**Available protein**  The amount of truly digestible protein in the feed less the amount of that protein required to equal the loss of nitrogen as metabolic fecal nitrogen (MFN) as:

Available protein (%) = apparent digestible crude protein (%) − 6.25 MFN (100/BV − 1)

**Biological value (BV)**  The percentage of the truly digestible nitrogen of feed protein that is retained and used for maintenance, growth, milk production, and pregnancy as:

$$BV = 100 \times \frac{1 - (F - MFN) - (U - EUN)}{1 - (F - MFN)}$$

$$= 100 \times \frac{NR + MFN + EUN}{1 - (F - MFN)}$$

where $1$ = feed intake of nitrogen, F = total fecal nitrogen, MFN = metabolic fecal nitrogen, U = total urinary nitrogen, EUN = endogenous urinary nitrogen, and NR = nitrogen retention, all in grams per day.

**British thermal unit**  The heat required to raise the temperature of 1 lb of water at its maximum density 1°F and is equal to about 232 calories.

**Calorie**  The amount of heat required to raise the temperature of 1g water through 1°C from 14.5 to 15.5°C. The calorie is equal to 4.1852 international joules.

**Centigrade temperature (°C)**  A scale with the ice point at 0° and the boiling point of water at 100°; now frequently called *Celsius temperature scale*. Centigrade is converted to Fahrenheit by the formula:

$$C = 5/9 \ (°F - 32)$$

**Crude metabolizable energy** The sum of the metabolizable energy values of the various components of the diet, each determined at the maintenance level of feeding, is the crude metabolizable energy. At higher levels of feeding, there is a decrease in the proportion of the gross energy of a food that is metabolizable, and the crude metabolizable energy of a diet needs to be corrected to give the intake of true metabolizable energy.

**Crude protein** Usually taken as the total nitrogen multiplied by 6.25, since protein on the average contains 16% nitrogen.

**Density** Concentration of a substance as measured by the mass per unit volume.

**Digestibility coefficient:** The same as percentage of apparent digestibility.

**Efficiency of utilization of metabolizable energy (K)** The increase in energy retention resulting from an increase of 100 kcal in the intake of metabolizable energy as:

$$K = 100 \times \frac{\text{increment in energy retention}}{\text{increment in intake of metabolizable energy}}$$

**Equilibrium, chemical** Occurs when a chemical reaction and its reverse reaction are taking place at equal velocities in such a manner that concentrations of reacting substances remain constant.

**Endogenous loss or excretion** Is a nutrient loss from tissues even though the nutrient is absent from the diet.

**Energy** The capability of doing work, quantitatively expressed as joules, calories, or ergs. An erg is the amount of energy expended when a force of 1 dyn acts through a distance of 1 cm.

**Energy retention** Determined as the heat of combustion of the feed intake minus the heats of combustion of feces, urine, and gaseous products of digestion, and heat production. This is equivalent to the metabolizable energy intake minus heat production. When energy is being stored in the body, energy retention is positive.

**Fahrenheit temperature (°F)** A scale with the ice point at 32° and the boiling point of water at 212°. Fahrenheit is converted to Celsius (or Centigrade) by the formula:

$$F = 9/5 \, (°C + 32)$$

**Feeding level** Amount of diet intake expressed as a multiple of the quantity of the same diet required for maintenance.

**Formula, chemical** A combination of symbols with subscripts representing the constitutents of a substance and their proportions by weight.

**Gas** Matter with neither shape nor volume and in which the molecules are practically unrestricted by cohesive forces.

**Gaseous products of digestion** Consist principally of methane, carbon dioxide, and hydrogen and, under some conditions, small amounts of hydrogen sulfide.

**Gross energy** In a diet, excreta, or other products, is equivalent to the heat of combustion.

**Heat** Energy transferred by a chemical reaction. Heat can be measured in terms of calories, joules, or ergs.

**Heat increment** The heat production increase that results from an increase in the intake of metabolizable energy.

**Heat of combustion** The amount of heat produced by the combustion of 1 molecular weight in grams of a substance.

**Heat of fermentation** Energy lost as heat as a result of fermentation in the rumen.

**International system of units** A metric system of units based on the kilogram, joule, meter, liter, and Centigrade temperature.

**Ion** An atom or group of atoms that are not electrically neutral, but carry a positive or negative charge. Positive ions occur when neutral atoms or molecules lose valence electrons, and negative ions occur when they gain electrons.

# Glossary

**Ionophores** Substances like sodium chloride, rumensin, and lasalocid, which exist only as ionic lattices in the pure crystalline form and, when dissolved in a appropriate solvent, give conductances that change according to some fractional power of the concentration. Their solutions have no neutral molecules that can dissociate, but they may have associated ions.

**Joule** A fundamental unit of energy in the International system of energy measurement; 1 joule = $10^7$ ergs = 0.239 calorie.

**Kilocalorie (kcal)** One thousand calories.

**Kilogram** The mass of a particular cylinder of platinum-iridium alloy, called the *International prototype kilogram,* preserved in a vault at Seures, France. It is equal to 2.205 lb.

**Liquid** A condition of matter in which the molecules are relatively free to change their positions with respect to each other, but are restricted by cohesive forces that maintain them in a relatively fixed volume.

**Logarithm** A number $m$ related to another number $p$ as $B^m = p$, where $B$ is larger than 1. Logarithms have the basic properties that $\log pq = \log p + \log q$ and $\log p^q = q \log p$. In systems commonly used, $B = 10$ (common logarithms) or $B = e = 2.71828$ (natural logarithms).

**Maintenance** The animal's requirement for a nutrient for normal vital body processes, including replacement of endogenous fecal and urinary losses and losses from the skin. It is sufficient so that the net gain or loss of the nutrient from the tissues is zero.

**Mass** Quantity of matter, and units of mass including the gram, which is 1/1000th the quantity of matter in the International prototype kilogram. In the United States, the standard mass is the avoirdupois pound and is defined as 1/2.20462 kg.

**Megacalorie (Mcal)** One million calories or 1000 kcal.

**Metabolic fecal nitrogen (MFN)** The nitrogen excreted in the feces other than the undigested feed nitrogen.

**Metabolizability of the gross energy of a food (Q)** The metabolizable energy of a food expressed as the percentage of its gross energy; $Q_m$ denotes the $Q$ value determined at the maintenance level of feeding.

**Metabolizable energy (ME)** In a feed, the heat of combustion of the feed minus the heats of combustion of the feces, urine, and gaseous products of digestion. At the maintenance level of feeding, ME by definition equals heat production.

**Mixtures** Two or more substances intermingled, with no constant percentage composition. Each component retains its essential original properties.

**Molecular weight** The sum of the atomic weights of all of the atoms in a molecule.

**Net requirement** The quantity of energy or another nutrient that should be absorbed in order to meet the animal's needs for maintenance, including replacement of endogenous losses and amounts needed for given rates of growth or for reproduction.

**Nitrogen retention** The net gain or loss of nitrogen by an animal; can be positive or negative. It is the intake of an animal minus the nitrogen lost in the feces and urine and from the animal's external surfaces.

**Oxidation** Loss of electrons by an element or substance. It is any process that increases the proportion of oxygen in a compound.

**Pound** A unit of mass equal in the United States to 0.4536 kg.

**Reduction** A gain in electrons by an element or substance. It is also defined as any process that increases the proportion of hydrogen in a compound.

**Relative humidity** The ratio of the amount of water vapor present in the atmosphere to the quantity that would saturate the atmosphere at the existing temperature.

**Requirement** The requirement for energy or other nutrient is the amount that must be supplied in the diet to meet the needs of the animal fed an otherwise adequate diet.

**Salt** A substance that yields ions other than hydrogen or hydroxyl ions and may be obtained by displacing the hydrogen of an acid by a metal.

**Specific gravity** The ratio of the mass of a body to the mass of an equal volume of water at 4°C or other specified temperature.

$$\text{Specific gravity of a beef carcass} = \frac{\text{wt of carcass in air, kg}}{\text{wt of carcass in air} - \text{wt of carcass in water}}$$

**Starch equivalent** The number of pounds of starch that in an adult steer would promote the accumulation of the same amount of fat as 100 lb of the feedstuff being tested.

**Temperature** The condition of a substance that determines the transfer of heat to or from other bodies. The customary unit of temperature is the degree Centigrade, which is 1/100th of the difference between the temperature of melting ice and that of boiling water under atmospheric pressure. The degree Fahrenheit is 1/180th of the same difference in temperature.

**Total digestible nutrients (TDN)** The TDN value of a diet is calculated from the percentages of apparently digestible nutrients as follows: TDN per 100 lb of feed = % of apparently digestible crude protein plus % of apparently digestible nitrogen-free extract + % of apparently digestible crude fiber plus 2.25 (% of apparently digestible crude fat). The Weende scheme of analysis is used in the above evaluation of a feedstuff.

**True digestibility** The amount of energy or nutrient in a feed minus the amount in the residue of the feed excreted in the feces, expressed as the percentage of the intake as:

$$\text{True digestibility} = \frac{\text{amount in feed} - (\text{amount in feces} - \text{metabolic fecal loss})}{\text{amount in feed}}$$

**Volume** The basic unit is the cubic centimeter (CC), or the volume of a cube whose edges are 1 cm in length. A liter is 1000 cc or 1000 ml, and a quart equals 980 ml.

**Voluntary intake** The intake of feed by an animal when provided an excess of the feed.

**Water requirement** The water from all sources, including free water and water in feedstuff, as well as water from metabolism, necessary to maintain the water balance.

**Weende scheme of analysis** A system of analysis in which the amounts of water, ash, crude protein, crude fiber, and fat are determined by standard methods, and the remainder is calculated by difference and is called the *nitrogen-free extract*.

**Weight** The product of the mass of a body and its acceleration due to gravity. A given mass has different weights at different altitudes.

**Weights and volumes**
  1 kilogram (kg) = 1000 grams (g) = 2.205 pounds (lb)
  1 pound (lb) = 453.6 grams (g) = 16 ounces (oz)
  1 ounce (oz) = 28.35 grams (g)
  1 gallon (gal) = 10.0 pounds (lb) = 4.536 kilograms (kg) = 4.536 liters

# Bibliography

Aberle, E. D., Reeves, E. S., Judge, M. D., Hunsley, R. E., and Perry, T. W. (1981). *J. Anim. Sci.* **52,** 757.
Abou Akkada, A. R., and Blackburn, T. H. (1963). *J. Gen. Microbiol.* **31,** 461.
Abou Akkada, A. R., and Howard, B. H. (1960). *Biochem. J.* **76,** 445.
Abou Akkada, A. R., Hobson, P. N., and Howard, B. H. (1959). *Biochem. J.* **73,** 44P.
Abrams, S. M., Klopfenstein, T. J., Stock, R. A., Britton, R. A., and Nelson, M. L. (1983). *J. Anim. Sci.* **57,** 729.
Adams, D. C., Galyean, M. L., Kiesling, H. E., Wallace, J. D., and Finkner, M. D. (1981). *J. Anim. Sci.* **53,** 870.
Adeeb, N. N., Wilson, G. F., and Campling, R. C. (1971). *Anim. Prod.* **13,** 391 (Abstr.)
Adolph, E. F. (1933). *Physiol. Rev.* **13,** 336.
Adriano, D. C., Pratt, P. F., and Bishop, S. E. (1971). *Soil Sci. Soc. Am. Proc.* **35,** 759.
Agricultural Research Council (1965). "Nutrient Requirements of Farm Animals, No. 2. Ruminants," Technical Reviews and Summaries. ARC, London.
Agricultural Research Council (1980). "The Nutrient Requirements of Ruminant Livestock," Technical Review by an Agricultural Research Council Working Party. ARC, Farnham Royal, England.
Ahrens, F. A. (1967). *Am. J. Vet. Res.* **28,** 1335.
Aimone, J. C., and Wagner, D. G. (1977). *J. Anim. Sci.* **44,** 1088.
Aines, G. E., Lamm, W. D., Webb, K. E., Jr., and Fontenot, J. P. (1982). *J. Anim. Sci.* **54,** 504.
Aitchison, T. E., Mertens, D. R., McGilliard, A. D., and Jacobson, N. L. (1976). *J. Dairy Sci.* **59,** 2056.
Allden, W. G. (1968). *Australian J. Agric. Res.* **19,** 997.
Allden, W. G. (1980). *In* "Grazing Animals" (F. H. W. Morley, ed.), p. 289. Elsevier, Amsterdam.
Allen, J. D., and Harrison, D. G. (1979). *Proc. Nutr. Soc.* **38,** 32A (abstr.).
Allison, M. J. (1970). Nitrogen Metabolism of Ruminal Micro-organisms. *In* "Physiology of Digestion and Metabolism in the Ruminant" (A. T. Phillipson, ed.), p. 456. Oriel Press, Ltd., Newcastle upon Tyne, England.
Allison, M. J., Bryant, M. P., and Doetsch, R. N. (1958). *Science* **128,** 474.
Allison, M. J., Dougherty, R. W., Bucklin, J. A., and Snyder, E. E. (1964). *Science* **144,** 54.
Al-Rabbat, M. F., Baldwin, R. L., and Weir, W. C. (1971). *J. Dairy Sci.* **54,** 1150.
Alsmeyer, R. H., Palmer, A. Z., Wallace, H. D., Shirley, R. L., and Cunha, T. J. (1955). *J. Anim. Sci.* **14,** 1226 (abstr.).
Altona, R. E. (1971). *J. S. Afr. Vet. Med. Assoc.* **42,** 45.
Altschul, A. M., Lyman, C. M., and Thurber, F. H. (1958). "Processed Plant Protein Foodstuffs." Academic Press, New York.

# Bibliography

American Association of Analytical Chemists (1975). "Official Methods of Analysis," 12th ed. AAOAC, Washington, D. C.
Ames, D. R., and Insley, L. W. (1975). *J. Anim. Sci.* **40,** 161.
Ammerman, C. B., and Block, S. S. (1964). *Agric. Food Chem.* **12,** 539.
Ammerman, C. B., van Walleghem, P. A., Easley, J. F., Arrington, L. R., and Shirley, R. L. (1963). *Proc. Fla. State Hortic. Soc.* **76,** 245.
Ammerman, C. B., Harms, R. H., Dennison, R. A., Arrington, L. R., and Loggins, P. E. (1965a). *Bull.—Fla., Agric. Exp. Stn,* **691.**
Ammerman, C. B., Hendrickson, R., Hall, G. M., Easley, J. F., and Loggins, P. E. (1965b). *Proc. Fla. State Hortic. Soc.* **78,** 307.
Ammerman, C. B., Waldroup, P. W., Arrington, L. R., Shirley, R. L., and Harms, R. H. (1966). *Agric. Food Chem.* **14,** 279.
Ammerman, C. B., Verde, G. J., Moore, J. E., Burns, W. C. and Chicco, C. F. (1972). *J. Anim. Sci.* **35,** 121.
Amos, H. E., and Evans, J. (1976). *J. Anim. Sci.* **43,** 861.
Amos H. E., Little, C. O., Ely, D. G., and Mitchell, G. E., Jr. (1971). *Can. J. Anim. Sci.* **51,** 51.
Amos, H. E., Burdick, D., and Huber, T. A. (1974). *J. Anim. Sci.* **38,** 702.
Anderson, G. D., Berger, L. L., and Fahey, G. C., Jr. (1981). *J. Anim. Sci.* **52,** 144.
Anderson, K. J. (1956). *Nature (London)* **177,** 96.
Anderson, M. J., Adams, D. C., Lamb, R. C., and Walters, J. L. (1979). *J. Dairy Sci.* **62,** 1098.
Anderson, P. C., and Anderson, K. L. (1977). *In* "Feed Composition, Animal Nutrient Requirements and Computerization of Diets" (P. W. Fonnesbeck, L. E. Harris, and L. C. Kearl, eds.), p. 306. Utah State University, Logan.
Andrews, R. P., Escuder-Volonte, J., Curran, M. K., and Holmes, W. (1972). *Anim. Prod.* **15,** 167.
Anthony, W. B. (1970). *J. Anim. Sci.* **30,** 274.
Anthony, W. B., Cunningham, J. R., Jr., and Harris, R. R. (1969). *In* "Cellulases and Their Applications" (Robert F. Gould, ed.), p. 315. Am. Chem. Soc., Washington, D.C.
Armentano, I. E., and Rakes, A. H. (1982). *J. Dairy Sci.* **65,** 390.
Armstrong, D. G. (1965). *In* "Physiology of Digestion in the Ruminant" (R. W. Dougherty, R. S. Allen, W. Burroughs, N. L. Jacobson, and A. D. McGilliard, eds.), p. 272. Butterworth, London.
Armstrong, D. G. (1969). *In* "Handbuch der Tierernahrung" (W. Lenkeit, K. Brierem, and E. Craseman, eds.), Vol. 1, p. 390. Parey, Berlin.
Armstrong, D. G., and Beever, D. E. (1969). *Proc. Nutr. Soc.* **28,** 121.
Armstrong, D. G., and Blaxter, K. L. (1957). *Br. J. Nutr.* **11,** 413.
Armstrong, D. G., and Blaxter, K. L. (1961). *Publ.—Eur. Assoc. Anim. Prod.* **10,** 187.
Armstrong, D. G., Blaxter, K. L., Graham, N. McC., and Wainman, F. W. (1958). *Br. J. Nutr.* **12,** 177.
Arndt, D. L., Day, D. L., and Hatfield, E. E. (1979). *J. Anim. Sci.* **8,** 157.
Arnett, W. D. (1963). Ph.D. Dissertation, Oklahoma State University, Stillwater.
Arthaud, V. H., Mandigo, R. W., Koch, R. M., and Kotula, A. W. (1977). *J. Anim. Sci.* **44,** 53.
Ash, R. W. (1956). *J. Physiol. (London)* **133,** 75P.
Ash, R. W. (1961). *J. Physiol. (London* **156,** 93.
Ashworth, A. (1969). *Br. J. Nutr.* **23,** 835.
Asplund, J. M., and Pfander, W. H. (1972). *J. Anim. Sci.* **35,** 1271.
Association of American Feed Control Officials (AAFCO) (1955). "Historical Sketch 1909-1955." P.O. Box 3160, College Station, Texas.
Aston, K., and Tayler, J. C. (1980). *Anim. Prod.* **31,** 243.
Atwal, A. S., Milligan, L. P., and Young, B. A. (1974). *Can. J. Anim. Sci.* **54,** 393.
Bae, D. H., Welch, J. G., and Smith, A. M. (1979). *J. Anim. Sci.* **49,** 1292.

## Bibliography

Bae, D. H., Welch, J. G., and Smith, A. M. (1981). *J. Anim. Sci.* **52,** 1371.
Baile, C. A. (1968). *Fed. Proc., Fed. Am. Soc. Exp. Biol.* **27**(6), 1361.
Bailey, C. B. (1961). *Br. J. Nutr.* **15,** 443.
Bailey, C. B., Hironaka, R., and Slen, S. B. (1962). *Can. J. Anim. Sci.* **42,** 1.
Bailey, R. W. (1958a). *J. Sci. Food Agric.* **9,** 743.
Bailey, R. W. (1958b). *J. Sci. Food Agric.* **9,** 748.
Baker, A. J. (1973). *J. Anim. Sci.* **36,** 768.
Baker, A. J., Mohaupt, A. A., and Spino, D. F. (1973). *J. Anim. Sci.* **37,** 179.
Baker, F. S., Jr., Palmer, A. Z., and Carpenter, J. W. (1973). *In* "Crossbreeding Beef Cattle" (M. Koger, T. J. Cunha, and A. C. Warnick, eds.), Ser 2, p. 277. Univ. of Florida Press, Gainesville.
Balch, C. C. (1950). *Br. J. Nutr.* **4,** 361.
Balch, C. C. (1971). *Br. J. Nutr.* **26,** 383.
Balch, C. C., and Campling, R. C. (1962). *Nutr. Abstr. Rev.* **32,** 669.
Balch, C. C., and Campling, R. C. (1965). *In* "Physiology and Digestion in the Ruminant" (R. W. Dougherty, R. S. Allen, W. Burroughs, N. L. Jacobson, and A. D. McGilliard, eds.), Butterworth, London. p. 108.
Balch, C. C., and Kelley, A. (1950). *Br. J. Nutr.* **4,** 395.
Balch, C. C., Kelley, A., and Heim, G. (1951). *Br. J. Nutr.* **5,** 207.
Balch, C. C., Balch, D. A., Johnson, V. W., and Turner, J. (1953). *Br. J. Nutr.* **7,** 212.
Balch, D. A., and Rowland, S. J. (1957). *Br. J. Nutr.* **11,** 288.
Baldwin, J. A., Hentges, J. R., Jr., Bagnall, L. O., and Shirley, R. L. (1975). *J. Anim. Sci.* **40,** 968.
Baldwin, R. L. (1968). *J. Dairy Sci.* **51,** 104.
Baldwin, R. L. (1970). *Am. J. Clin. Nutr.* **23,** 1508.
Baldwin, R. L., and Denham, S. C. (1979). *J. Anim. Sci.* **49,** 1631.
Baldwin, R. L., and Milligan, L. P. (1964). *Biochim. Biophys. Acta* **92,** 421.
Baldwin, R. L., Wood, W. A., and Emery, R. S. (1962). *J. Bacteriol.* **83,** 907.
Baldwin, R. L., Wood, W. A., and Emery, R. S. (1963). *J. Bacteriol.* **85,** 1346.
Baldwin, R. L., Wood, W. A., and Emery, R. S. (1965). *Biochim. Biophys. Acta* **97,** 202.
Baldwin, R. L., Koong, L. J., and Ulyatt, M. J. (1977). *Agric. Syst.* **2,** 255.
Bales, G. L., Kellog, D. W., and Miller, D. D. (1979). *J. Anim. Sci.* **49,** 1324.
Ballard, F. J., Hanson, R. W., and Kronfeld, D. S. (1969). *Fed. Proc., Fed. Am. Soc. Exp. Biol.* **28,** 218.
Banks, W., Clapperton, J. L., Ferrie, M. E., and Wilson, A. G. (1976). *J. Dairy Res.* **43,** 213.
Barcroft, J., McAnally, R. A., and Phillipson, A. T. (1944). *J. Exp. Biol.* **20,** 120.
Barnes, R. F., Muller, L. D., Bauman, L. F., and Colenbrander, V. F. (1971). *J. Anim. Sci.* **33,** 881.
Barnett, A. J. G. (1954). "Silage Fermentation." Academic Press, New York.
Barnicoat, C. R. (1945). *N. Z. J. Sci. Techol. Sect. A.* **27,** 202.
Barry, T. N. (1976a). *Proc. Nutr. Soc.* **35,** 221.
Barry, T. N. (1976b). *J. Agric. Sci.* **86,** 379.
Barth, K. M., Corrick, J. A., Shumway, P. E., and Coleman, S. W. (1974). *J. Anim. Sci.* **38,** 687.
Bartley, E. E. (1965). *J. Dairy Sci.* **48,** 102.
Bartley, E. E., and Cannon, C. Y. (1947). *J. Dairy Sci.* **30,** 555 (abstr.).
Bartley, E. E., and Meyer, R. M. (1967). *J. Anim. Sci.* **29,** 913 (abstr.).
Bartley, E. E., and Yadava, I. S. (1961). *J. Anim. Sci.* **20,** 648.
Bartley, E. E., Lippke, H., Pfost, H. B., Nijweide, R. J., Jacobson, N. L., and Meyer, R. M. (1965). *J. Dairy Sci.* **48,** 1657.
Bartley, E. E., Farmer, E. L., Pfost, H. B., and Dayton, A. D. (1968). *J. Dairy Sci.* **51,** 706.
Bartley, E. E., Barr, G. W., and Mickelsen, R. (1975a). *J. Anim. Sci.* **41,** 752.

Bartley, E. E., Meyer, R. M., and Fina, L. R. (1975b). *In* "Digestion and Metabolism in the Ruminant" (I. W. McDonald and A. C. I. Warner, eds.), p. 551. University of New England Publishing Unit, Armidale, N.S.W., Australia.
Bartley, E. E., Davidovich, A. D., Barr, G. W., Griffel, G. W., Dayton, A. D., Deyoe, C. W., and Bechtle, R. M. (1976). *J. Anim. Sci.* **43**, 835.
Bartley, E. E., Ibbetson, R. W., Chyba, L. J., and Dayton, A. D. (1978). *J. Anim. Sci.* **47**, 791.
Bartley, E. E., Herod, E. L., Bechtle, R. M., Sapienza, D. A., Brent, B. E., and Davidovich, A. (1979). *J. Anim. Sci.* **49**, 1066.
Bartley, E. E., Meyer, R. M., and Call, E. P. (1981). *J. Anim. Sci.* **52**, 1150.
Barton, J. S., Bull, L. S., and Hemken, R. W. (1971). *J. Anim. Sci.* **33**, 682.
Bass, J. M., Fishwick, G., and Parkins, J. J. (1981). *Anim. Prod.* **33**, 15.
Bassett, J. M. (1978). *Proc. Nutr. Soc.* **37**, 273.
Bastie, A. M. (1957). *Arch. Sci. Physiol.* **11**, 87.
Bauchop, T. (1967). *J. Bacteriol.* **94**, 171.
Bauchop, T., and Elsden, S. R. (1960). *J. Gen. Microbiol.* **23**, 457.
Beames, R. M. (1959). *Queensl. J. Agric. Sci.* **16**, 223.
Beames, R. M. (1960). *Proc. Aust. Soc. Anim. Prod.* **3**, 86.
Beaudouin, J., Shirley, R. L., and Hammell, D. L. (1980). *J. Anim. Sci.* **50**, 572.
Becker, B. A., Campbell, J. R., and Martz, F. A. (1975). *J. Dairy Sci.* **58**, 1677.
Becker, D. E., Adams, C. R., Terrill, S. W., and Meade, R. J. (1953). *J. Anim. Sci.* **12**, 107.
Becker, E. R., and Husing, T. S. (1929). *Proc. Natl. Acad. Sci. U.S.A.* **15**, 684.
Becker, R. B., and Arnold, P. T. D. (1951). *Circ.—Fla. Agric. Exp. Stn.* **S-40**.
Bedrak, E., Warnick, A. C., Hentges, J. F., and Cunha, T. J. (1969). *Bull.—Fla., Agric. Exp. Stn.* **678**.
Beede, D. K., and Farlin, S. D. (1977a). *J. Anim. Sci.* **45**, 385.
Beede, D. K., and Farlin, S. D. (1977b). *J. Anim. Sci.* **45**, 393.
Beeson, W. M. (1975). *Proc. Conf.—Distill. Feed Res. Counc.* **30**, 14.
Beeson, W. M., and Chen, M. C. (1976). *Proc. Conf.—Distill. Feed Res. Counc.* **31**, 8.
Beeson, W. M., and Perry, T. W. (1958). *J. Anim. Sci.* **17**, 368.
Beever, D. E., Coelho da Silva, J. F., and Armstrong, D. G. (1970). *Proc. Nutr. Soc.* **29**, 43A.
Beever, D. E., Coehlo da Silva, J. F., Prescott, J. H. D., and Armstrong, D. G. (1972). *Br. J. Nutr.* **28**, 347.
Beever, D. E., Thomson, D. J., Cammell, S. B., and Harrison, D. G. (1977). *J. Agric. Sci.* **88**, 61.
Bell, F. R. (1984). *J. Anim. Sci.* **59**, 1369.
Bellingham, F., and Bernstein, M. (1973). *J. Appl. Bacteriol.* **36**, 183.
Belyea, R. L., Martz, F. A., McIlroy, W., and Keene, K. E. (1979). *J. Anim. Sci.* **49**, 1281.
Bement, R. E. (1969). *J. Range Manage.* **22**, 83.
Bensadoun, A., Paladines, O. L., and Reid, J. T. (1962). *J. Dairy Sci.* **45**, 1203.
Bergen, W. G., and Yokoyama, M. T. (1977). *J. Anim. Sci.* **45**, 573.
Bergen, W. G., and Bates, D. B. (1984). *J. Anim. Sci.* **58**, 1465.
Bergen, W. G., Purser, D. B., and Cline, J. H. (1968a). *J. Anim. Sci.* **27**, 1497.
Bergen, W. G., Purser, D. B., and Cline, J. H. (1968b). *J. Dairy Sci.* **51**, 1698.
Bergen, W. G., Cash, E. H., and Henderson, H. E. (1974). *J. Anim. Sci.* **39**, 629.
Bergen, W. G., Black, J. R., and Fox, D. G. (1979). *Feedstuffs* **51**(8), 25.
Berger, L. L., Anderson, G. D., and Fahey, G. C. (1981a). *J. Anim. Sci.* **52**, 138.
Berger, L. L., Ricke, S. C., and Fahey, G. C., Jr. (1981b). *J. Anim. Sci.* **53**, 1440.
Bergman, E. N. (1973). *Cornell Vet.* **63**, 341.
Bergman, E. N., and Hogue, D. E. (1967). *Am. J. Physiol.* **213**, 1378.
Bergman, E. N., Roe, W. E., and Kon, K. (1966). *Am. J. Physiol.* **211**, 793.
Bertrand, J. E., and Dunavin, L. S. (1975). *J. Anim. Sci.* **41**, 1206.

Bertrand, J. E., Lutrick, M. C., and Dunavin, L. S. (1975). *Bull.–Fla. Agric. Exp. Stn.* **779** (Jay).
Bertrand, J. E., Lutrick, M. C., Breland, H. L., and West, R. L. (1980). *J. Anim. Sci.* **50,** 35.
Bertrand, J. E., Lutrick, M. C., Edds, G. T., and West, R. L. (1981). *J. Anim. Sci.* **53,** 146.
Bhalerao, V. R., Endres, J., and Kummerrow, F. A. (1961). *J. Dairy Sci.* **44,** 1283.
Bhattacharya, A. N., and Warner, R. G. (1968). *J. Dairy Sci.* **51,** 1481.
Bianca, W. (1965). *J. Dairy Res.* **32,** 291.
Bickerstaffe, R., Annison, E. F., and Linzell, J. L. (1974). *J. Agric. Sci.* **82,** 71.
Biddle, G. N., and Evans, J. L. (1973). *J. Anim. Sci.* **36,** 123.
Biddle, G. N., Evans, J. L., and Trout, J. R. (1975). *J. Nutr.* **105,** 1584.
Bines, J. A., and Davey, A. W. F. (1970). *Br. J. Nutr.* **24,** 1013.
Bines, J. A., and Hart, J. C. (1978). *Proc. Nutr. Soc.* **37,** 281.
Bishop, R. B. (1971). *J. Dairy Sci.* **54,** 1240 (abstr.).
Bishop, R. B., and Murphy, W. D., Jr. (1972). *J. Dairy Sci.* **55,** 711 (abstr.).
Black, A. L., Egan, A. R., Anand, R. S., and Chapman, T. E. (1968). "Isotope Studies on the Nitrogen Chain," p. 247. IAEA, Vienna.
Black, W. H., Quesenberry, J. R., and Baker, A. L. (1938). *U.S. Dep. Agric., Tech. Bull.* **603**.
Blackburn, T. H., and Hobson, P. N. (1960). *J. Gen. Microbiol.* **22,** 282.
Blair-West, J. R., Bott, E., Boyd, G. W., Coghlan, J. P., Denton, D. A., Goding, J. R., Weller, S., Wintour, M., and Wright, R. D. (1965). *In* "Physiology of Digestion in the Ruminant" (D. W. Daugherty, R. S. Allen, W. Burroughs, N. L. Jacobson, and A. D. McGilliard, eds.), p. 198. Butterworth, London.
Blaxter, K. L. (1962). "The Energy Metabolism of Ruminants." Thomas, Springfield, Illinois.
Blaxter, K. L., and Clapperton, J. L. (1965). *Br. J. Nutr.* **19,** 511.
Blaxter, K. L., and Graham, N. McC. (1956). *J. Agric. Sci.* **47,** 207.
Blaxter, K. L., and Wilson, R. S. (1962). *Anim. Prod.* **4,** 351.
Blaxter, K. L., and Wilson, R. S. (1963). *Anim. Prod.* **5,** 27.
Blaxter, K. L., Reineke, E. P., Carampton, E. W., and Petersen, W. E. (1949). *J. Anim. Sci.* **8,** 307.
Blaxter, K. L., Hutcheson, M. K., Robertson, J. M., and Wilson, A. L. (1952). *Br. J. Nutr.* **6,** i (abstr.).
Blaxter, K. L., Graham, N. McC., and Wainman, F. W. (1956). *Br. J. Nutr.* **10,** 69.
Block, E., Kilmer, L. H., and Muller, L. D. (1981). *J. Anim. Sci.* **52,** 1164.
Boda, J. M., Cupps, P. T., Colvin, H., and Cole, H. H. (1956). *J. Am. Vet. Med. Assoc.* **128,** 532.
Bohman, V. R., and Torell, C. (1956). *J. Anim. Sci.* **15,** 1089.
Bohman, V. R., Trimberger, G. W., Loosli, J. K., and Turk, K. L. (1954). *J. Dairy Sci.* **37,** 284.
Bohman, V. R., Wade, M. A., and Hunter, J. E. (1957). *J. Anim. Sci.* **16,** 833.
Bohman, V. R., Melendy, H., and Wade, M. A. (1962). *J. Anim. Sci.* **20,** 553.
Boling, J. A., Bradley, N. W., and Campbell, L. D. (1977). *J. Anim. Sci.* **44,** 867.
Bolsen, K. K., Woods, W., and Klopfenstein, T. (1973). *J. Anim. Sci.* **36,** 1186.
Bolsen, K. K., Berger, L. L., Conway, K. L., and Riley, J. G. (1976). *J. Anim. Sci.* **42,** 185.
Bond, J., and McDowell, R. E. (1972). *J. Anim. Sci.* **35,** 820.
Bond, J., and Oltjen, R. R. (1973a). *J. Anim. Sci.* **37,** 141.
Bond, J., and Oltjen, R. R. (1973b). *J. Anim. Sci.* **37,** 1040.
Bond, J. and Rumsey, T. S. (1973). *J. Anim. Sci.* **37,** 593.
Bond, J., and Wiltbank, J. N. (1970). *J. Anim. Sci.* **30,** 438.
Bond, J., Everson, D. O., Gutierrez, J., and Warwick, E. J. (1962). *J. Anim. Sci.* **21,** 728.
Bond, J., Rumsey, T. M., and Weinland, B. T. (1976). *J. Anim. Sci.* **43,** 873.
Bond, T. E., Ittner, N. R., and Heitman, H. (1952). *Agric. Eng.* **33,** 148.
Bond, T. E., Kelly, C. F., and Heitman, H., Jr. (1958). *J. Hered.* **49,** 75.
Borchers, R. (1965). *J. Anim. Sci.* **24,** 1033.

Bothast, R. J., Adams, G. H., Hatfield, E. E., and Lancaster, E. B. (1975). *J. Dairy Sci.* **58**, 386.
Bouchard, R., and Conrad, H. R. (1973). *J. Dairy Sci.* **56**, 665 (abstr.).
Bowen, J. M. (1962). *Am. J. Vet. Res.* **23**, 685.
Boyd, C. E. (1968). *Hyacinth Control J.* **7**, 26.
Bradley, N. W., Jones, B. M., Jr., Mitchell, G. E., Jr., and Little, C. O. (1966). *J. Anim. Sci.* **25**, 480.
Bratzler, J. W., and Forbes, E. B. (1940). *J. Nutr.* **19**, 611.
Bray, A. C., and Hemsley, R. V. (1969). *Aust. J. Agric. Res.* **20**, 759.
Brazle, F. K., and Harbers, L. H. (1977). *J. Anim. Sci.* **46**, 506.
Breirem, K., and Ulvesli, O. (1960). *Herb. Abstr.* **30**, 1.
Brent, B. E., Adepoju, A., and Portela, F. (1971). *J. Anim. Sci.* **32**, 794.
Brethour J. F., and Duitsman, W. W. (1966). *Kan., Agric. Exp. Stn., Bull.* **492**, 40.
Breur, L. H., Riewe, M. E., Smith, J. D., Dohme, C. K., Taylor, T. M., and Harper, J. C. (1967). *Proc. Annu. Tex. Nutr. Conf.* **22**, 215.
Briggs, P. K., Hogan, J. P., and Reid, R. L. (1957). *Aust. J. Agric. Res.* **8**, 674.
Briggs, P. K., McBarron, E. O., Grainger, T. E., and Franklin, M. C. (1960). *Proc. Int. Grassl. Congr., 8th, 1960*, p. 579.
Britt, D. G., and Huber, J. T. (1975). *J. Dairy Sci.* **58**, 1666.
Broadbent, P. J. (1976). *Anim. Prod.* **23**, 165.
Broadbent, P. J., Ball, C., and Dodsworth, T. L. (1976). *Anim. Prod.* **23**, 341.
Brobeck, J. R. (1960). *Recent Prog. Horm. Res.* **16**, 439.
Broberg, G. (1957). *Nord. Veterinaermed.* **9**, 942.
Brod, D. L., Bolsen, K. K., and Brent, B. E. (1982). *J. Anim. Sci.* **54**, 179.
Broderick, G. A. (1975). *In* "Protein Nutritional Quality of Foods and Feeds" (I. M. Friedman, ed.), Vol. 1, Part II, p. 211. Dekker, New York.
Broderick, G. A., and Balthrop, J. E. (1979). *J. Anim. Sci.* **49**, 1101.
Broderick, G. A., and Craig, W. M. (1980). *J. Nutr.* **110**, 2381.
Broderick, G. A., Kang-Meznarich, J. H., and Craig, W. M. (1981). *J. Dairy Sci.* **64**, 1731.
Brody, S. (1945). "Bioenergetics and Growth." Van Neostrand- Rheinhold, Princeton, New Jersey.
Brommelsiek, W. A., Shirley, R. L., Bertrand, J. E., and Palmer, A. Z. (1979). *J. Anim. Sci.* **48**, 1475.
Brookes, I. M., Owens, F. N., Isaacs, J., Brown, R. E., and Garrigus, U. S. (1972). *J. Anim. Sci.* **35**, 877.
Brooks, C. C., Garner, G. B., Gehrke, C. W., Muhrer, M. E., and Pfander, W. H. (1954). *J. Anim. Sci.* **13**, 758.
Brouwer, E. (1965). *Publ.—Eur. Assoc. Anim. Prod.* **11**, 441.
Brown, C. J. (1961). *Bull.—Arkansas, Agric. Exp. Stn.* **641**.
Brown, H., Elliston, N. G., McAskill, J. W., Muenster, O. A., and Tonkinson, L. V. (1973). *J. Anim. Sci.* **37**, 1085.
Brown, H., Bing, R. F., Grueter, H. P., McAskill, J. R., Cooley, C. O., and Rathmacher, R. P. (1975). *J. Anim. Sci.* **40**, 207.
Brown, L. D. (1966). *J. Dairy Sci.* **49**, 223.
Bryant, H. T., Blazer, R. E., Hammes, R. L., and Fontenot, J. P. (1970). *J. Anim. Sci.* **30**, 153.
Bryant, M. P. (1956). *J. Bacteriol.* **72**, 162.
Bryant, M. P. (1959). *Bacteriol. Rev.* **23**, 125.
Bryant, M. P. (1963). *J. Anim. Sci.* **22**, 801.
Bryant, M. P. (1965). Rumen methanogenic bacteria. *In* "Physiology of Digestion in the Ruminant" (R. W. Daugherty, R. S. Allen, W. Burroughs, N. L. Jacobson, and A. D. McGilliard, eds.), p. 411. Butterworth, London.

Bryant, M. P., and Burkey, L. A. (1953). *J. Dairy Sci.* **36,** 205.
Bryant, M. P., and Doetsch, R. N. (1954). *Science* **120,** 944.
Bryant, M. P., and Robinson, I. M. (1961a). *Appl. Microbiol.* **9,** 91.
Bryant, M. P., and Robinson, I. M. (1961b). *Appl. Microbiol.* **9,** 96.
Bryant, M. P., and Robinson, I. M. (1961c). *J. Dairy Sci.* **44,** 1446.
Bryant, M. P., and Robinson, I. M. (1962). *J. Bacteriol.* **84,** 605.
Bryant, M. P., Small, N., Bouwa, C., and Robinson, I. M. (1958a). *J. Dairy Sci.* **41,** 1747.
Bryant, M. P., Small, N., Bouwa, C., and Chu, H. (1958b). *J. Bacteriol.* **76,** 15.
Bryant, M. P., Robinson, I. M., and Chu, H. (1959). *J. Dairy Sci.* **42,** 1831.
Buchanan-Smith, J. G., Bannister, W., Durham, R. M., and Curl, S. E. (1964). *J. Anim. Sci.* **23,** 902 (abstr.).
Buchanan-Smith, J. G., Totusek, R., and Tillman, A. D. (1968). *J. Anim. Sci.* **27,** 525.
Buchanan-Smith, J. G., Macleod, G. K., and Mowat, D. N. (1974). *J. Anim. Sci.* **38,** 133.
Buchman, D. T., Shirley, R. L., and Killinger, G. D. (1968a). *Q. J. Fla. Acad. Sci.* **31,** 143.
Buchman, D. T., Shirley, R. L., and Killinger, G. B. (1968b). *Proc.—Soil Crop Sci. Soc. Fla.* **28,** 209.
Bucholtz, H. F., and Bergen, W. G. (1973). *Appl. Microbiol.* **25,** 504.
Bull, L. S., and Vandersall, J. H. (1973). *J. Dairy Sci.* **56,** 106.
Bull, L. S., Reid, J. T., and Johnson, D. E. (1970). *J. Nutr.* **100,** 262.
Bull, R. C., Olson, D. P., Stosjek, M. J., and Ross, R. H. (1978). *J. Anim. Sci.* **47** Suppl. 1, 193.
Burghardi, S. R., Goodrich, R. D., and Meiske, J. C. (1980). *J. Anim. Sci.* **50,** 729.
Burgos, A., and Olsen, H. H. (1970). *J. Dairy Sci.* **53,** 467 (abstr.).
Burkhardt, J. D., Embry, L. B., and Luther, R. M. (1969). *J. Anim. Sci.* **29,** 153 (Abstr.).
Burns, R. E. (1963). *Ga., Agric. Exp. Stn., Tech. Bull.* [N.S.] **32.**
Burns, R. H., Johnston, A., Hamilton, J. W., McColloch, R. J., Duncan, W. E., and Fisk, H. G. (1964). *J. Anim. Sci.* **23,** 5.
Burns, W. C., Shirley, R. L., Koger, M., Cunha, T. J., Easley, J. F., and Chapman, H. L., Jr. (1973). *Proc.—Soil Crop Sci. Soc. Fla.* **33,** 45.
Burris, M. J., and Priode, B. M. (1957). The value of apple pomace as a roughage for wintering beef cattle. *Va., Agric. Exp. Stn., Res. Rep.* **12.**
Burris, W. R., Bradley, N. W., and Boling, J. A. (1974a). *J. Anim. Sci.* **38,** 200.
Burris, W. R., Boling, J. A., Bradley, N. W., and Ludwick, R. L. (1974b). *J. Anim. Sci.* **39,** 818.
Burris, W. R., Bradley, N. W., and Boling, J. A. (1975). *J. Anim. Sci.* **40,** 714.
Burris, W. R., Boling, J. A., Bradley, N. W., and Young, A. W. (1976). *J. Anim. Sci.* **42,** 699.
Burroughs, W., and Trenkle, A. (1969). Initial experiment with methioninehydroxy analog added to an all-urea supplement for finishing heifer calves. *Iowa State Univ., Agric. Home Econ. Exp. Stn., Anim. Sci. Leaflet.* **R-122.**
Burroughs, W., Frank, N. A., Gerlaugh, P., and Bethke, R. M. (1950). *J. Nutr.* **40,** 9.
Burroughs, W., Woods, W., Ewing, S. A., Greig, J. and Theurer, B. (1960). *J. Anim. Sci.* **19,** 458.
Burroughs, W., Geasler, M., and Fleming, B. (1973). *Beef,* May 9, 8.
Burroughs, W., Nelson, D. K., and Mertens, D. R. (1975a). *J. Anim. Sci.* **41,** 933.
Burroughs, W., Nelson, D. K., and Mertens, D. R. (1975b). *J. Dairy Sci.* **58,** 611.
Burt, A. W. A. (1966). *J. Agric. Sci.* **66,** 131.
Burton, G. W. (1970). *J. Anim. Sci.* **30,** 143.
Burton, J. H., Anderson, M., and Reid, J. T. (1974). *Br. J. Nutr.* **32,** 515.
Butcher, J. E. (1966). *J. Anim. Sci.* **26,** 590 (abstr.).
Butterworth, M. H. (1967). *Nutr. Abstr. Rev.* **37,** 349.
Byers, F. M. (1980a). *J. Anim. Sci.* **50,** 1127.
Byers, F. M. (1980b). *J. Anim. Sci.* **51,** 158.

Byers, F. M., and Moxon, A. L. (1980). *J. Anim. Sci.* **50,** 1136.
Byers, F. M., Johnson, D. E., and Matsushima, J. K. (1976). *Publ.—Eur. Assoc. Anim. Prod.* **19,** 253.
Byers, J. H., and Ormiston, E. E. (1964). *J. Dairy Sci.* **47,** 707 (abstr.).
Cahill, V. R. (1964). *Proc. Recip. Meat Conf.* **17,** 35.
Caldwell, D. R., and Bryant, M. P. (1966). *Appl. Microbiol.* **14,** 794.
Campbell, T. C., Loosli, J. K., Warner, R. G., and Tasaki, I. (1963). *J. Anim. Sci.* **22,** 139.
Campbell, T. W., Bartley, E. E., Bechtle, R. M., and Dayton, A. D. (1976). *J. Dairy Sci.* **59,** 1452.
Campling, R. C. (1966). *J. Dairy Res.* **33,** 13.
Campling, R. C., and Freer, M. (1966). *Br. J. Nutr.* **20,** 229.
Campling, R. C., Freer, M., and Balch, C. C. (1961). *Br. J. Nutr.* **15,** 531.
Cardon, B. P. (1953). *J. Anim. Sci.* **12,** 536.
Carroll, E. J., and Hungate, R. E. (1955). *Arch. Biochem. Biophys.* **56,** 525.
Caswell, L. F., Webb, K. E., Jr., and Fontenot, J. P. (1977). *J. Anim. Sci.* **44,** 803.
Chace, L. E., Wangsness, P. J., and Baumgardt, B. R. (1976). *J. Dairy Sci.* **59,** 1923.
Chalupa, W. (1975). *J. Dairy Sci.* **58,** 1198.
Chalupa, W. (1976). *J. Anim. Sci.* **43,** 828.
Chalupa, W. (1977). *J. Anim. Sci.* **46,** 585.
Chalupa, W., and Montgomery, A. (1979). *J. Anim. Sci.* **48,** 393.
Chalupa, W., Oltjen, R. R., Slyter, L. L., and Dinius, D. A. (1971). *J. Anim. Sci.* **33,** 278 (abstr.).
Chalupa, W., Corbett, W. and Brethour, J. R. (1980). *J. Anim. Sci.* **51,** 170.
Chalupa, W., Patterson, J. A., Parish, R. C. and Chow, A. W. (1983). *J. Anim Sci.* **57,** 195.
Chamberlain, D. G., Thomas, P. C., and Wilson, A. G. (1976). *J. Sci. Food Agric.* **27,** 231.
Chambers, M. I., and Synge, R. L. M. (1954). *J. Agric. Sci.* **44,** 263.
Chandler, P. T. (1970). Improving protein nutrition of ruminants. *VA. Feed Conv. Nutr. Conf.*, p. 1.
Chandler, P. T., and Jahn, E. (1973). *J. Dairy Sci.* **56,** 666 (abstr.).
Channon, H. J., Jenkins, G. N., and Smith, J. A. B. (1937). *Biochem. J.* **31,** 41.
Chapman, H. L., and Palmer, A. Z. (1972). *Circ.—Fla., Agric. Exp. Sta.* **S-216.**
Chapman, H. L., Haines, C. E., Crockett, J. R., and Kidder, R. W. (1958). *Fla., Agric. Exp. Stn., Dep. Anim. Sci. Mimeo Rep.* **59-3.**
Chapman, H. L., Shirley, R. L., Palmer, A. Z., and Carpenter, J. W. (1971). *Bull.—Fla., Agric. Exp. Stn.* **748.**
Chapman, H. L., Jr., Kidder, R. W., and Plank, S. W. (1953). *Bull.—Fla., Agric. Exp. Stn.* **531.**
Chapman, H. L., Jr., Kidder, R. W., Koger, M., Crockett, J. R., and McPherson, W. K. (1965). *Fla., Agric. Exp. Stn., Bull.* **701.**
Chapman, H. L., Jr., Beardsley, D. W., Cunha, T. J., and McPherson, W. K. (1967). *Fla., Agric. Exp. Stn., Bull.* **719.**
Chapman, H. L., Jr., Ammerman, C. B., Baker, F. S., Jr., Hentges, J. F., Hayes, B. W., and Cunha, T. J. (1972). *Bull.—Fla., Agric. Exp. Stn.* **751.**
Chavez, M. I., Shirley, R. L., and Easley, J. F. (1975). *Proc.—Soil Crop Sci. Soc. Fla.* **35,** 74.
Chen, M. C., Beeson, W. M., and Perry, T. W. (1976). *J. Anim. Sci.* **43,** 1280.
Chen, M. C., Ammerman, C. B., Henry, P. R., Palmer, A. Z., and Long, S. K. (1981). *J. Anim. Sci.* **53,** 253.
Chigaru, P. R. N., and Topps, J. H. (1981). *Anim. Prod.* **32,** 95.
Christiansen, W. C., Woods, W., and Burroughs, W. (1964). *J. Anim. Sci.* **23,** 984.
Church, D. C. (1970). *In* "Digestive Physiology and Nutrition of Ruminants" (D. C. Church, ed.), Vol 1 p. 85. Oregon State Univ. Book Stores, Corvallis.
Church, D. C., Daugherty, D. A., and Kennick, W. H. (1982). *J. Anim. Sci.* **54,** 337.
Clanton, D. C. (1978). *J. Anim. Sci.* **47,** 765.
Clanton, D. C., and England, M. E. (1980). *J. Anim. Sci.* **51,** 539.

Clanton, D. C., Rothlisberger, J. A., Baker, G. N., and Ingalls, J. E. (1966). "Beef Cattle Progress Report." Anim. Sci. Dep., University of Nebraska, Lincoln.
Clanton, D. C., England, M. E., and Parrott, J. C., III (1981). *J. Anim. Sci.* **53,** 873.
Clapperton, J. L. (1964). *Br. J. Nutr.* **18,** 47.
Clapperton, J. L. (1977). *Anim. Prod.* **24,** 169.
Clark, J. H. (1975). *In* "Protein Nutritional Quality of Foods and Feeds" (I. M. Friedman, ed.), Vol. 1, Part II, p. 261. Dekker, New York.
Clark, J. L., Pfander, W. H., Bloomfield, R. A., Krause, G. F., and Thompson, G. B. (1970). *J. Anim. Sci.* **31,** 961.
Clark, J. L., Hedrick, H. B., and Thompson, G. B. (1976). *J. Anim. Sci.* **42,** 352.
Clark, R., Barrett, E. L., and Kellerman, J. H. (1965). *J. S. Afr. Vet. Med. Assoc.* **36,** 79.
Clark, W. S. (1979). *J. Dairy Sci.* **62,** 96.
Clarke, R. T. J. (1977). *In* "Microbial Ecology of the Gut" (R. T. J. Clarke and T. Bauchop, eds.), p. 23. Academic Press, London.
Clarke, S. D., and Dyer, I. A. (1973). *J. Anim. Sci.* **37,** 1022.
Clemens, E. T., and Johnson, R. R. (1973a). *J. Anim. Sci.* **37,** 1027.
Clemens, E. T., and Johnson, R. R. (1973b). *J. Nutr.* **103,** 1406.
Clemens, E. T., and Johnson, R. R. (1974). *J. Anim. Sci.* **39,** 937.
Cody, R. E., Jr., Morrill, J. L., and Hibbs, C. M. (1968). *J. Dairy Sci.* **51,** 952 (Abstr.).
Cody, R. E., Jr., Morrill, J. L., and Hibbs, C. M. (1972). *J. Anim. Sci.* **35,** 460.
Cole, H. H., and Kleiber, M. (1945). *Am. J. Vet. Res.* **6,** 188.
Cole, N. A., and Hutcheson, D. P. (1981). *J. Anim. Sci.* **53,** 907.
Cole, N. A., and Hutcheson, D. P. (1985). *J. Anim. Sci.* **60,** 772.
Cole, N. A., Johnson, R. R., and Owens, F. N. (1976a). *J. Anim. Sci.* **43,** 483.
Cole, N. A., Johnson, R. R., and Owens, F. N. (1976b). *J. Anim Sci.* **43,** 490.
Cole, N. A., Johnson, R. R., Owens, F. N., and Males, J. R. (1979c). *J. Anim. Sci.* **43,** 497.
Coleman, G. S. (1960). *J. Gen. Microbiol.* **22,** 555.
Coleman, G. S. (1964). *J. Gen. Microbiol.* **37,** 209.
Coleman, G. S. (1969). *J. Gen. Microbiol.* **57,** 81.
Coleman, G. S. (1972). *J. Gen. Microbiol.* **71,** 117.
Coleman, G. S. (1975). *In* "Digestion and Metabolism in the Ruminant" (I. W. McDonald and A. C. I. Warner, eds.), p. 149. University of New England Publishing Unit, Armidale, N.S.W., Australia.
Coleman, G. S., Davies, J. I., and Cash, M. A. (1972). *J. Gen Microbiol.* **73,** 509.
Coleman, S. W., and Barth, K. M. (1974). *J. Anim. Sci.* **39,** 408.
Coleman, S. W., and Barth, K. M. (1977). *J. Anim. Sci.* **45,** 1180.
Coleman, S. W., Pate, F. M., and Beardsley, D. W. (1976). *J. Anim. Sci.* **42,** 27.
Colovos, N. F., Hotter, J. B., Koes, R. M., Urban, W. E., Jr., and Davis, H. A. (1970). *J. Anim. Sci.* **30,** 819.
Conrad, H. R., and Hibbs, J. W. (1961). *Ohio Farm Home Res.* **46,** 13.
Cook, C. W., and Harris, L. E. (1968). *Bull.—Utah Agric. Exp. Stn.* **475.**
Cook, L. J., and Scott, T. W. (1970). *Proc. Aust. Biochem. Soc.* **3,** 93.
Cook, R. J., and Fox, D. G. (1977). *Mich. State Univ. Res. Rep.* **328,** 96.
Coombe, J. B. (1959). *J. Aust. Inst. Agric. Sci.* **25,** 299.
Coop, I. E., and Blakely, R. L. (1949). *N. Z. J. Sci. Technol., Sect. A.* **30** 277.
Coop, I. E., and Blakely, R. L. (1950). *N. Z. J. Sci. Technol., Sect. A* **31** (5), 44.
Coppock, C. E., Peplowski, M. A., and Lake, G. B. (1976). *J. Dairy Sci.* **59,** 1152.
Cornman, A. W., Lamm, W. D., Webb, K. E., Jr., and Fontenot, J. P. (1981). *J. Anim. Sci.* **52,** 1233.
Crabtree, J. R., and Williams, G. L. (1971). *Anim. Prod.* **13,** 83.

Crane, T. D. (1973). *Vet. Bull.* **43,** 165.
Crawford, R. J., Jr., Hoover, W. H., Sniffen, C. J., and Crooker, B. A. (1978). *J. Anim. Sci.* **46,** 1768.
Crickenberger, R. G., Henderson, H. E., Reddy, C. A., and Magee, W. T. (1977). *J. Anim. Sci.* **45,** 566.
Crickenberger, R. G., Henderson, H. E., and Reddy, C. A. (1981). *J. Anim. Sci.* **52,** 677.
Cromwell, G. L., Pickett, R. A., and Beeson, W. M. (1967). *J. Anim. Sci.* **26,** 1325.
Crooker, B. A. (1978). M. S. Thesis, University of Maine, Orono.
Cross, D. L., and Jenny, B. F. (1976). *J. Dairy Sci.* **59,** 919.
Cross, D. L., Boling, J. A., and Bradley, N. W. (1973). *J. Anim. Sci.* **36,** 982.
Cross, D. L., Ludwick, R. L., Boling, J. A., and Bradley, N. W. (1974). *J. Anim. Sci.* **38,** 404.
Cross, D. L., Skelley, G. C., Thompson, C. S., and Jenny, B. F. (1978). *J. Anim. Sci.* **47,** 544.
Cross, H. R., and Dinius, D. A. (1978). *J. Anim. Sci.* **47,** 1265.
Cuitun, L. L., Hale, W. H., Theurer, B. Dryden, F. D., and Marchello, J. A. (1975). *J. Anim. Sci.* **40,** 691.
Cullison, A. E., Park, C. S., Wheeler, W. E., McCampbell, H. C., and Warren, E. P. (1973). *J. Anim. Sci.* **37,** 852.
Cullison, A. E., McCampbell, H. C., Cunningham, A. C., Lowrey, R. S., Warren, E. P., McLendon, B. D., and Sherwood, D. H. (1976). *J. Anim. Sci.* **42,** 219.
Cunningham, M. D., Martz, F. A., and Merilan, C. P. (1964). *J. Dairy Sci.* **47,** 382.
Cuthbertson, D. P., and Chalmers, M. I. (1959). *Biochem. J.* **46,** XVII (abstr.).
Czerkawski, J. W. (1971). The inhibition of methane production in the rumen. *Hannah Res. Inst. Rep.,* pp. 40.
Czerkawski, J. W. (1972). *Proc. Nutr. Soc.* **31,** 141.
Czerkawski, J. W. (1974). *Publ.—Eur. Assoc. Anim. Prod.* **14,** 95.
Czerkawski, J. W. (1978). *J. Dairy Sci.* **61,** 1261.
Czerkawski, J. W., and Breckenridge, G. (1975). *Br. J. Nutr.* **34,** 447.
Dain, J. A., Neal, A. L., and Dougherty, R. W. (1955). *J. Anim. Sci.* **14,** 930.
Dale, H. E., Stewart, R. E., and Brody, S. (1954). *Cornell Vet.* **44,** 368.
Daniel, J. W., Fuquay, J. W., McGee, W. H., Brown, W. H., and Cardwell, J. T. (1973). *J. Dairy Sci.* **56,** 306 (abstr.).
Danielli, J. F., Hitchcock, M. W. S., Marshall, R. A., and Phillipson, A. T. (1945). *J. Exp. Biol.* **22,** 75.
Daniels, L. B., Campbell, J. R., and Martz, F. A. (1970). *J. Dairy Sci.* **52,** 908.
Danner, M. L., Fox, D. G., and Black, J. R. (1980). *J. Anim. Sci.* **50,** 394.
Dartt, R. M., Boling, J. A., and Bradley, N. W. (1978). *J. Anim. Sci.* **46,** 345.
Daugherty, D. A., and Church, D. C. (1982). *J. Anim. Sci.* **54,** 345.
Davidovich, A., Bartley, E. E., Milliken, G. A., Dayton, A. D., Deyoe, C. W., and Bechtle, R. M. (1977). *J. Anim. Sci.* **45,** 1397.
Davidson, K. L., and Woods, W. (1960). *J. Anim. Sci.* **19,** 54.
Davidson, K. L., and Woods, W. (1963). *J. Anim. Sci.* **22,** 27.
Davis, A. V., and Merilan, C. P. (1960). *J. Dairy Sci.* **43,** 871 (abstr.).
Davis, C. L., Brown, R. E., and Beitz, D. C. (1964). *J. Dairy Sci.* **47,** 1217.
Davis, G. V., and Erhart, A. B. (1976). *J. Anim. Sci.* **43,** 1.
Davis, L. E., Westfall, B. A., and Dale, H. E. (1965). *Am. J. Vet. Res.* **26,** 1403.
Dehority, B. A. (1973). *Fed. Proc., Fed. Am. Soc. Exp. Biol.* **32,** 1819.
Dehority, B. A., and Johnson, R. R. (1961). *J. Dairy Sci.* **44,** 2242.
Deinum, B., van Es, A. J. H., and van Soest, P. J. (1968). *Neth. J. Agric. Sci.* **16,** 217.
Della-Fera, M. A., and Baile, C. A. (1984). *J. Anim. Sci.* **59,** 1362.
Demeyer, D. I., and Van Nevel, C. J. (1975). *In* "Digestion and Metabolism in the Ruminant" (I.

W. McDonald and A. C. I. Warner, eds.), p. 366. University of New England Publishing Unit, Armidale, N.S.W., Australia.
Dennis, S. M., Nagaraja, T. G., and Bartley, E. E. (1981a). *J. Dairy Sci.* **64**, 2350.
Dennis, S. M., Nagaraja, T. G., and Bartley, E. E. (1981b). *J. Anim. Sci.* **52**, 418.
Denton, D. A. (1957). *Q. J. Exp. Physiol. Cogn. Med. Sci.* **42**, 72.
Deutscher, G. H., and Whiteman, J. V. (1971). *J. Anim. Sci.* **33**, 337.
Dick, A. T., Dann, A. T., and Bull, L. B. (1963). *Nature (London)* **197**, 207.
Dinius, D. A., and Baile, C. A. (1977). *J. Anim. Sci.* **45**, 147.
Dinius, D. A., and Bond, J. (1975). *J. Anim. Sci.* **41**, 629.
Dinius, D. A., and Oltjen, R. R. (1971a). *J. Anim. Sci.* **33**, 312 (abstr.).
Dinius, D. A., and Oltjen, R. R. (1971b). *J. Anim. Sci.* **33**, 1344.
Dinius, D. A., and Oltjen, R. R. (1972). *J. Anim. Sci.* **34**, 137.
Dinius, D. A., and Williams, E. E. (1975). *J. Anim. Sci.* **41**, 1170.
Dinius, D. A., Oltjen, R. R., and Satter, L. D. (1974). *J. Anim. Sci.* **38**, 887.
Dinius, D. A., Simpson, M. E., and March, P. B. (1976). *J. Anim. Sci.* **42**, 229.
Dinius, D. A., Goering, H. K., Oltjen, R. R., and Cross, H. R. (1978). *J. Anim. Sci.* **46**, 761.
Dirksen, G. (1970). *In* "Physiology of Digestion and Metabolism in the Ruminant" (A. T. Phillipson, ed.), p. 612. Oriel Press, Ltd., Newcastle upon Tyne, England.
Diven, R. H., Reed, R. E., and Pistor, W. J. (1964). *Ann. N. Y. Acad. Sci.* **111**, 638.
Dobson, A., and Phillipson, A. T. (1958). *J. Physiol. (London)* **140**, 94.
Dobson, M. E. (1959). *Aust. Vet. J.* **35**, 225.
Domermuth, W. F., Veum, T. L., Alexander, M. A., Hedrick, H. B., and Clark, J. L. (1973). *J. Anim. Sci.* **37**, 259 (abstr.).
Dougherty, R. W., and Cello, R. M. (1949). *Cornell Vet.* **39**, 403.
Dougherty, R. W., and Habel, R. E. (1955). *Cornell Vet.* **45**, 459.
Drapala, W. J., Raymond, L. C., and Crampton, E. W. (1947). *Sci. Agric.* **27**, 36.
Drews, J. E., Moody, N. W., Hays, V. W., Speer, V. C., and Evan, R. C. (1969). *J. Nutr.* **97**, 537.
Dronawat, N. S., Stanley, R. W., Cobb, E., and Morita, K. (1966). *J. Dairy Sci.* **49**, 28.
Drori, D., Levy, D., Folman, Y., and Holzer, Z. (1974). *J. Anim. Sci.* **38**, 654.
Drude, R. E., Escano, J. R., and Russoff, L. L. (1971). *J. Dairy Sci.* **54**, 773 (abstr.).
Dryden, F. D., and Marchello, J. A. (1973). *J. Anim. Sci.* **37**, 33.
Dryden, F. D., Marchello, J. A., Cuitun, L. L., and Hale, W. H. (1975). *J. Anim. Sci.* **40**, 697.
Dunlop, R. H. (1972). *Adv. Vet. Sci. Comp. Med.* **16**, 259.
Durand, M., Kumaresan, A., Beumatin, P., Dumay, C., and Gueguen, L. (1976). *Tracer Stud. Non-Protein Nitrogen Ruminants 3, Proc. Res. Co-Ord. Meet., 1976*, p. 27.
DuToit, P. J., Malan, A. J., and Groenewald, J. W. (1934). *Onderstepoort J. Vet. Sci.* **2**, 565.
Dyer, I. A. (1969). *Anim. Nutr. Health.* **24**, 11.
Dyer, I. A., Ensminger, M. E., and Blue, R. L. (1957). *J. Anim. Sci.* **16**, 828.
Dyer, I. A., Koes, R. M., Herlugson, M. L., Ojikutu, L. B., Preston, R. L., Zimmer, P., and DeLay, R. (1980). *J. Anim. Sci.* **51**, 843.
Eadie, J. M., and Mann, S. O. (1970). *In* "Physiology of Digestion and Metabolism in the Ruminant" (A. T. Phillipson, ed.), p. 335. Oriel Press Ltd., Newcastle upon Tyne, England.
Eadie, J. M., Hyldgaard-Jensen, J., Mann, S. O., Reid, R. S., and Whitelaw, F. G. (1970). *Br. J. Nutr.* **24**, 157.
Earley, R., Vatthauer, R., Paulson, W., and Peshel, D. (1977). Effect of nutritional level on performance of young crossbred heifers. *Lancaster Cow-Calf Day Rep.*, p. 47.
Easley, J. F., and Shirley, R. L. (1974). *Hyacinth Control J.* **12**, 82.
Edwards, R. L., Tove, S. B., Blumer, R. N., and Barrick, E. R. (1961). *J. Anim. Sci.* **20**, 712.
Egan, A. R. (1976). *Rev. Rural Sci.* **2**, 27.
Egan, A. R., and Black, A. L. (1968). *J. Nutr.* **96**, 450.

Egan, A. R., and Moir, R. J. (1965). *Aust. J. Agric. Res.* **16,** 437.
Egan, A. R., Moller, F., and Black, A. L. (1970). *J. Nutr.* **100,** 419.
Egan, A. R., Walker, D. J., Nader, C. J., and Storer, G. (1975). *Aust. J. Agric. Res.* **26,** 909.
Elam, C. J. (1976). *J. Anim. Sci.* **43,** 898.
Elam, C. J., and Davis, R. E. (1962a). *J. Anim. Sci.* **21,** 327.
Elam, C. J., and Davis, R. E. (1962b). *J. Anim. Sci.* **21,** 568.
Elam, C. J., Gutierrez, J., and Davis, R. E. (1960). *J. Anim. Sci.* **19,** 1089.
Elliott, R. C. (1967). *J. Agric. Sci.* **69,** 375.
Ellis, G. J., Matrone, G., and Maynard, L. A. (1946). *J. Anim. Sci.* **5,** 285.
Ellis, W. C. (1968). *J. Agric. Food Chem.* **16,** 220.
Elsden, S. R. (1946). *Biochem. J.* **40,** 252.
El-Shazly, K. (1952). *Biochem. J.* **51,** 640.
El-Shazly, K., Dehority, B. A., and Johnson, R. R. (1961). *J. Anim. Sci.* **20,** 268.
Ely, D. G., Little, C. O., Woolfolk, P. G., and Mitchell, G. E., Jr. (1967). *J. Nutr.* **91,** 314.
Emerick, R. J. (1974). *Fed. Proc., Fed. Am. Soc. Exp. Biol.* **33**(5), 1183.
Emery, R. S. (1978). *J. Dairy Sci.* **61,** 825.
Emery, R. S., and Brown, L. D. (1961). *J. Dairy Sci.* **44,** 1899.
Emery, R. S., Smith, C. K., Grimes, R. M., Huffman, C. F., and Duncan, C. W. (1960). *J. Dairy Sci.* **43,** 76.
Emery, R. S., Brown, L. D., and Thomas, J. W. (1964). *J. Dairy Sci.* **47,** 1325.
Ensminger, M. E., and Olentine, C. G., Jr. (1978). "Feeds and Nutrition—Complete First Edition." Ensminger Publ. Co., Clovis, California.
Ensor, W. L., Shaw, J. C., and Tellechea, H. F. (1959). *J. Dairy Sci.* **42,** 189.
Erdman, R. A., Bull, L. S., and Hemken, R. W. (1981). *J. Dairy Sci.* **64,** 1679.
Erwin, E. S., Dyer, I. A., and Ensminger, M. E. (1956). *J. Anim. Sci.* **15,** 717.
Erwin, E. S., Sterner, W., and Marco, G. J. (1963). *J. Am. Oil Chem. Soc.* **40,** 344.
Esplin, G., Hale, W. H., Hubert, F., Jr., and Taylor, B. (1963). *J. Anim. Sci.* **22,** 695.
Essig, H. W., and Shawver, C. B. (1968). *J. Anim. Sci.* **27,** 1669.
Essig, H. W., Rogillio, C. E., Hagan, F., and Drapala, W. J. (1972). *J. Anim. Sci.* **34,** 653.
Evans, E., Moran, E. T., Jr., and Walker, J. P. (1978). *J. Anim. Sci.* **46,** 520.
Evans, J. L., and Davis, G. K. (1966). *J. Anim. Sci.* **25,** 1014.
Evans, R. A., Axford, R. F. E., and Offer, N. W. (1975). *Proc. Nutr. Soc.* **34,** 65A (abstr.).
Evans, R. J., Bandemer, S. L., Davidson, J. A., and Schaible, P. J. (1958). *Poult. Sci.* **36,** 798.
Eversole, D. E., Bergen, W. G., Merkel, R. A., Magee, W. T., and Harpster, H. W. (1981). *J. Anim. Sci.* **53,** 91.
Ewan, R. C., Hatfield, E. E., and Garrigus, U. S. (1958). *J. Anim. Sci.* **17,** 298.
Faichney, G. J., Davis, H. L., Scott, T. W., and Cook, L. J. (1972). *Aust. J. Biol. Sci.* **25,** 205.
Farlin, S., and Guyer, P. Q. (1974). Protein levels for finishing cattle. *Nebr. Agric. Exp. Stn., Beef Cattle Rep.* **EC74-218,** 10.
Feist, W. C., Baker, A. J., and Tarkow, H. (1970). *J. Anim. Sci.* **30,** 832.
Fenderson, C. L., and Bergen, W. O. (1976). *J. Anim. Sci.* **42,** 1323.
Fergusan, K. A. (1975). *In* "Digestion and Metabolism in the Ruminant" (I. W. McDonald and A. C. I. Warner, eds.), p. 448. University of New England, Armidale, N.S.W., Australia.
Fernandez, J. A., Coppock, C. E., and Schake, L. M. (1982). *J. Dairy Sci.* **65,** 242.
Ferrell, C. L., Garrett, W. N., Hinman, N., and Grichting, G. (1976). *J. Anim. Sci.* **42,** 937.
Fick, K. R., Ammerman, C. B., McGowan, C. H., Loggins, P. E., and Cornell, J. A. (1973). *J. Anim. Sci.* **36,** 137.
Field, R. A. (1971). *J. Anim. Sci.* **32,** 849.
Fifield, R. B., and Johnson, R. J. (1978). *J. Anim. Sci.* **47,** 577.
Findley, J. D. (1958). *Proc. Nutr. Soc.* **17,** 186.

Flachowsky, G. (1975). *Arch. Tierernaehr* **25**, 139.
Flatt, W. P., Warner, R. G., and Loosli, J. K. (1958). *J. Dairy Sci.* **41**, 1593.
Flatt, W. P., Moe, P. W., Oltjen, R. R., Putnam, P. A., and Hooven, N. W. (1969), Energy metabolism studies with dairy cows receiving purified diets. *Symp. Energy Metab. 4th, 1969,* 12:109.
Flatt, W. P., Moe, P. W., and van Es, A. J. H. (1969). Calorimetric studies as a basis for feeding standards for lactating dairy cows. *Proc. World Conf. Anim. Prod., 2nd, 1969,* p. 339. (Blaxter, K. L., Kielanowski, J. and Thorbek, A, eds.) Oriel Press, Newcastle upon Tyne.
Fleck, A. T., Schalles, R. R., and Kiracofe, G. H. (1980). *J. Anim. Sci.* **51**, 816.
Folman, Y., Drori, D., Holzer, Z., and Levy, D. (1974). *J. Anim. Sci.* **39**, 788.
Fonnesbeck, P. V., Harris, L. E., and Kearl, L. C. (1971). *Proc., Annu. Meet.—Am. Soc. Anim. Sci., West. Sect.* **22**, 161.
Fonnesbeck, P. V., Kearl, L. C., and Harris, L. E. (1975). *J. Anim. Sci.* **40**, 1150.
Fonseca, H. A. (1957). M. S. Thesis, University of Florida, Gainesville.
Fontenot, J. P., Bovard, K. P., Oltjen, R. R., Rumsey, T. S., and Priode, B. M. (1977). *J. Anim. Sci.* **45**, 513.
Foote, L. E., Girouard, R. E., Johnston, J. E., Rainey, J., Brown, P. B., and Willis, W. H. (1968). *J. Dairy Sci.* **51**, 584.
Forbes, R. M., and Garrigus, W. P. (1950). *J. Anim. Sci.* **9**, 531.
Fordyce, J., and Kay, M. (1974). *Anim. Prod.* **18**, 105.
Forero, O., Owens, F. N., and Lusby, K. S. (1980). *J. Anim. Sci.* **50**, 532.
Forrest, W. W., and Walker, D. J. (1965). *J. Bacteriol.* **89**, 1448.
Forsberg, C. W., and Lam, K. (1977). *Appl. Environ. Microbiol.* **33**, 528.
Fortin, A., Reid, J. T., Maiga, A. M., Sim, D. W., and Wellington, G. H. (1980a). *J. Anim. Sci.* **51**, 331.
Fortin, A., Reid, J. T., Maiga, A. M., Sim, D. W., and Wellington, G. H. (1980b). *J. Anim. Sci.* **51**, 1288.
Fortin, A., Reid, J. T., Maiga, A. M., Sim, D. W., and Wellington, G. H. (1981). *J. Anim. Sci.* **53**, 982.
Fox, D. G., and Black, J. R. (1977). *Mich. State Univ. Res. Rep.* **328**, 141.
Fox, D. G. and Black, J. R. (1984). *J. Anim. Sci.* **58**, 725.
Fox, D. G., Klosterman, E. W., Newland, H. W., and Johnson, R. R. (1970). *J. Anim. Sci.* **30**, 303.
Fox, D. G., Johnson, R. R., Preston, R. L., Dockerty, T. R., and Klosterman, E. W. (1972). *J. Anim. Sci.* **34**, 310.
Fox, D. G., Crickenberger, R. G., Bergen, W. G., and Black, J. R. (1977). *Mich. State Univ. Res. Rep.* **328**, 141.
Frahm, R. R., Walters, L. E., and McLellan, C. R. (1971). *J. Anim. Sci.* **32**, 463.
Franklin, M. C., and Reid, R. L. (1944). *Aust. Vet. J.* **20**, 332.
Freitag, R. R., Theurer, B. and Hale, W. H. (1970). *J. Anim. Sci.* **31**, 434.
Fritschel, P. R., Satter, L. D., Baker, A. J., McGovern, J. N., Vatthauer, R. J., and Millett, M. A. (1976). *J. Anim. Sci.* **42**, 1513.
Fulghum, R. S., and Moore, W. E. C. (1963). *J. Bacteriol.* **85**, 808.
Fuller, J. R., and Johnson, D. E. (1981). *J. Anim. Sci.* **53**, 1574.
Fuquay, J. W. (1981). *J. Anim. Sci.* **52**, 164.
Fuquay, J. W., McGee, W. H., and Custer, E. W. (1974). *J. Dairy Sci.,* **57**, 132 (Abstr.).
Gaillard, B. D. E. (1962). *J. Agric. Sci.* **59**, 369.
Galgan, M. W., and Schneider, B. H. (1951). *Wash., Agric. Exp. Stn., Stn. Circ.* **166**.
Gall, L. S., and Huhtanen, C. N. (1951). *J. Dairy Sci.* **34**, 353.
Galloway, D. (1975). *Pulp Pap.* **49**, 104.
Galyean, M. L., and Chabot, R. C. (1981). *J. Anim. Sci.* **52**, 1197.

Galyean, M. L., Wagner, D. G., and Johnson, R. R. (1976). *J. Anim. Sci.* **43**, 1088.
Galyean, M. L., Wagner, D. G., and Owens, F. N. (1979). *J. Anim. Sci.* **49**, 199.
Galyean, M. L., Lee, R. W., and Hubbert, M. E. (1981a). *J. Anim. Sci.* **53**, 7.
Galyean, M. L., Wagner, D. G., and Owens, F. N. (1981b). *J. Dairy Sci.* **64**, 1804.
Garrett, W. N. (1970). *J. Anim. Sci.* **31**, 242.
Garrett, W. N. (1971). *J. Anim. Sci.* **32**, 451.
Garrett, W. N. (1976). "Thirteenth California Feeders Day Report." University of California, Davis. p. 21.
Garrett, W. N. (1979). *J. Anim. Sci.* **49**, 1403.
Garrett, W. N., and Hinman, N. (1969). *J. Anim. Sci.* **28**, 1.
Garrett, W. N., Lofgreen, G. P., and Hull, J. L. (1966). "California Feeders Day Report." p. 25.
Garrett, W. N., Yang, Y. T., Dunkley, W. L., and Smith, L. M. (1976). *J. Anim. Sci.* **42**, 1522.
Garton, G. A., Hobson, P. N., and Lough, A. K. (1958). *Nature (London)* **182**, 1511.
Geay, Y. (1984). *J. Animal Sci.* **58**, 766.
Gengler, W. R., Martz, F. A., and Johnson, H. D. (1970). *J. Dairy Sci.* **53**, 434.
Gharib, F. H., Goodrich, R. D., Meiske, J. C., and El Serafy, A. M. (1975). *J. Anim. Sci.* **40**, 727.
Ghorban, K. Z., Knox, K. L., and Ward, G. M. (1966). *J. Dairy Sci.* **49**, 1515.
Gibbs, H. O. (1966). M. S. Thesis, Texas A & M University Library, College Station.
Gibson, J. P. (1981). *Anim. Prod.* **32**, 275.
Gil, J. W., and King, K. W. (1957). *Agric. Food Chem.* **5**, 363.
Gil, L. A., Shirley, R. L., and Moore, J. E. (1973a). *J. Anim. Sci.* **37**, 159.
Gil, L. A., Shirley, R. L., and Moore, J. E. (1973b). *J. Dairy Sci.* **56**, 757.
Gil, L. A., Shirley, R. L., Moore, J. E., and Easley, J. F. (1973c). *Proc. Soc. Exp. Biol. Med.* **142**, 670.
Gilchrist, F. M. C., and Clark, R. (1957). *S. Afr. Vet. Med. Assoc.* **28**, 299.
Gill, D. R., Martin, J. R., and Lake, R. (1976). *J. Anim. Sci.* **43**, 363.
Gillingham, J. T., Shirer, M. M., Starnes, J. J., Page, N. R., and McClain, E. F. (1969). *Agron. J.* **61**, 727.
Glen, B. P., Ely, D. G., and Boling, J. A. (1977). *J. Anim. Sci.* **45**, 871.
Glewen, M. J., and Young, A. W. (1982). *J. Anim. Sci.* **54**, 713.
Glimp, H. A., Karr, M. R., Little, C. O., Woolfolk, P. G., Mitchell, G. E., Jr., and Hudson, L. W. (1967). *J. Anim. Sci.* **26**, 858.
Godfrey, N. W. (1961). *J. Agric. Sci.* **57**, 173.
Goering, H. K. (1976). *J. Anim. Sci.* **43**, 869.
Goering, H. K., and van Soest, P. J. (1970). *U.S., Dep. Agric., Agric. Handb.* **379**.
Golding, E. J., Moore, J. E., Franke, D. E., and Ruelke, O. C. (1976a). *J. Anim. Sci.* **42**, 710.
Golding, E. J., Moore, J. E., Franke, D. E., and Ruelke, O. C. (1976b). *J. Anim. Sci.* **42**, 717.
Goode, L., Barrick, E. R., and Tungman, D. F. (1955). *Proc. Assoc. South. Agric. Work.* **52**, 64.
Goodrich, R. B., Garrett, J. E., Gast, D. R., Kirick, M. A., Larsen, B. A., and Meiske, J. C. (1984). *J. Anim. Sci.* **58**, 1484.
Gordon, J. G. (1958). *J. Agric. Sci.* **51**, 78.
Gossett, W. H., Perry, T. W., Mohler, M. T., Plumlee, M. P., and Beeson, W. M. (1962). *J. Anim. Sci.* **21**, 248.
Gottschalk, A., and Graham, E. R. B. (1958). *Z. Naturforsch.* **13**, 821.
Graf, G. C., and Petersen, W. E. (1953). *J. Dairy Sci.* **36**, 1036.
Graham, N. McC. (1964). *Aust. J. Agric. Res.* **15**, 127.
Gray, F. V., Pilgrim, A. F., Rodda, H. J., and Weller, R. A. (1952). *J. Exp. Biol.* **29**, 57.
Gray, F. V., Pilgrim, A. F., and Weller, R. A. (1954). *J. Exp. Biol.* **31**, 49.
Griel, L. C., Jr., Patton, R. A., McCarthy, R. D., and Chandler, P. T. (1968). *J. Dairy Sci.* **51**, 1866.

Gutierrez, G. G., Schake, L. M., and Byers, F. M. (1982). *J. Anim. Sci.* **54,** 863.
Gutierrez, J. (1955). *Biochem. J.* **60,** 516.
Gutierrez, J., and Davis, R. E. (1959). *J. Protozool.* **6,** 222.
Gutierrez, J., and Hungate, R. E. (1957). *Science* **126,** 511.
Gutierrez, J., Davis, R. E., Lindahl, I. L., and Warnick, E. J. (1959). *Appl. Microbiol.* **7,** 16.
Gutierrez, J., Davis, R. E., and Lindahl, I. L. (1961). *Appl. Microbiol.* **9,** 209.
Haaland, G. L., Matsushima, J. K., Johnson, D. E., and Ward, G. M. (1981). *J. Anim. Sci.* **52,** 696.
Hadjipanayiotou, M., and Louca, A. (1976). *Anim. Prod.* **23,** 129.
Haecker, T. L. (1920). *Stn. Bull.—Minn., Agric. Exp. Stn.* 193.
Hagemeister, H., Kaufmann, W., and Pfeffer, E. (1976). *In* "Protein Metabolism and Nutrition" (D. J. A. Cole, K. N. Boorman, P. J. Buttery, D. Lewis, R. J. Neale, and H. Swan, eds.), Butterworth, London.
Hahn, G. (1981). *J. Anim. Sci.* **52,** 175.
Haines, C. E., Chapman, H. L., Jr., and Kidder, R. W. (1959). *Fla. Everglades Mimeo Rep.* **59,** 13.
Haines, C. E., Chapman, H. L., Jr., Kidder, R. W., and Greene, R. E. L. (1965). *Bull.—Fla., Agric. Exp. Stn.* **693.**
Hale, W. H. (1963). *Feedstuffs* **35**(20), 84.
Hale, W. H. (1970). *In* "Sorghum Production and Utilization" (J. S. Wall and W. M. Ross, eds.), p. 507. AVI Publishing Co., Westport, Connecticut.
Hale, W. H. (1978). Feedlot nutrition for beef cattle in warm weather. *Proc. Fla. Nutr. Conf.*, p. 61.
Hale, W. H., and King, R. P. (1955). *Proc. Soc. Exp. Biol. Med.* **89,** 112.
Hale, W. H., and Prouty, F. L. (1980). *Anim. Sci. Res. Rep., Univ. Ariz.*, p. 5.
Hale, W. H., and Theurer, C. B. (1972). *In* "Digestive Physiology and Nutrition of Ruminants" (D. C. Church, ed.) Vol. 3, p. 47. Oregon State University, Corvallis.
Hale, W. H., Sherman, W. C., Reynolds, W. M., and Appel, P. P. (1959). *J. Anim. Sci.* **18,** 1522 (abstr.).
Hale, W. H., Cuitun, L., Saba, W. J., Taylor, B. and Theurer, B. (1966a). *J. Anim. Sci.* **25,** 392.
Hale, W. H., Theurer, C. B., Prouty, F., Muntifering, R., Shell, L. and Felix, S. (1980). *Anim. Sci. Res. Rep., Univ. Ariz.* p. 1
Hall, G. A. B., Absher, C. W., Totusek, R., and Tillman, A. D. (1968). *J. Anim. Sci.* **27,** 165.
Ham, W. E., and Sandstedt, R. M. (1944). *J. Biol. Chem.* **154,** 505.
Hamernik, D. L., Males, J. R., Gaskins, C. T., and Reeves, J. J. (1985). *J. Anim. Sci.* **60,** 358.
Hamlin, L. J., and Hungate, R. E. (1956). *J. Bacteriol.* **72,** 548.
Hankins, O. G., and Howe, P. E. (1946). *U.S., Dep. Agric., Tech. Bull.* **926.**
Hansen, T. L., and Klopfenstein, T. (1979). *J. Anim. Sci.* **48,** 474.
Harbers, L. H. (1975). *J. Anim. Sci.* **41,** 1496.
Hardison, W. A., Reid, J. T., Martin, C. M., and Woolfolk, P. G. (1954). *J. Dairy Sci.* **37,** 89.
Harkins, J., Edwards, R. A., and McDonald, P. (1974). *Anim. Prod.* **19,** 141.
Harmeyer, H., Holler, H., Martens, H., and von Grabe, C. (1976). *Trace Stud. Non-Protein Nitrogen Ruminants 3, Proc. Res. Co-Ord. Meet., 1976*, p. 69.
Harmon, B. W., Fontenot, J. P., and Webb, K. E., Jr. (1975a). *J. Anim. Sci.* **40,** 144.
Harmon, B. W., Fontenot, J. P., and Webb, K. E., Jr. (1975b). *J. Anim. Sci.* **40,** 156.
Harmon, D. L., Brink, D. R., Britton, R. A., and Steele, R. T. (1981). *Proc., Annu. Meet.—Am. Soc. Anim. Sci., Raleigh, N.,* Abstr. No. 679.
Harpster, H. W., Fox, D. G., and Magee, W. T. (1976). *Mich. Agric. Exp. Stn., Res. Rep.* **328,** 42.
Harpster, H. W., Long, T. A., and Wilson, L. L. (1978). *J. Anim. Sci.* **46,** 238.
Harris, B. (1975). "Dairy Information Sheet," DY 73–11. University of Florida, Gainesville.
Harris, J. M., Cash, E. H., Wilson, L. L., and Stricklin, W. R. (1979). *J. Anim. Sci.* **49,** 613.
Harris, J. M., Shirley, R. L., and Palmer, A. Z. (1982). *J. Anim. Sci.* **55,** 1293.
Harris, L. E., and Mitchell, H. H. (1941). *J. Nutr.* **22,** 167.

Harris, L. E., Asplund, J. M., and Crampton, E. W. (1968). *Bull.—Utah Agric. Exp. Stn.* **479**.
Harrison, D. G., Beever, D. E., Thomson, D. J., and Osbourn, D. E. (1975). *J. Agric. Sci.* **85**, 93.
Harrison, D. G., Beever, D. E., Thomson, D. G., and Osbourn, D. E. (1976). *J. Sci. Food. Agric.* **27**, 617.
Hart, E. B., Bohstedt, G., Deobald, H. J., and Wegner, M. I. (1939). *J. Dairy Sci.* **22**, 785.
Hart, I. C., Bines, J. A., and Mordant, S. V. (1979). *J. Dairy Sci.* **62**, 270.
Hart, M. R., Walker, H. G., Jr., Graham, R. P., Hanni, P. J., Brown, A. H., and Kohler, G. O. (1981). *J. Anim. Sci.* **51**, 402.
Hartnell, G. F., and Satter, L. D. (1975). *J. Anim. Sci.* **41**, 403 (abstr.).
Hartnell, G. F., and Satter, L. D. (1978). *J. Anim. Sci.* **47**, 935.
Harvey, R. W., Burns, J. C., Blumer, T. N., and Linnerud, A. C. (1975). *J. Anim. Sci.* **41**, 740.
Hatch, C. F., and Beeson, W. M. (1972). *J. Anim. Sci.* **35**, 854.
Hatch, C. F., Perry, T. W., Mohler, M. T., and Beeson, W. M. (1972). *J. Anim. Sci.* **34**, 483.
Hatfield, E. E., and Smith, G. S. (1963). *J. Anim. Sci.* **22**, 1122 (abstr.).
Hatfield, E. E., Forbes, R. M., Neumann, A. L., and Garrigus, U. S. (1955). *J. Anim. Sci.* **14**, 1206 (abstr.).
Hatfield, E. E., Garrigus, U. S., Forbes, R. M., Neumann, A. L., and Gaither, W. (1959). *J. Anim. Sci.* **18**, 1208.
Hazzard, D. G., Kesler, E. M., Arnott, D. R., and Patton, S. (1958). *J. Dairy Sci.* **41**, 1439.
Heald, P. J. (1951). *Br. J. Nutr.* **5**, 84.
Heald, P. J., and Oxford, A. E. (1953). *Biochem. J.* **53**, 506.
Heath, M. E., and Plumlee, M. P. (1971). Using corn stalks for pasture. "Field Day Report. p 1 Purdue University, West Lafayette, Indiana.
Hedrick, H. B., Meyer, W. E., Alexander, M. A., Zobrisky, S. E., and Naumann, H. D. (1962). *J. Anim. Sci.* **21**, 362.
Heinemann, W. W., and Dyer, I. A. (1972). *Bull.—Wash. Agric. Exp. Stn.* **747**.
Heinemann, W. W., and van Keuren, R. W. (1956). *J. Anim. Sci.* **15**, 1097.
Heinemann, W. W., Hanks, E. M., and Young, D. C. (1978). *J. Anim. Sci.* **47**, 34.
Helmer, L. G., Bartley, E. E., and Meyer, R. M. (1965). *J. Dairy Sci.* **48**, 575.
Helmer, L. G., Bartley, E. E., Deyoe, C. W., Meyer, R. M., and Pfost, H. B. (1970a). *J. Dairy Sci.* **53**, 330.
Helmer, L. G., Bartley, E. E., and Deyoe, C. W. (1970b). *J. Dairy Sci.* **53**, 883.
Henderickx, H. K. (1961). *Arch. Int. Physiol. Biochim.* **69**, 449.
Henderickx, H. K., and Demeyer, D. I. (1967). *Naturwissenschaften* **14**, 369.
Henke, L. A. (1952). *Hawaii, Agric. Exp. Stn., Biannu. Rep., 1950–1952*, p. 30.
Hennessy, D. W., Williamson, P. J., Lowe, R. F., and Baigent, D. R. (1981). *J. Agric. Sci.* **96**, 205.
Hennig, A., Schuler, D., Freytag, H. H., Voigt, C., Gruhn, K., and Jeroch, H. (1972). *Jahrb. Tierernaehr. Futterung* **8**, 226.
Hentges, J. F., Jr., Moore, J. E., Palmer, A. Z., and Carpenter, J. W. (1966). *Bull.—Fla. Agric. Exp. Stn.* **708**.
Heydén, S. (1961). *Lantbrukshoegsk. Ann.* **27**, 273.
Hidiroglou, M., Ivan, M., and Jenkins, K. J. (1977). *J. Dairy Sci.* **60**, 1905.
Hill, J. R., and Godley, W. C. (1974). *J. Anim. Sci.* **39**, 156 (abstr.).
Hilton, J. H., Wilbur, J. W., and Hauge, S. M. (1932). *J. Dairy Sci.* **15**, 277.
Hinks, C. E., and Henderson, A. R. (1977). *Anim. Prod.* **25**, 53.
Hinks, C. E., and Whittemore, C. T. (1976). *Anim. Prod.* **22**, 415.
Hinman, D. D., and Johnson, R. R. (1974). *J. Anim. Sci.* **39**, 958.
Hird, F. J. R., and Wiedemann, M. J. (1964). *Biochem. J.* **92**, 585.
Hironaka, R. (1969). *Can. J. Anim. Sci.* **49**, 181.

# Bibliography

Hironaka, R., and Peters, H. F. (1969). *Can. J. Anim. Sci.* **49**, 323.
Hironaka, R., Miltimore, J. E., McArthur, J. M., McGregor, D. R., and Smith, E. S. (1973). *Can. J. Anim. Sci.* **53**, 75.
Hoar, D. W., Embry, L. B., King, H. R., and Emerick, R. J. (1968a). *J. Anim. Sci.* **27**, 556.
Hoar, D. W., Embry, L. B., and Emerick, R. J. (1968b). *J. Anim. Sci.* **27**, 1727.
Hobson, P. N. (1965). *J. Gen. Microbiol.* **38**, 167.
Hobson, P. N., and MacPherson, M. (1952). *Biochem. J.* **52**, 671.
Hobson, P. N., and Mann, S. O. (1961). *J. Gen. Microbiol.* **25**, 227.
Hobson, P. N., Mann, S. O., and Oxford, A. E. (1958). *J. Gen. Microbiol.* **19**, 462.
Hodges, E. M., Kirk, W. G., Peacock, F. M., Jones, D. W., Davis, G. K., and Neller, J. R. (1964). *Bull.—Fla., Agric. Exp. Stn.* **686**.
Hogan, J. P. (1961). *Aust. J. Biol. Sci.* **14**, 448.
Hogan, J. P. (1975). *J. Dairy Sci.* **58**, 1164.
Hogan, J. P., and Weston, R. H. (1970). *In* "Physiology of Digestion and Metabolism in the Ruminant" (A. T. Phillipson, ed.), p. 474 Oriel Press, Ltd., Newcastle upon Tyne, England.
Hogan, J. P., and Weston, R. H. (1971). *Aust. J. Agric. Res.* **22**, 951.
Holloway, J. W., and Totusek, R. (1973a). *J. Anim. Sci.* **37**, 800.
Holloway, J. W., and Totusek, R. (1973b). *J. Anim. Sci.* **37**, 807.
Holloway, J. W., Stephens, D. F., Whiteman, J. V., and Totusek, R. (1975). *J. Anim. Sci.* **40**, 114.
Holmes, J. H. G., Drennan, M. J., and Garrett, W. N. (1970). *J. Anim. Sci.* **31**, 409.
Holzer, Z., Tagari, H., Levy, D., and Volcani, R. (1976). *Anim. Prod.* **22**, 41.
Hopper, T. H. (1944). *J. Agric. Res.* **68**, 239.
Horn, F. P., Telford, J. P., McCroskey, J. E., Stephens, D. F., Whiteman, J. V., and Totusek, R. (1979). *J. Anim. Sci.* **49**, 1051.
Horn, G. W., Gordon, J. L., Prigge, E. C., and Owens, F. N. (1979). *J. Anim. Sci.* **48**, 683.
Horn, G. W., Mader, T. L., Armbruster, S. L., and Frahm, R. R. (1981). *J. Anim. Sci.* **52**, 447.
Horton, G. M. J. (1980a). *Anim. Prod.* **30**, 441.
Horton, G. M. J. (1980b). *J. Anim. Sci.* **50**, 1160.
Horton, G. M. J., and Holmes, W. (1974). *J. Br. Grassl. Soc.* **29**, 73.
Horton, G. M. J., and Holmes, W. (1975). *J. Anim. Sci.* **40**, 706.
Horton, G. M. J., and Holmes, W. (1976). *Anim. Prod.* **22**, 419.
Horton, G. M. J., and Holmes, W. (1978). *J. Anim. Sci.* **46**, 297.
Horton, G. M. J., and Nicholson, H. H. (1980). *Can. J. Anim. Sci.* **60**, 919.
Horton, G. M. J., and Nicholson, H. H. (1981). *J. Anim. Sci.* **52**, 1143.
Horton, G. M. J., Keeler, E. H., and Bassendowski, K. A. (1981). *Anim. Prod.* **32**, 267.
Howard, A., and Lawrie, R. A. (1957). *CSIRO, Spec. Rep.* **64**.
Howard, B. H. (1961). *Proc. Nutr. Soc.* **20**, 29 (abstr.).
Huber, J. T. (1970). *J. Anim. Sci.* **31**, 244 (abstr.).
Huber, J. T. (1976). *J. Anim. Sci.* **43**, 902.
Huber, J. T., and Kung, L. (1981). *J. Dairy Sci.* **64**, 1170.
Huber, J. T., Graf, G. C., and Engel, R. W. (1965). *J. Dairy Sci.* **48**, 1121.
Hudson, L. W., Glimp, H. A., Little, C. O., and Woolfolk, P. G. (1970). *J. Anim. Sci.* **30**, 609.
Huffman, D. L., Palmer, A. Z., Carpenter, J. W., and Shirley, R. L. (1969). *J. Anim. Sci.* **28**, 443.
Huffman, M. P., and Self, H. L. (1972). *J. Anim. Sci.* **35**, 871.
Hume, I. D. (1970). *Aust. J. Agric. Res.* **21**, 305.
Hume, I. D. (1974). *Aust. J. Agric. Res.* **25**, 155.
Hume, I. D., and Bird, P. R. (1970). *Aust. J. Agric. Res.* **21**, 315.
Hume, I. D., Moir, R. J., and Somers, M. (1970). *Aust. J. Agric. Res.* **21**, 283.
Hungate, R. E. (1943). *Biol. Bull. (Woods Hole, Mass.)* **84**, 157.
Hungate, R. E. (1957). *Can. J. Microbiol.* **3**, 289.

Hungate, R. E. (1966). "The Rumen and Its Microbes." Academic Press, New York.
Hungate, R. E., Daugherty, R. W., Bryant, M. P., and Cello, R. M. (1952). *Cornell Vet.* **42**, 423.
Hungate, R. E., Reichl, J., and Prins, R. (1970). *Appl. Microbiol.* **22**, 1104.
Hunt, C. H., Hershberger, T. V., and Cline, J. H. (1954). *J. Anim. Sci.* **13**, 570.
Hunter, G. D., and Millson, G. C. (1964). *Res. Vet. Sci.* **5**, 1.
Hunter, R. A., and Seibert, B. D. (1980). *Aust. J. Agric. Res.* **31**, 1037.
Husted, W. T., Mehen, S., Hale, W. H., Little, M., and Theurer, B. (1968). *J. Anim. Sci.* **27**, 531.
Hutjens, M. F., and Schultz, L. H. (1971). *J. Dairy Sci.* **54**, 1637.
Hutcheson, D. P., Cole, N. A., and McLaren, J. B. (1984). *J. Anim. Sci.* **58**, 700.
Hutson, F. M., Fuquay, J. W., Cardwell, J. T., and Brown, W. H. (1971). *J. Dairy Sci.* **54**, 459 (abstr.).
Irwin, L. N., Tucker, R. E., Mitchell, G. E., Jr., and Schelling, G. T. (1972). *J. Anim. Sci.* **35**, 267.
Irwin, L. N., Mitchell, G. E., Jr., Tucker, R. E., and Schelling, G. T. (1979). *J. Anim. Sci.* **48**, 367.
Isaacs, J., and Owens, F. N. (1972). *J. Anim. Sci.* **35**, 267.
Isaacson, H. R., Hinds, F. C., Bryant, M. P., and Ownes, F. N. (1975). *J. Dairy Sci.* **58**, 1645.
Ittner, N. R., Bond, T. E., and Kelly, C. F. (1955). *J. Anim. Sci.* **14**, 818.
Iwata, H., Kobayashi, K., Sato, Y., and Watanabe, Y. (1959). *Kyushu Univ. Fac. Agric. Sci., Bull.* **17**, 69.
Jacobs, N. J., and Wolin, M. J. (1963). *Biochim. Biophys. Acta* **69**, 29.
Jahn, E., and Chandler, P. T. (1976). *J. Anim. Sci.* **42**, 724.
Jahn, E., Chandler, P. T., and Kelly, R. F. (1976). *J. Anim. Sci.* **42**, 736.
Jainudeen, M. R., Hansel, W., and Davidson, K. L. (1964). *J. Dairy Sci.* **47**, 1382.
Jakobsen, P. E., Sorensen, P. H., and Larsen, H. (1957). *Acta Agric. Scand.* **7**, 103.
James, A. T., and Martin, A. J. P. (1952). *Biochem. J.* **50**, 679.
James, L. F., Allison, M. J., and Littledike, E. T. (1975). *In* "Digestion and Metabolism in the Ruminant" (I. W. McDonald and A. C. I. Warner, ed.). University of New England Publishing Unit, Armidale, N.S.W., Australia.
Jarvis, B. D. W. (1968). *Appl. Microbiol.* **16**, 714.
Jayasuriya, G. C. N., and Hungate, R. E. (1959). *Arch. Biochem. Biophys.* **82**, 274.
Jaysainghe, J. B. (1961). *Ceylon Vet. J.* **9**, 135.
Jensen, H. L., and Kourmaran, K. A. (1956). *Ann. Inst. Pasteur (Paris), Suppl.* **3**, 184.
Jensen, H. L., and Schroder, M. (1965). *J. Appl. Bacteriol.* **28**, 473.
Jensen, R., Connell, W. E., and Deen, A. W. (1954). *Am. J. Vet. Res.* **15**, 425.
Jesse, G. W., Thompson, G. B., Clark, J. L., Hedrick, H. B., and Weimer, K. G. (1976a). *J. Anim. Sci.* **43**, 418.
Jesse, G. W., Thompson, G. B., and Clark, J. L. (1976b). *J. Anim. Sci.* **43**, 1044.
Jesse, G. W., Thompson, G. B., Clark, J. L., Weimer, K. G., and Hutcheson, D. P. (1976c). *J. Anim. Sci.* **43**, 1049.
Jimenez, A. A., Perry, T. W., Pickett, R. A., and Beeson, W. M. (1963). *J. Anim. Sci.* **22**, 471.
Joanning, S. W., Johnson, D. E., and Barry, B. P. (1981). *J. Anim. Sci.* **53**, 1095.
Johns, A. T. (1953). *N. Z. J. Sci. Technol. Sect. A* **35**, 262.
Johnson, B. C., Hamilton, T. S., Mitchell, H. H., and Robinson, W. B. (1942). *J. Anim. Sci.* **1**, 236.
Johnson, D. E., Matsushima, J. K., and Knox, K. L. (1968). *J. Anim. Sci.* **27**, 1431.
Johnson, H. D., and Yeck, R. G. (1964). *Res. Bull.—Mo., Agric. Exp. Stn.* **865**.
Johnson, R. J., and Dyer, I. A. (1966). *J. Anim. Sci.* **25**, 903.
Johnson, R. R., and McClure, K. E. (1973). *J. Anim. Sci.* **36**, 397.
Johnson, R. R., Dehority, B. A., and Bentley, D. G. (1958). *J. Anim. Sci.* **17**, 841.
Johnson, W. H., Goodrich, R. D., and Meiske, J. C. (1971). *J. Anim. Sci.* **32**, 778.

Johnston, R. P., Kesler, E. M., and McCarthy, R. D. (1961). *J. Dairy Sci.* **44,** 331.
Johnstone-Wallace, D. B., and Kennedy, K. (1944). *Proc. N. Z. Soc. Anim. Prod.* **31,** 92.
Jones, S. D. M., Price, M. A., and Berg, R. T. (1978). *J. Anim. Sci.* **46,** 1151.
Jordan, R. M., and Croom, H. G. (1957). *J. Anim. Sci.* **16,** 118.
Jordan, W. A., Lister, E. E., and Rowlands, G. J. (1968). *Can. J. Anim. Sci.* **48,** 145.
Jorgensen, N. A., Santini, F. J., and Crowley, J. W. (1981). *Proc.—Cornell Nutr. Conf. Feed Manuf., 1981,* p. 24.
Joyce, J. P., and Blaxter, K. L. (1964). *Br. J. Nutr.* **18,** 5.
Joyner, A. E., and Baldwin, R. L. (1966). *J. Bacteriol.* **92,** 1321.
Joyner, A. E., Kesler, E. M., and Holter, J. B. (1963). *J. Dairy Sci.* **46,** 1108.
Juhász, B., and Szegedi, B. (1968). *Acta Vet. Acad. Sci. Hung.* **18,** 63.
Kamstra, L. D., Ronning, D., Walker, H. G., Kohler, G. O., and Wyman, O. (1980). *J. Anim. Sci.* **50,** 153.
Kandatsu, M., Matsumoto, T., Kazama, Y., Kikuno, K., and Ichinose, Y. (1955). *J. Agric. Chem. Soc. Jpn.* **29,** 759.
Karr, M. R., Garrigus, U. S., Hatfield, E. E., and Norton, H. W. (1965a). *J. Anim. Sci.* **24,** 459.
Karr, M. R., Garrigus, U. S., Hatfield, E. E., Norton, H. W., and Doane, B. B. (1965b). *J. Anim. Sci.* **24,** 469.
Karr, M. R., Little, C. O., and Mitchell, G. E., Jr. (1966). *J. Anim. Sci.* **25,** 652.
Kartchner, R. J. (1981). *J. Anim. Sci.* **51,** 432.
Kartchner, R. J., Rittenhouse, L. R., and Raleigh, R. J. (1979). *J. Anim. Sci.* **48,** 425.
Kassell, B., and Laskowski, M. (1956). *J. Biol. Chem.* **219,** 203.
Kay, R. N. B., and Hobson, P. N. (1963). *J. Dairy Res.* **30,** 261.
Kearl, L. C., Frischknecht, N. C., and Harris, L. E. (1971). *Proc., Annu. Meet.—Am. Soc. Anim. Sci., West. Sect.* **22,** 63.
Keating, E. K., Saba, W. J., Hale, W. H., and Taylor, B. (1965). *J. Anim. Sci.* **24,** 1080.
Keener, H. A., Colovos, N. F., and Eckberg, R. B. (1957). *N. H., Agric. Exp. Stn. Bull.* **438.**
Keith, E. A., and Daniels, L. B. (1976). *J. Anim. Sci.* **42,** 888.
Keith, E. A., Colenbrander, V. R., Perry, T. W., and Bauman, L. F. (1981). *J. Anim. Sci.* **52,** 8.
Kellems, R. O., Wayman, O., Nguyen, A. H., Nolan, J. C., Jr., Campbell, C. M., Carpenter, J. R., and Ho-a, E. B. (1979). *J. Anim. Sci.* **48,** 1040.
Kelley, R. O., Martz, F. A., and Johnson, H. D. (1967). *J. Dairy Sci.* **50,** 531.
Kelly, C. F., Bond, T. E., and Ittner, N. R. (1955). *Agric. Eng.* **36,** 173.
Kelly, N. C., Thomas, P. C., and Chamberlain, D. G. (1978). *Proc. Nutr. Soc.* **37,** 34A (abstr.).
Kennedy, P. M., and Milligan, L. P. (1978). *Br. J. Nutr.* **39,** 105.
Kennedy, P. M., Christopherson, R. J., and Milligan, L. P. (1976). *Br. J. Nutr.* **36,** 231.
Kezar, W. W., and Church, D. C. (1979). *J. Anim. Sci.* **49,** 1396.
Kiddle, P., Marshall, R. A., and Phillipson, A. T. (1957). *J. Physiol. (London)* **113,** 307.
Kienholz, E. W., Ward, G. M., and Johnson, D. E. (1976). *J. Anim. Sci.* **43,** 230 (Abstr.).
Kiesling, H. E., McCroskey, J. E., and Wagner, D. G. (1973). *J. Anim. Sci.* **37,** 790.
King, K. W. (1956). *Va., Agric. Exp. Stn., Tech. Bull.* **127,**
King, K. W., and Vessal, M. I. (1969). *In* "Enzymes of the Cellulose Complex Symposium: Cellulases and Their Applications" (R. E. Gould, ed.), p. 7. Am. Chem. Soc., Washington, D.C.
Kirk, W. G., and Davis, G. K. (1954). *Bull.—Fla., Agric. Exp. Stn.* **538.**
Kirk, W. G., Peacock, F. M., Hodges, E. M., and Jones, D. W. (1958). *Bull.—Fla., Agric. Exp. Stn.* **603.**
Kirk, W. G., Chapman, H. L., Jr., Peacock, F. M., and Davis, G. K. (1969). *Bull.—Fla. Agric. Exp. Stn.* **614A.**

Kistner, A., Gouws, L., and Gilchrist, F. M. C. (1962). *J. Agric. Sci.* **59**, 85.
Kleiber, M. (1975). "The Fire of Life," p. 257. Robert E. Krieger Publishing Co., Huntington, New York.
Klopfenstein, T. J., and Abrams, S. M. (1981). Distillers grains as a naturally protected protein for ruminants. *Nebr. Beef Cattle Rep.* p 1.
Klosterman, E. W., and Parker, C. F. (1976). Effect of size, breed and sex on feed efficiency of beef cattle. *Ohio, Agr. Exp. Stn., Res. Bull.* **1088**.
Klosterman, E. W., Parker, C. F., Bishop, R. B., and Cahill, V. R. (1972). *Ohio Agric. Summ.* **63**, 1.
Knox, K. L., and Handley, T. M. (1973). *J. Anim. Sci.* **37**, 190.
Koenig, S. E., Hatfield, E. E., and Spears, J. W. (1978). *J. Anim. Sci.* **46**, 490.
Koers, W. C., Britton, R., Klopfenstein, T. J., and Woods, W. R. (1976). *J. Anim. Sci.* **43**, 684.
Koffman, M. (1937). *Lantbrukshoegsk. Ann.* **5**, 201.
Kotb, A. R., and Luckey, T. D. (1972). *Nutr. Abstr. Rev.* **42**, 813.
Kowalczyk, J., Orskov, E. R., Robinson, J. J., and Stewart, C. S. (1977). *Br. J. Nutr.* **37**, 251.
Kraft, H. W. (1963). *Proc. S. Afr. Soc. Anim. Prod.* **2**, 43.
Krause, V., Klopfenstein, T., Guyer, P. Q., and Tolman, W. (1980). *J. Anim. Sci.* **50**, 216.
Kraybill, H. F., Bitter, H. L., and Hankins, O. G. (1952). *J. Appl. Physiol.* **4**, 575.
Kress, D. D., Hauser, E. R., and Chapman, A. B. (1969). *J. Anim. Sci.* **29**, 373.
Kretschmer, A. E. (1958). *Agron. J.* **50**, 314.
Kriss, M. (1930). *J. Agric. Res.* **40**, 283.
Kriss, M. (1943). *J. Anim. Sci.* **2**, 63.
Krogh, N. (1961). *Acta Vet. Scand.* **2**, 357.
Kromann, R. P. (1967). *J. Anim. Sci.* **26**, 1131.
Kropp, J. R., Johnson, R. R., Males, J. R., and Owens, F. N. (1977a). *J. Anim. Sci.* **45**, 844.
Kropp, J. R., Johnson, R. R., Males, J. R., and Owens, F. N. (1977b). *J. Anim. Sci.* **46**, 837.
Krzywaneck, F. W. (1929). *Arch. Gen. Physiol.* **222**, 89.
Kuc, J., and Nelson, O. E. (1964). *Arch. Biochem. Biophys.* **105**, 103.
Kung, L., Jr., and Stanley, R. W. (1982). *J. Anim. Sci.* **54**, 689.
Kunitz, M. (1945). *Science* **101**, 668.
Kunitz, M. (1947). *J. Gen. Physiol.* **30**, 291.
Kunkle, W. E., Fetter, A. W., and Preston, R. L. (1976). *J. Anim. Sci.* **42**, 1263.
Kyker, G. C. (1962). *In* "Mineral Metabolism" (C. L. Comar and F. Bonner, eds.), Vol. 2, Part 2B, p. 499. Academic Press, New York.
Labosky, P., Dick, J. W., and Cross, D. L. (1977). *Poult. Sci.* **56**, 2064.
Ladd, J. N. (1959). *Biochem. J.* **71**, 16.
Ladd, J. N., and Walker, D. J. (1959). *Biochem. J.* **71**, 364.
Lake, R. P., Hildebrand, R. L., Clanton, D. C., and Jones, L. E. (1974). *J. Anim. Sci.* **39**, 827.
Lamm, W. D., Ward, J. K., and White, G. C. (1977). *J. Anim. Sci.* **45**, 1231.
Land, H., and Virtanen, A. I. (1959). *Acta Chem. Scand.* **13**, 489.
Langlands, J. P. (1966). *Anim. Prod.* **8**, 253.
Lanigan, G. W. (1970). *Aust. J. Agric. Res.* **21**, 633.
Larsen, H. J., Stoddard, G. E., Jacobson, N. L., and Allen, R. S. (1956). *J. Anim. Sci.* **15**, 473.
Lassiter, J. W., Hamdy, M. K., and Buranamanas, P. (1963). *J. Anim. Sci.* **22**, 335.
Laster, D. B., Glimp, H. A., and Gregory, K. E. (1973). *J. Anim. Sci.* **36**, 734.
Leathem, J. H. (1961). *J. Anim. Sci., Suppl.* **25**, 68.
Leaver, J. D. (1973). *Anim. Prod.* **17**, 43.
Lechtenberg, V. L., Muller, L. D., Bauman, L. F., Rhykerd, C. L., and Barnes, R. F. (1972). *Agron. J.* **64**, 657.
Ledger, H. P., Gilliver, B., and Robb, J. M. (1973). *J. Agric. Sci.* **80**, 381.

Lee, R. W., Galyean, M. L., Lofgreen, G. P., Leighton, E. A., and Hubbert, M. E. (1981). *Proc., Annu. Meet. Am. Soc. Anim. Sci., Raleigh, N.C.,* Abstr. No. 700.
Leitch, I., and Thompson, J. S. (1944). *Nutr. Abstr. Rev.* **14,** 197.
Lemenager, R. P., Owens, F. N., Lusby, K. S., and Totusek, R. (1978a). *J. Anim. Sci.* **47,** 247.
Lemenager, R. P., Owens, F. N., Schockey, B. J., Lusby, K. S., and Totusek, R. (1978b). *J. Anim. Sci.* **47,** 255.
Lemenager, R. P., Owens, F. N., and Totusek, R. (1978c). *J. Anim. Sci.* **47,** 262.
Lemenager, R. P., Smith, W. H., Martin, T. C., Singleton, W. L., and Hodges, J. R. (1980). *J. Anim. Sci.* **51,** 837.
Lemenager, R. P., Martin, T. G., Stewart, T. S., and Perry, T. W. (1981). *J. Anim. Sci.* **53,** 26.
Leng, R. A. (1970a). *Adv. Vet. Sci. Comp. Med.* **14,** 209.
Leng, R. A. (1970b). *In* "Physiology of Digestion and Metabolism in the Ruminant" (A. T. Phillipson, ed.), p. 406. Oriel Press, Ltd., Newcastle upon Tyne, England.
Leng, R. A., Steel, J. W., and Luick, J. R. (1967). *Biochem. J.* **103,** 785.
Leu, B. M., Hoffman, M. P., and Self, H. L. (1977). *J. Anim. Sci.* **44,** 717.
Levy, D., Amir, S., Holzer, Z., and Neumark, H. (1972). *Anim. Prod.* **15,** 157.
Lewis, D., and McDonald, I. W. (1958). *J. Agric. Sci.* **51,** 108.
Lewis, J. K., Gartner, F. R., and Nesvold, J. (1964). *Proc., Annu. Meet.— Am. Soc. Anim. Sci., West. Sect.* **15,** 59.
Lewis, T. R., and Emery, R. S. (1962). *J. Dairy Sci.* **45,** 765.
Lichtenwalner, R. E., Fontenot, J. P., and Webb, K. E., Jr. (1971). *Res. Div. Rep.—Va. Polytech. Inst. State Univ.* **125,** 117.
Liebholz, J. (1972). *Aust. J. Agric. Res.* **23,** 1073.
Liener, I. E., and Pallansch, M. J. (1952). *J. Biol. Chem.* **197,** 29.
Lindahl, I. L. (1959). *In* "Techniques and Procedures Used in Animal Production" (C. E. Terrill, ed.), p. 173. Am. Soc. Anim. Prod., Madison, Wisconsin.
Lindahl, I. L., Davis, R. E., Jacobson, D. R., and Shaw, J. C. (1957). *J. Anim. Sci.* **16,** 165.
Lindsay, J. A., and Davies, H. L. (1981). *Anim. Prod.* **32,** 85.
Lindsay, J. R., and Hogan, J. P. (1972). *Aust. J. Agric. Res.* **23,** 321.
Linn, J. G., Staba, E. J., Goodrich, R. D., Meiske, J. C., and Otterby, D. E. (1975a). *J. Anim. Sci.* **41,** 601.
Linn, J. G., Goodrich, R. D., Otterby, D. E., Meiske, J. C., and Staba, E. J. (1975b). *J. Anim. Sci.* **41,** 610.
Lippke, H. (1975). *J. Dairy Sci.* **58,** 1860.
Lippke, H., Vetter, R. L., and Jacobson, N. L. (1969). *J. Anim. Sci.* **28,** 819.
Lippke, H., Vetter, R. L., and Jacobson, N. L. (1970). *J. Anim. Sci.* **31,** 1195.
Lister, E. E., Jordan, W. A., Wauthy, J. W., Comeau, J. E., and Pigden, W. J. (1972). *Can. J. Anim. Sci.* **52,** 671.
Little, C. O., Burroughs, W., and Woods, W. (1963). *J. Anim. Sci.* **22,** 358.
Little, C. O., Mitchell, G. E., Jr., and Potter, G. D. (1968). *J. Anim. Sci.* **27,** 1722.
Little, C. O., Bradley, N. W., and Mitchell, G. E., Jr. (1969). *J. Anim. Sci.* **28,** 135 (abstr.).
Loerch, S. C., and Berger, L. L. (1981). *J. Anim. Sci.* **53,** 1198.
Lofgreen, G. P. (1965). *Publ.—Eur. Assoc. Anim. Prod.* **11,** 309.
Lofgreen, G. P., and Christiansen, A. C. (1962). *J. Anim. Sci.* **21,** 262.
Lofgreen, G. P., and Garrett, W. N. (1968). *J. Anim. Sci.* **27,** 793.
Lofgreen, G. P., and Otagaki, K. K. (1960). *J. Anim. Sci.* **19,** 392.
Lofgreen, G. P., Loosli, J. K., and Maynard, L. A. (1947). *J. Anim. Sci.* **6,** 343.
Lofgreen, G. P., Bath, D. L., and Strong, H. T. (1963). *J. Anim. Sci.* **22,** 598.
Lofgreen, G. P., Givens, R. L., Morrison, S. R., and Bond, T. E. (1975a). *J. Anim. Sci.* **40,** 223.
Lofgreen, G. P., Dunbar, J. R., Addis, D. G., and Clark, J. G. (1975b). *J. Anim. Sci.* **41,** 1256.

Lofgreen, G. P., Stinocher, L. H., and Kiesling, H. E. (1980). *J. Anim. Sci.* **50**, 590.
Lofgreen, G. P., El Tayeb, A. E., and Kiesling, H. E. (1981). *J. Anim. Sci.* **52**, 959.
Lohman, T. G., Breidenstein, B. C., Twardock, A. R., Smith, G. W., and Norton, H. W. (1966). *J. Anim. Sci.* **25**, 1218.
Loosli, J. K., and Harris, L. E. (1945). *J. Anim. Sci.* **4**, 435.
Loosli, J. K., and McDonald, I. W. (1968). *FAO Agric. Stud.* **75**.
Loveday, H. D., and Dikeman, M. E. (1980). *J. Anim. Sci.* **51**, 78.
Lowman, B. G., and Knight, D. W. (1970). *Anim. Prod.* **12**, 525.
Lucas, D. M., Fontenot, J. P., and Webb, K. E., Jr. (1975). *J. Anim. Sci.* **41**, 1480.
Lutwak-Mann, C. (1958). *Vitam. Horm. (N.Y.)* **16**, 35.
Lyford, S. J., Jr., Smart, W. W. G., Jr., and Matrone, G. (1963). *J. Nutr.* **79**, 105.
Lyman, C. M., Kuiken, K. A., and Hale, F. (1956). *J. Agric. Food Chem.* **4**, 1008.
Lyman, C. M., Baliga, B. P., and Slay, M. W. (1959). *Arch. Biochem. Biophys.* **84**, 486.
Lyttleton, J. W. (1960). *N. Z. J. Agric. Res.* **3**, 63.
McAllan, A. B., and Smith, R. H. (1974). *Proc. Nutr. Soc.* **33**, 41A (abstr.).
McAllan, A. B., and Smith, R. H. (1976). *Br. J. Nutr.* **36**, 511.
McArthur, J. M., and Miltimore, J. E. (1961). *Can. J. Anim. Sci.* **41**, 187.
McCartor, M. M., and Smith, G. C. (1978). *J. Anim. Sci.* **47**, 270.
McCartor, M. M., Randel, R. D., and Carroll, L. H. (1979). *J. Anim. Sci.* **48**, 488.
McCarty, R. D., Porter, G. A., and Griel, L. C., Jr. (1968). *J. Dairy Sci.* **51**, 459.
McCarthy, F. D., Hawkins, D. R., and Bergen, W. G. (1985). *J. Anim. Sci.* **60**, 781.
McCaskey, T. A., and Anthony, W. B. (1979). *J. Anim. Sci.* **48**, 163.
McClure, T. J. (1961). *N. Z. Vet. J.* **9**, 107.
McCullough, T. A. (1969). *Anim. Prod.* **11**, 145.
McCullough, M. E. (1972). *J. Anim. Sci.* **34**, 127.
McCullough, M. E. (1973). "Optimum Feeding of Dairy Animals for Meat and Milk," 2nd ed. Univ. of Georgia Press, Athens.
McCullough, M. E., and Neville, W. E. (1972). *Feedstuffs* **44**, 27.
McDonald, I. W. (1952). *Biochem. J.* **51**, 86.
McDonald, I. W. (1954). *Biochem. J.* **56**, 120.
McDonald, M. A., and Bell, J. M. (1958). *Can. J. Anim. Sci.* **38**, 160.
McDonald, P., and Whittenbury, R. (1973). *In* "Chemistry and Biochemistry of Herbage" (G. W. Butler and R. W. Bailey, eds.), Vol. 3, p. 33. Academic Press, London.
McDonald, P., Sterling, A. C., Henderson, A. R., Dewar, W. A., Stark, G. H., Davie, W. G., Macpherson, H. T., Reid, A. M., and Slater, J. (1960). "Studies on Ensilage," Tech. Bull. No. 24. Edinburgh School of Agriculture.
McDowell, R. E., Moody, E. G., van Soest, P. J., Lehmann, R. P., and Ford, G. L. (1969). *J. Dairy Sci.* **52**, 188.
Macfarlane, W. V. (1965). *In* "Studies in Physiology," p. 191. Springer-Verlag, Berlin and New York.
McGuire, R. L., Bradley, N. W., and Little, C. O. (1966). *J. Anim. Sci.* **25**, 185.
MacKenzie, H. I., and Altona, R. E. (1964). *J. S. Afr. Vet. Med. Assoc.* **35**, 301.
McLaren, G. A., Anderson, G. C., Welch, J. A., Campbell, C. D., and Smith, G. A. (1959). *J. Anim. Sci.* **18**, 1319.
McLaren, G. A., Anderson, G. C., and Barth, K. M. (1965). *J. Anim. Sci.* **24**, 231.
Macleod, G. K., Yu, Y., and Schaeffer, L. R. (1977). *J. Dairy Sci.* **60**, 726.
McManus, W. R. (1961). *Nature (London)* **192**, 1161.
McMeniman, N. P., and Armstrong, D. G. (1977). *Anim. Feed Sci. Technol.* **2**, 255.
McNaught, M. L., Owen, E. C., Henry, K. M., and Kon, A. K. (1954). *Biochem. J.* **56**, 151.
McNeill, J. W., Potter, G. D., and Riggs, J. K. (1971). *J. Anim. Sci.* **33**, 1371.

MacRae, J. C. (1976) *Rev. Rural Sci.* **2,** 93.
Maeng, W. J., and Baldwin, R. L. (1976a). *J. Dairy Sci.* **59,** 636.
Maeng, W. J., and Baldwin, R. L. (1976b). *J. Dairy Sci.* **59,** 643.
Maeng, W. J., and Baldwin, R. L. (1976c). *J. Dairy Sci.* **59,** 648.
Maeng, W. J., van Nevel, C. J., Baldwin, R. L., and Morris, J. G. (1976). *J. Dairy Sci.* **59,** 68.
Mahadevan, S., Sauer, F., and Erfle, J. D. (1976). *J. Anim. Sci.* **42,** 745.
Mahadevan, S., Erfle, J. D., and Sauer, F. D. (1979). *Feedstuffs* **51**(51), 20.
Mahadevan, S., Erfle, J. D., and Sauer, F. D. (1980). *J. Anim. Sci.* **50,** 723.
Mangan, J. L. (1959). *N. Z. J. Agric. Res.* **2,** 47.
Mangan, J. L. (1972). *Br. J. Nutr.* **27,** 261.
Mangus, W. L., and Brinks, J. S. (1971). *J. Anim. Sci.* **32,** 17.
Mann, S. O. (1970). *J. Appl. Bacteriol.* **33,** 403.
Marchello, J. A., Wooten, R. A., Roubicek, C. B., Swingle, R. S., and Dryden, F. D. (1979). *J. Anim. Sci.* **48,** 831.
Margolin, S. (1930). *Biol. Bull. (Woods Hole, Mass.)* **59,** 301.
Markoff, J. (1913). *Biochem. J.* **57,** 1.
Marlowe, T. J., Mast, C. C., and Schalles, R. R. (1955). *J. Anim. Sci.* **24,** 494.
Marshall, S. P., and van Horn, H. H. (1975). *J. Dairy Sci.* **58,** 896.
Martin, J. E., Arrington, L. R., Moore, J. E., Ammerman, C. B., Davis, G. K., and Shirley, R. L. (1964). *J. Nutr.* **83,** 60.
Martin, L. C., Ammerman, C. B., Burns, W. C., and Koger, M. (1976). *J. Anim. Sci.* **42,** 21.
Martin, T. G., Lane, G. T., Mudge, M. D., and Albright, J. L. (1978a). *J. Dairy Sci.* **61,** 1151.
Martin, T. G., Perry, T. W., Beeson, W. M., and Mohler, M. T. (1978b). *J. Anim. Sci.* **47,** 29.
Martin, T. G., Lemenager, R. P., Srinivasan, G., and Alenda, A. (1981). *J. Anim. Sci.* **53,** 33.
Martinez, A., and Church, D. C. (1970). *J. Anim. Sci.* **31,** 982.
Martinsson, K., and Lindell, L. (1981). *Swed. J. Agric. Res.* **11,** 23.
Mason, V. C., and Palmer, R. (1971). *J. Agric. Sci.* **76,** 567.
Masonite Corporation (1977). Anon. Masonex Product Sheet, Rep. MC877-3B, Chicago, Illinois.
Masson, M. J., and Phillipson, A. T. (1952). *J. Physiol. (London)* **116,** 98.
Mata-Hernandez, A., Marchello, J. A., Roubicek, C. B., Ochoa, M. F., Bennett, J. A., and Gorman, W. D. (1981). *J. Anim. Sci.* **53,** 1246.
Mather, R. E., and Apgar, W. P. (1956). *J. Dairy Sci.* **39,** 938 (abstr.).
Mathison, G. W., and Milligan, L. R. (1971). *Br. J. Nutr.* **25,** 351.
Mathison, G. W., Hartin, R. T., and Beck, B. E. (1981). *Can. J. Anim. Sci.* **61,** 375.
Matsumoto, T. (1961). *Tohoku J. Agric. Res.* **12,** 213.
Matsushima, J. K., and Montgomery, R. L. (1967). *Col., Agric. Exp. Stn., Colo. Farm Home Res.* **17,** 4.
Matsushima, J. K., and Stenquist, N. J. (1967). *J. Anim. Sci.* **26,** 925 (abstr.).
Maust, L. E., McDowell, R. E., and Hooven, N. W. (1972). *J. Dairy Sci.* **55,** 1133.
Maxson, W. E., Shirley, R. L., Bertrand, J. E., and Palmer, A. Z. (1973). *J. Anim. Sci.* **37,** 1451.
Mayes, R. W., and Orskov, E. R. (1974). *Br. J. Nutr.* **32,** 143.
Meacham, T. N., Warnick, A. C., Cunha, T. J., Hentges, J. F., and Shirley, R. L. (1964). *J. Anim. Sci.* **23,** 380.
Meacham, T. N., Bovard, K. P., Priode, B. M., and Fontenot, J. P. (1970). *J. Anim. Sci.* **31,** 428.
Mead, S. W., Cole, H. H., and Regan, W. M. (1944). *J. Dairy Sci.* **27,** 779.
Mehrez, A. Z., Orskov, E. R., and McDonald, I. (1977). *Br. J. Nutr.* **38,** 437.
Meiske, J. C., van Arsdell, W. J., Luecke, R. W., and Hoefer, J. A. (1955). *J. Anim. Sci.* **14,** 941.
Meiske, J. C., Goodrich, R. D., and Owens, F. N. (1968). "Beef Cattle Feeders Day Report," B-112. University of Minnesota, St. Paul.
Meiske, J. C., Goodrich, R. D., and Owens, F. N. (1969). *J. Anim. Sci.* **29,** 165 (abstr.).

Mendel, V. E., and Boda, J. M. (1961). *J. Dairy Sci.* **44,** 1881.
Merricks, D. L., and Salsbury, R. L. (1976). *J. Anim. Sci.* **42,** 955.
Mertens, D. R. (1977). *Proc. Ga. Nutr. Conf. Feed Ind.,* p. 30.
Mertens, D. R., Campbell, J. R., Martz, F. A., and Hildebrand, E. S. (1971a). *J. Dairy Sci.* **54,** 667.
Mertens, D. R., Martz, F. A., and Campbell, J. R. (1971b). *J. Dairy Sci.* **54,** 931.
Mertz, E. T., Bates, L. S., and Nelson, O. E. (1964). *Science* **145,** 279.
Mertz, E. T., Veron, O. A., Bates, L. S., and Nelson, O. E. (1965). *Science* **148,** 1741.
Meyer, J. H., and Clawson, W. J. (1964). *J. Anim. Sci.* **23,** 214.
Meyer, J. H., Weir, W. C., and Smith, J. D. (1955). *J. Anim. Sci.* **14,** 160.
Meyer, J. H., Hull, J. L., Weitkamp, W. H., and Bonilla, S. (1965). *J. Anim. Sci.* **24,** 29.
Meyer, R. M., and Bartley, E. E. (1971). *J. Anim. Sci.* **33,** 1018.
Meyer, R. M., and Bartley, E. E. (1972a). U.S. Patent 3, 686, 416.
Meyer, R. M., and Bartley, E. E. (1972b). *J. Anim. Sci.* **34,** 234.
Meyer, R. M., Helmer, L. G., and Bartley, E. E. (1965). *J. Dairy Sci.* **48,** 503.
Meyer, R. M., Bartley, E. E., and Barr, G. W. (1973). *J. Anim. Sci.* **37,** 351 (Abstr.).
Miller, E. L. (1973). *Proc. Nutr. Soc.* **32,** 79.
Miller, J. K., Perry, S. C., Chandler, P. T., and Cragle, R. G. (1967). *J. Dairy Sci.* **50,** 355.
Miller, R. W., Hemken, R. W., Waldo, D. R., Okamoto, M., and Moore, L. A. (1965). *J. Dairy Sci.* **48,** 1455.
Millet, M. A., Baker, A. J., Feist, W. C., Mellenberger, R. W., and Satter, L. D. (1970). *J. Anim. Sci.* **31,** 781.
Millet, M. A., Baker, A. J., Satter, L. D., McGovern, J. N., and Dinius, D. A. (1973). *J. Anim. Sci.* **37,** 599.
Milligan, J. D., and Christison, G. I. (1974). *Can. J. Anim. Sci.* **54,** 605.
Milligan, L. P. (1971). *Fed. Proc., Fed. Am. Soc. Exp. Biol.* **30,** 1454.
Milne, J. A., Maxwell, T. J., and Souter, W. (1981). *Anim. Prod.* **32,** 185.
Minson, D. J. (1963). *J. Br. Grassl. Soc.* **18,** 39.
Minson, D. J. (1976). *Rev. Rural Sci.* **2,** 37.
Minson, D. J. (1981). *Anim. Feed Sci. Technol.* **6,** 223.
Mishra, B. D., Bartley, E. E., Fina, L. R., and Claydon, T. J. (1967). *J. Anim. Sci.* **26,** 606.
Mishra, B. D., Bartley, E. E., Fina, L. R., and Bryant, M. P. (1968). *J. Anim. Sci.* **27,** 1651.
Mishra, M. F., Martz, F. A., Stanley, R., Johnson, H. D., Campbell, J. R., and Hildebrand, E. (1969). *J. Anim. Sci.* **29,** 166 (abstr.).
Miyazaki, A. K., Okamoto, K., Tsude, E., Kawashima, R., and Uesaka, S. (1974). *Nippon Chikusan Gakkai Ho* **45,** 183, *Chem. Abstr.* **81,** 12232.
Mizwicki, K. L., Owens, F. N., Poling, K., and Burnett, G. (1980). *J. Anim. Sci.* **51,** 698.
Moe, P. W., and Tyrrell, H. F. (1971). *J. Dairy Sci.* **55,** 480.
Moe, P. W., and Tyrrell, H. F. (1972). *J. Anim. Sci.* **35,** 271.
Moe, P. W., and Tyrrell, H. F. (1973). *J. Anim. Sci.* **37,** 183.
Moe, P. W., and Tyrrell, H. F. (1979). *J. Dairy Sci.* **62,** 1583.
Moe, P. W., Reid, J. T., and Tyrrell, H. F. (1965). *J. Dairy Sci.* **48,** 1053.
Moe, P. W., Flatt, W. P., and Tyrrell, H. F. (1972). *J. Dairy Sci.* **55,** 945.
Moir, R. J., and Harris, L. E. (1962). *J. Nutr.* **77,** 285.
Moir, R. J., and Somers, M. (1956). *Aust. J. Agric. Res.* **8,** 253.
Moir, R. J., Somers, M., and Bray, A. C. (1967). *Sulfur Inst. J.* **3,** 15.
Monteith, J. L., and Mount, L. E. (1974). "Heat Loss from Man and Animals." Butterworth, London.
Montgomery, M. J., and Baumgardt, B. R. (1965). *J. Dairy Sci.* **48,** 569.

Moore, J. E., and Mott, G. O. (1973). *In* "Anti-quality Components of Forages" (A. G. Matches, ed.), CSSA Spec. Publ. No. 4, p. 53. Crop Sci. Soc. Am., Madison, Wisconsin.
Moore, L. A. (1964). *J. Anim. Sci.* **23,** 230.
Morris, M. P., and Garcia-Rivera, J. (1955). *J. Dairy Sci.* **38,** 1169.
Morrison, F. B. (1959). "Feeds and Feeding," 22nd ed. Morrison Publishing Co., Clinton, Iowa.
Morrison, S. R., Givens, R. L., and Lofgreen, G. P. (1973). *J. Anim. Sci.* **36,** 428.
Moseley, W. M., McCartor, M. M., and Randel, R. D. (1977). *J. Anim. Sci.* **45,** 961.
Mott, G. O., Quinn, L. R., Bissehoff, W. V. A., and du Rocha, G. L. (1967). *Pesqui. Agropecu. Bras.* **2,** 9.
Mott, G. O., Rhykerd, C. L., Tayler, R. W., Perry, T. W., and Huber, D. A. (1968). *ASA Spec. Publ.* **13,** 94.
Moulton, C. R. (1923). *J. Biol. Chem.* **57,** 79.
Moulton, C. R., Trowbridge, P. F., and Haigh, L. D. (1922). *Res. Bull.—Mo., Agric. Exp. Stn.* **55.**
Mowat, D. N., and Deelstra, K. (1972). *J. Anim. Sci.* **34,** 332.
Mowat, D. N., McCaughey, P., and Macleod, G. K. (1981). *Can. J. Anim. Sci.* **61,** 703.
Mueller, R. G., and Harris, G. A. (1967). *J. Range Manage.* **20,** 67.
Muir, L. A., and Barreto, A. (1979). *J. Anim. Sci.* **48,** 468.
Muir, L. A., Rickes, E. L., Duquette, P. F., and Smith, G. E. (1981). *J. Anim. Sci.* **52,** 635.
Mulholland, J. G., Coombe, J. B., and McManus, W. R. (1976). *Aust. J. Agric. Res.* **27,** 139.
Mullenax, C. H., Keeler, R. F., and Allison, M. J. (1966). *Am. J. Vet. Res.* **27,** 857.
Muller, L. D., Lechtenberg, V. L., Bauman, L. F., Barnes, R. F., and Rhykerd, C. L. (1972). *J. Anim. Sci.* **35,** 883.
Munro, H. N. (1964). *In* "Mammalian Protein Metabolism" (H. N. Munro and J. B. Allison, eds.), Vol. 1, p. 38. Academic Press, New York.
Muntifering, R. B., Theurer, B., Swingle, R. S., and Hale, W. H. (1980). *J. Anim. Sci.* **50,** 930.
Naga, M A , and Harmeyer, J. H. (1975). *J. Anim. Sci.* **40,** 374.
Nagaraja, T. G., Bartley, E. E., Fina, L. R., Anthony, H. D., and Bechtle, R. M. (1978a). *J. Anim. Sci.* **47,** 226.
Nagaraja, T. G., Bartley, E. E., Fina, L. R., and Anthony, H. D. (1978b). *J. Anim. Sci.* **47,** 1329.
Nagaraja, T. G., Avery, T. B., Bartley, E. E., Galitzer, S. J., and Dayton, A. D. (1981). *J. Anim. Sci.* **53,** 206.
Nagaraja, T. G., Avery, T. B., Bartley, E. E., Roof, S. K., and Dayton, A. D. (1982). *J. Anim. Sci.* **54,** 649.
Nakamura, K., Kanegasaki, S., and Takahashi, H. (1971). *J. Gen. Appl. Microbiol.* **17,** 13.
National Academy of Sciences (1972). "Water Quality Criteria." Natl. Acad. Sci., Environ. Prot. Agency, Washington, D.C.
National Research Council (1958). "Composition of Cereal Grains and Forages." Publ. No. 585. NAS—NRC, Washington, D.C.
National Research Council (1970). "Nutrient Requirements of Domestic Animals - Nutrient Requirements of Beef Cattle," 4th ed. NAS—NRC, Washington, D.C.
National Research Council (1971a). "Nutrient Requirements of Domestic Animals - Nutrient Requirements of Dairy Cattle," 4th ed. NAS—NRC, Washington, D.C.
National Research Council (1971b). "Atlas of Nutrition Data of U.S. and Canadian Feeds." NAS—NRC, Washington, D.C.
National Research Council (1984). "Nutrients and Toxic Substances in Water for Livestock and Poultry." NAS—NRC, Washington, D.C.
National Research Council (1975). "Nutrient Requirements of Domestic Animals - Nutrient Requirements of Sheep," 5th ed. NAS—NRC, Washington, D.C.
National Research Council (1976a). "Nutrient Requirements of Domestic Animals - Nutrient Requirements of Beef Cattle," 5th ed. NAS—NRC, Washington, D.C.

National Research Council (1976b). "Urea and Other Nonprotein Nitrogen Compounds in Animal Nutrition." NAS—NRC, Washington, D.C.

National Research Council (1978). "Nutrient Requirements of Domestic Animals - Nutrient Requirements of Dairy Cattle," 5th ed. NAS—NRC, Washington, D.C.

National Research Council (1981). "Nutritional Energetics of Domestic Animals and Glossary of Energy Terms." NAS—NRC, Washington, D.C.

Nehring, K., Beyer, K. M., and Huffman, B. (1971). Futtermitteltabellenwerk VEB Deutscher Landwirtschaltsverlag, Berlin.

Nelson, D. B., and Campbell, W. D. (1954). *Okla., Agric. Exp. Stn. [Misc. Publ.] MP* **MP-34**.

Netemeyer, D. T., Bush, L. J., and Owens, F. N. (1980). *J. Dairy Sci.* **63**, 574.

Neville, W. E., Jr. (1974). *J. Anim. Sci.* **38**, 681.

Neville, W. E., Jr., Baird, D. M., and Sell, D. E. (1960). *J. Anim. Sci.* **19**, 1223.

Neville, W. E., Jr., Hellwig, R. E., Ritter, R. J., III, and McCormick, W. C. (1977). *J. Anim. Sci.* **44**, 687.

Newson, J. R., Totusek, R., Renbarger, R., Nelson, E. C., Franks, L., Nauhaus, V., and Basler, W. (1968). *Okla., Agric. Exp. Stn. [Misc. Publ.] MP* **MP-80**.

Newton, G. L. (1980). *Proc. Ga. Nutr. Conf., 1980*, p. 102.

Newton, G. L., and Utley, P. R. (1978). *J. Anim. Sci.* **47**, 1338.

Newton, G. L., Utley, P. R., Ritter, R. J., and McCormick, W. C. (1977). *J. Anim. Sci.* **44**, 447.

Nichols, J. R., Zeigler, J. H., White, J. M., Kesler, E. M., and Watkins, J. L. (1964). *J. Dairy Sci.* **47**, 179.

Nicholson, J. W. G., Loosli, J. K., and Warner, R. G. (1960). *J. Anim. Sci.* **19**, 1071.

Nicholson, J. W. G., Cunningham, H. M., and Friend, D. W. (1962). *Can. J. Anim. Sci.* **42**, 75.

Nikolic, J. A., and Jovanovic, M. (1973). *J. Agric. Sci.* **81**, 1.

Nishihara, H., Shoji, K., and Hori, M. (1965). *Biken J.* **8**, 23.

Nisizawa, K., and Pigman, W. (1960). *Biochem. J.* **75**, 293.

Nocek, J. E., Herbein, J. H., and Polan, C. E. (1980). *J. Nutr.* **110**, 2355.

Nolan, J. V., and Stachiw, S. (1979). *Br. J. Nutr.* **42**, 63.

Noland, P. R., Ford, B. F., and Ray, M. L. (1955). *J. Anim. Sci.* **14**, 860.

Noller, C. H., White, J. L., and Wheeler, W. E. (1980). *J. Dairy Sci.* **63**, 1947.

Nordin, M., and Campling, R. C. (1976). *Anim. Prod.* **23**, 305.

Norris, K. H., Barnes, R. F., Moore, J. E., and Shenk, J. S. (1976). *J. Anim. Sci.* **43**, 889.

Nour, A. M., Abou Akkada, A. R., El-Shazly, K., Naga, M. A., Borhami, B. E., and Abaza, M. A. (1979). *J. Anim. Sci.* **49**, 1300.

Nugent, J. H. A., and Mangan, J. L. (1978). *Proc. Nutr. Soc.* **37**, 48A.

O'Donovan, P. B., and Conway, A. (1968). *J. Br. Grassl. Soc.* **23**, 228.

Offer, N. W., Asford, R. F. E., and Evans, R. A. (1978). *Br. J. Nutr.* **40**, 35.

Ogilvie, B. M., McClymont, G. L., and Shorland, F. B. (1961). *Nature (London)* **190**, 725.

Okorie, A. U., Buttery, P. J., and Lewis, D. (1977). *Proc. Nutr. Soc.* **36**, 38A (Abstr.).

Olbrich, S. E., Martz, F. A., Vogt, J. R., and Hildebrand, E. S. (1971). *J. Anim. Sci.* **33**, 899.

Oliphant, J. M. (1974). *Anim. Prod.* **18**, 211.

Olson, T. M. (1925). *S. D., Agric. Exp. Stn. [Bull.]* **B 215**.

Oltjen, R. R. (1969). *J. Anim. Sci.* **28**, 673.

Oltjen, R. R. (1972). *Proc. Mont. Nutr. Conf.* **23**, 30.

Oltjen, R. R., and Bolsen, K. K. (1980). *J. Anim. Sci.* **51**, 958.

Oltjen, R. R., and Davis, R. E. (1965). *J. Anim. Sci.* **24**, 198.

Oltjen, R. R., and Dinius, D. A. (1976). *J. Anim. Sci.* **43**, 201.

Oltjen, R. R., and Williams, P. P. (1974). *J. Anim. Sci.* **38**, 915.

Oltjen, R. R., Smith, E. F., Koch, B. A., and Baker, F. H. (1959). *J. Anim. Sci.* **18**, 1196.

Oltjen, R. R., Putnam, P. A., Williams, E. E., Jr., and Davis, R. E. (1966). *J. Anim. Sci.* **25**, 1000.

Oltjen, R. R., Burns, W. C., and Ammerman, C. B. (1974). *J. Anim. Sci.* **38**, 975.
Oltjen, R. R., Rumsey, T. S., Fontenot, J. P., Bovard, K. P., and Priode, B. M. (1977). *J. Anim. Sci.* **46**, 532.
Orskov, E. R. (1972). *World Congr. Anim. Feed. [Gen. Rep.]* **2**, 267.
Orskov, E. R. (1982). "Protein Nutrition in Ruminants." Academic Press, New York.
Orskov, E. R., and Fraser, C. (1970). *Proc. Nutr. Soc.* **29**, 31A.
Orskov, E. R., and Greenhaugh, J. F. (1977). *J. Agric. Sci.* **89**, 253.
Orskov, E. R., and McDonald, I. (1970). *Publ.—Eur. Assoc. Anim. Prod.* **13**, 131.
Orskov, E. R., and Oltjen, R. R. (1967). *J. Nutr.* **93**, 222.
Orskov, E. R., Fraser, C., and Kay, R. N. B. (1969). *Br. J. Nutr.* **23**, 217.
Orskov, E. R., Fraser, C., and McDonald, I. (1971a). *Br. J. Nutr.* **26**, 477.
Orskov, E. R., Mayes, R. W., and Penn, A. (1971b). *Proc. Nutr. Soc.* **30**, 43A.
Orskov, E. R., Grubb, D. A., Webster, A. J. F., and Smith, J. S. (1978). *Proc. Br. Nutr. Soc.* **37**, 67A (abstr.).
Osborne, T. B., and Mendel, L. B. (1916). *Am. J. Physiol.* **40**, 16.
Osborne, T. B., and Mendel, L. B. (1917). *J. Biol. Chem.* **32**, 369.
Osbourn, D. F., Beever, D. E., and Thomson, D. J. (1976). *Proc. Nutr. Soc.* **35**, 191.
Osbourn, D. F., Terry, R. A., Spooner, M. C., and Tetlow, R. M. (1981). *Anim. Feed. Sci. Technol.* **6**, 387.
Osman, H. E., Abou Akkada, A. R., and Agabawi, K. A. (1970a). *Anim. Prod.* **12**, 267.
Osman, H. E., Theurer, B., Hale, W. H., and Mehen, S. M. (1970b). *J. Nutr.* **100**, 1133.
Otagaki, K. K., Lofgreen, G. P., Cobb, E., and Dull, G. D. (1961). *J. Dairy Sci.* **44**, 491.
Otchere, E. O., McGilliard, A. D., and Young, J. W. (1974). *J. Dairy Sci.* **57**, 1189.
Owens, F. N., and Isaacson, H. R. (1977). *Fed. Proc., Fed. Am. Soc. Exp. Biol.* **36**, 198.
Owens, F. N., Meiske, J. C., and Goodrich, R. D. (1967). "Minnesota Beef Cattle Feeders Day," Rep. B-87. University of Minnesota, St. Paul.
Owens, F. N., Meiske, J. C., and Goodrich, R. D. (1970a). *J. Anim. Sci.* **30**, 455.
Owens, F. N., Meiske, J. C., and Goodrich, R. D. (1970b). *J. Anim. Sci.* **30**, 462.
Owens, F. N., Lusby, K. S., Mizwicki, K., and Forero, O. (1980). *J. Anim. Sci.* **50**, 527.
Palmquist, D. L. (1978). *Proc., Conf.—Distill. Feed Res. Counc.* **33**, 12.
Palmquist, D. L., and Moser, E. A. (1981). *J. Dairy Sci.* **64**, 1664.
Paquay, R., De Baere, R., and Lousse, A. (1972). *Br. J. Nutr.* **27**, 27.
Parrott, J. C., Mehen, S., Hale, W. H., Little, M., and Theurer, B. (1969). *J. Anim. Sci.* **28**, 425.
Parthasarathy, D., and Phillipson, A. T. (1953). *J. Physiol. (London)* **121**, 452.
Pate, F. M. (1981). *J. Anim. Sci.* **53**, 881.
Pate, F. M., and Crockett, J. R. (1978). *Bull.—Fla., Agric., Exp. Stn.* **799**.
Pate, F. M., Haines, C. E., Greene, R. E. L., Palmer, A. Z., Carpenter, J. W., and Beardsley, D. W. (1972). *Bull.—Fla., Agric. Exp. Stn.* **752**.
Payne, W. J. (1970). *Ann. Rev. Microbiol.* **24**, 17.
Peacock, F. M., and Kirk, W. G. (1959). *Bull.—Fla., Agric. Exp. Stn.* **616**.
Peacock, F. M., McCaleb, J. E., Hodges, E. M., and Kirk, W. G. (1961). *Bull.—Fla., Agric. Exp. Stn.* **635**.
Peacock, F. M., Kirk, W. G., Hodges, E. M., Palmer, A. Z., and Carpenter, J. W. (1964). *Bull.—Fla., Agric. Exp. Stn.* **667**.
Pearson, A. M., Wallace, H. D., and Hentges, J. F., Jr. (1954). *Bull.—Fla., Agric. Exp. Stn.* **553**.
Pearson, R. M., and Smith, J. A. B. (1943). *Biochem. J.* **37**, 153.
Peavy, A. H., III, Harris, B., Jr., van Horn, H. H., and Wilcox, C. J. (1980). *J. Dairy Sci.* **63**, 405.
Pendlum, L. C., Boling, J. A., and Bradley, N. W. (1976). *J. Anim. Sci.* **43**, 1307.
Pendlum, L. C., Boling, J. A., and Bradley, N. W. (1977). *J. Anim. Sci.* **44**, 18.
Pendlum, L. C., Boling, J. A., and Bradley, N. W. (1978). *J. Anim. Sci.* **46**, 535.

Pendlum, L. C., Boling, J. A., and Bradley, N. W. (1980). *J. Anim. Sci.* **50,** 29.
Pennington, R. J. (1954). *Biochem. J.* **56,** 410.
Pennington, R. J., and Pfander, W. H. (1957). *Biochem. J.* **65,** 109.
Pennington, R. J., and Sutherland, T. M. (1956). *Biochem. J.* **63,** 353.
Perry, K. D., and Briggs, C. A. E. (1955). *J. Pathol. Bacteriol.* **70,** 546.
Perry, T. W. (1964). *Feedstuffs* **36**(46), 26.
Perry, T. W., and Stewart, T. S. (1979). *J. Anim. Sci.* **48,** 900.
Perry, T. W., Beeson, W. M., and Mohler, M. T. (1960). High protein-urea supplements with and without lysine for fattening steer calves. *Purdue Univ., Agric. Exp. Stn., Mimeo* **AS-277.**
Perry, T. W., Beeson, W. M., Robinson, D. M., and Mohler, M. T. (1967). *Res. Prog. Rep.—Indiana, Agric. Exp. Stn.* **303.**
Perry, T. W., Beeson, W. M., DaBell, W. R., Kohler, G. O., and Gough, F. A. (1974). *J. Anim. Sci.* **39,** 1158.
Perry, T. W., Beeson, W. M., and Mohler, M. T. (1976a). *J. Anim. Sci.* **42,** 761.
Perry, T. W., Beeson, W. M., Mohler, M. T., and Baugh, E. (1976b). *J. Anim. Sci.* **43,** 945.
Petchey, A. M., and Broadbent, P. J. (1980). *Anim. Prod.* **31,** 251.
Peter, A. P., Driedger, A., Hatfield, E. E., Peterson, L. A., Hixon, D. L., and Garrigus, U. S. (1970). *J. Anim. Sci.* **31,** 250 (abstr.).
Peterson, C. B. (1970). *Publ.—Eur. Assoc. Anim. Prod.* **13,** 205.
Pew, J. C., and Weyna, P. (1962). *Tappi* **45,** 247.
Pfander, W. H., and Phillipson, A. T. (1953). *J. Physiol. (London)* **122,** 102.
Pfander, W. H., Ross, C. V., Preston, R. L., Vipperman, P. E., and Tyree, W. (1964). Molasses vs Masonex. *Univ. Mo., Spec. Rep. Bull.* **38.**
Phar, P. A., Bradley, N. W., Little, C. O., and Cundiff, L. V. (1970). *J. Anim. Sci.* **30,** 589.
Phillipson, A. T. (1952). *Br. J. Nutr.* **6,** 190.
Phillipson, A. T., and Mangan, J. L. (1959). *N. Z. J. Agric. Res.* **2,** 990.
Phillipson, A. T., Dobson, M. J., Blackburn, T. H., and Brown, M. (1962). *Br. J. Nutr.* **16,** 151.
Pickard, D. W., and Lamming, G. E. (1968). *Anim. Prod.* **10,** 223.
Pilgrim, A. F., Gray, F. V., Weller, R. A., and Belling, C. B. (1970). *Br. J. Nutr.* **24,** 589.
Pinney, D. O., Pope, L. S., and Stephens, D. F. (1962). Effects of alternate high and low winter feed levels on growth and reproduction of replacement heifers. *Okla., Agric. Exp. Stn., Livest. Feeders Day Rep.* **MP-67.**
Pinney, D. O., Stephens, D. F., and Pope, L. S. (1972). *J. Anim. Sci.* **34,** 1067.
Pinzon, F. J., and Wing, J. M. (1976). *J. Dairy Sci.* **59,** 1100.
Pitzen, D. F. (1974). Ph.D. Dissertation, Iowa State University, Ames.
Polan, C. E., Chandler, P. T., and Miller, C. N. (1970). *J. Dairy Sci.* **53,** 607.
Poos, M. I. (1981). *Feedstuffs* **53**(26), 19.
Poos. M. I., Bull, L. S., and Hemken, R. W. (1979). *J. Anim. Sci.* **49,** 1417.
Porter, J. W. G. (1961). *In* "Digestive Physiology and Nutrition in the Ruminant" (D. Lewis, ed.), P. 226. Butterworth, London.
Pothoven, M. A., and Beitz, D. C. (1975). *J. Nutr.* **105,** 1055.
Potter, E. L., Cooley, C. O., Richardson, L. F., Raun, A. P., and Rathmacher, R. P. (1976a). *J. Anim. Sci.* **43,** 665.
Potter, E. L., Raun, A. P., Cooley, C. O., Rathmacher, R. P., and Richardson, L. F. (1976b). *J. Anim. Sci.* **43,** 678.
Potter, G. D., Little, C. O., and Mitchell, G. E., Jr. (1969). *J. Anim. Sci.* **28,** 711.
Prasad, D. A., Morrill, J. L., Melton, S. L., Dayton, A. D., Arnett, D. W., and Pfost, H. B. (1975). *J. Anim. Sci.* **41,** 578.
Prescott, J. D. H., and Lamming, G. E. (1964). *J. Agric. Sci.* **63,** 341.
Prescott, J. M., Williams, W. T., and Ragland, R. S. (1959). *Proc. Soc. Exp. Biol. Med.* **102,** 490.

Preston, R. L. (1975). *J. Anim. Sci.* **41,** 622.
Preston, R. L. (1980). *Feedstuffs* **52**(35), 49.
Preston, R. L., and Cahill, V. R. (1972). *Ohio, Agric. Res. Dev. Cent., Res. Summ.* **63.**
Preston, R. L., and Cahill, V. R. (1973). *Ohio, Agric. Res. Dev. Cent., Res. Summ.* **68.**
Preston, R. L., and Cahill, V. R. (1974). *Ohio, Agric. Res. Dev. Cent., Res. Summ.* **77.**
Preston, R. L., Vance, R. D., Althouse, P. G., and Cahill, V. R. (1972). *Ohio, Agric. Res. Dev. Cent., Res. Summ.* **63,** 55.
Preston, R. L., Vance, R. D., Cahill, V. R., and Kock, S. W. (1974). *J. Anim. Sci.* **38,** 47.
Preston, R. L., Cahill, V. R., and Parker, C. F. (1975). Supplemental protein withdrawal in steer calves fed high concentrate rations. Ohio Beef Day and Cattlemen's Roundup Report, p. 1.
Preston, T. R., and Willis, M. B. (1970). "Intensive Beef Production," 1st ed. Pergamon, Oxford.
Preston, T. R., Willis, M. B., and Elias, A. (1970). *Anim. Prod.* **12,** 457.
Prigge, E. C., Johnson, R. R., Owens, F. N., and Williams, D. E. (1976). *J. Anim. Sci.* **43,** 705.
Prigge, E. C., Galyean, M. L., Owens, F. N., Wagner, D. G., and Johnson, R. R. (1978). *J. Anim. Sci.* **46,** 249.
Prigge, E. C., Varga, G. A., Vicini, J. L., and Reid, R. L. (1981). *J. Anim. Sci.* **53,** 1629.
Prins, R. A. (1977). *In* "Microbial Ecology of the Gut" (R. T. J. Clarke and T. Bauchop, eds.), p. 73. Academic Press, London.
Pullar, J. D., and Webster, A. J. F. (1972). *Proc. Br. Nutr. Soc.* **32,** 19A (abstr.).
Purser, D. B. (1970). *J. Anim. Sci.* **30,** 988.
Purser, D. B., and Moir, R. J. (1959). *Aust. J. Agric. Res.* **10,** 555.
Putnam, P. A., Lehmann, R., and Davis, R. E. (1966). *J. Anim. Sci.* **25,** 817.
Quinn, L. Y., Burroughs, W., and Christiansen, W. C. (1962). *Appl. Microbiol.* **10,** 583.
Rackis, J. J. (1965). *Fed. Proc., Fed. Am. Soc. Exp. Biol.* **24,** 1488.
Ragsdale, A. C., Brody, S., Thompson, H. J., and Worstell, D. M. (1948). *Res. Bull.—Mo., Agric. Exp. Stn.* **425.**
Raleigh, R. J., Horner, H. A., and Phillips, R. L. (1970). *Proc., Annu. Meet.—Am. Soc. Anim. Sci., West. Sect.* **21,** 81.
Rangnekar, D. V., Badve, V. C., Kharat, S. T., Sobale, B. N., and Joshi, A. L. (1982). *Anim. Feed. Sci. Technol.* **7,** 61.
Rattray, P. V., and Garrett, W. N. (1971). *J. Anim. Sci.* **33,** 298 (abstr.).
Rattray, P. V., Garrett, W. N., East, N. E., and Hinman, N. (1974). *J. Anim. Sci.* **38,** 383.
Raun, A. P., Cooley, C. O., Potter, E. L., Rathmacher, R. P., and Richardson, L. F. (1976). *J. Anim. Sci.* **43,** 670.
Raun, N. S., Renbarger, R., Stables, G., and Pope, L. S. (1965). *J. Anim. Sci.* **24,** 279 (abstr.).
Ray, D. E., and Roubicek, C. B. (1971). *J. Anim. Sci.* **33,** 72.
Raymond, W. F. (1969). *Adv. Agron.* **21,** 1.
Redd, T. L., Boling, J. A., Bradley, N. W., and Ely, D. G. (1975). *J. Anim. Sci.* **40,** 567.
Reddy, C. A., Henderson, H. E., and Erdman, M. D. (1976). *Appl. Environ. Microbiol.* **32,** 769.
Reid, C. S. W., Clarke, R. T. J., Cockrem, F. R. M., Jones, W. T., McIntosh, J. T., and Wright, D. E. (1975). *In* "Digestion and Metabolism in the Ruminant" (I. W. McDonald and A. C. I. Warner, eds.), p. 524. University of New England Publishing Unit, Armidale, N.S.W., Australia.
Reid, J. T., and White, O. D. (1978). *In* "New Protein Foods" (A. M. Altschul and H. L. Wilcke, eds.), Vol. 3, p. 116. Academic Press, New York.
Reid, J. T., Wellington, G. H., and Dunn, H. O. (1955). *J. Dairy Sci.* **38,** 1344.
Reid, J. T., Smith, A. M., and Anderson, M. J. (1958). Difference in requirements for maintenance of dairy cattle between pasture and barn feeding conditions. *Proc. Cornell Nutr. Conf.*, p. 88.
Reid, J. T., Moe, P. W., and Tyrrell, H. F. (1966). *J. Dairy Sci.* **49,** 215.
Reid, J. T., White, O. D., Anrique, R., and Fortin, A. (1980). *J. Anim. Sci.* **51,** 1393.

Reid, R. L., Baker, B. S., and Vona, L. C. (1984). *J. Anim. Sci.* **59,** 1403.
Reis, P. J., and Schinckel, P. G. (1963). *Aust. J. Biol. Sci.* **16,** 218.
Reiser, R., and Fu, H. C. (1962). *J. Nutr.* **76,** 215.
Renton, A. R., and Forbes, T. J. (1973). *Anim. Prod.* **16,** 173.
Rhodes, R. C., McCartor, M. M., and Randel, R. D. (1978). *J. Anim. Sci.* **46,** 769.
Ribeiro, J. M. de C. R., Brockway, J. M., and Webster, A. J. F. (1977). *Anim. Prod.* **25,** 107.
Richardson, C. R., and Hatfield, E. E. (1978). *J. Anim. Sci.* **46,** 740.
Richardson, C. R., Beville, R. N., Ratcliff, R. K., and Albin, R. C. (1981). *J. Anim. Sci.* **53,** 557.
Richter, M. F., Shirley, R. L., and Palmer, A. Z. (1980). *J. Anim. Sci.* **50,** 207.
Ricke, S. C., Berger, L. L., van der Aar, P. J., and Fahey, G. C., Jr. (1984). *J. Anim. Sci.* **58,** 194.
Riddett, W., Campbell, I. L., McDowell, F. H., and Cow, G. A. (1941). *N. Z. J. Sci. Technol.* **23,** 80.
Riggs, J. K. (1958). *J. Anim. Sci.* **17,** 981.
Riggs, J. K. (1965). *Proc. Annu. Tex. Nutr. Conf.* p **20.**
Riggs, J. K., and McGinty, D. D. (1970). *J. Anim. Sci.* **31,** 991.
Riggs, J. K., Sorenson, J. W., Jr., Adame, J. L., and Schake, L. M. (1970). *J. Anim. Sci.* **30,** 634.
Rittenhouse, L. R., Clanton, D. C., and Streeter, C. L. (1970). *J. Anim. Sci.* **31,** 1215.
Roberts, W. K., and McKirdy, J. A. (1964). *J. Anim. Sci.* **23,** 682.
Roberts, W. K., and St. Omer, V. V. (1965). *J. Anim. Sci.* **24,** 902 (abstr.).
Robison, W. L. (1930). *Ohio, Agric. Exp. Stn., Res. Bull.* **452.**
Robles, A. Y., Martz, F. A., Belyea, R. L., and Warren, W. P. (1981). *J. Anim. Sci.* **52,** 1417.
Rodgers, J. A., Marks, B. C., Davis, C. L., and Clark, J. H. (1979). *J. Dairy Sci.* **62,** 1599.
Roffler, R. E., Schwab, C. G., and Satter, L. D. (1976). *J. Dairy Sci.* **59,** 80.
Roman-Ponce, H., van Horn, H. H., Marshall, S. P., Wilcox, C. J., and Randell, P. F. (1975). *J. Dairy Sci.* **58,** 1320.
Roman-Ponce, H., Thatcher, W. W., Buffington, D. E., Wilcox, C. J., and van Horn, H. H. (1977). *J. Dairy Sci.* **60,** 424.
Rose, A. L. (1941). *Aust. Vet. J.* **17,** 211.
Rouquette, F. M., and Carpenter, Z. L. (1981). *J. Anim. Sci.* **53,** 892.
Rouquette, F. M., Jr., Griffin, J. L., Randel, R. D., and Carroll, L. H. (1980). *J. Anim. Sci.* **51,** 251.
Rumsey, T. S. (1978). *J. Anim. Sci.* **47,** 967.
Rumsey, T. S., Tyrrell, H. F., Dinius, D. A., Moe, P. W., and Cross, H. R. (1977). *J. Anim. Sci.* **45,** Suppl. 1, 264.
Rumsey, T. S., Kern, D. L., and Slyter, L. L. (1979). *J. Anim. Sci.* **48,** 1202.
Rumsey, T. S. (1982). *J. Anim. Sci.* **54,** 1030.
Rush, I. G., and Totusek, R. (1976). *J. Anim. Sci.* **42,** 497.
Russell, J. B., and Hespell, R. B. (1981). *J. Dairy Sci.* **64,** 1153.
Russell, J. B., Bottje, W. G., and Cotta, M. A. (1981). *J. Anim. Sci.* **53,** 242.
Russell, J. R., Young, A. W., and Jorgensen, N. A. (1980). *J. Anim. Sci.* **51,** 996.
Russell, J. R., Young, A. W., and Jorgensen, N. A. (1981). *J. Anim. Sci.* **52,** 1170.
Ryan, R. K. (1964). *Am. J. Vet. Res.* **25,** 646.
Salsbury, R. L., Hoefer, J. A., and Luecke, R. W. (1961). *J. Anim. Sci.* **20,** 569.
Salsbury, R. L., Marvin, D. K., Woodmansee, C. W., and Henlein, G. F. W. (1971). *J. Dairy Sci.* **54,** 390.
Samford, R. A., Riggs, J. K., Rooney, L. W., Patter, G. D., and Coon, J. (1971). "Beef Cattle Research in Texas," PR-2964. Texas A & M University, College Station.
Sander, E. G., Warner, R. G., Harrison, H. N., and Loosli, J. K. (1959). *J. Dairy Sci.* **42,** 1600.
Santos, K. A., Stern, M. D., and Satter, L. D. (1984). *J. Anim. Sci.* **58,** 244.

Satter, L. D., and Esdale, W. J. (1968). *Appl. Microbiol.* **16**, 680.
Satter, L. D. and Roffler, R. E. (1974). *British J. Nutrition* **32**, 199.
Satter, L. D., and Roffler, R. E. (1975). *J. Dairy Sci.* **58**, 1219.
Satter, L. D., and Roffler, R. E. (1976). Relation between ruminal ammonia and non-protein nitrogen utilization by ruminants. *Proc. Int. Atomic Energy Agency,* p. 119. IAEA, Vienna.
Satter, L. D., and Slyter, L. L. (1974). *Br. J. Nutr.* **32**, 199.
Satter, L. D., Whitlow, L. W., and Beardsley, G. L. (1977). *Proc., Conf.—Distill. Feed Res. Counc.* **32**, 63.
Satter, L. D., Whitlow, L. W., and Santos, K. A. (1979). *Proc. Conf.—Distill. Feed Res. Counc.* **34**, 77.
Saubidet, C. L., and Verde, L. S. (1976). *Anim. Prod.* **22**, 61.
Schaible, P. J., Moore, L. A., and Moore, J. M. (1934). *Science* **79**, 372.
Schaibly, G. E., and Wing, J. M. (1974). *J. Anim. Sci.* **38**, 697.
Schake, L. M., and Riggs, J. K. (1975). *J. Anim. Sci.* **40**, 561.
Schake, L. M., Riggs, J. K., and Butler, O. D. (1972). *J. Anim. Sci.* **34**, 926.
Schalk, A. F., and Amadon, R. S. (1928). *N. D. Agric. Exp. Stn., Bull.* **216**.
Schanbacher, B. D. (1984). *J. Anim. Sci.* **59**, 1621.
Scheidy, S. F., Acord, C., Bartley, E. E., Brethour, J. R., Clark, J., Heinemann, W. W., Thomas, O. O., and Curtin, L. V. (1972). *J. Anim. Sci.* **35**, 262 (abstr.).
Scheifinger, C., Russell, N., and Chalupa, W. (1976). *J. Anim. Sci.* **43**, 821.
Schelling, G. T., and Hatfield, E. E. (1967). *J. Anim. Sci.* **26**, 1484 (abstr.).
Schelling, G. T., and Hatfield, E. E. (1968). *J. Nutr.* **96**, 319.
Schiemann, R., Nehring, K., Hoffman, L., Jentsch, W., and Chudy, A. (1971). "Emergetische Futterbewertung und Energienormenen." VEB Deutscher Landwirtschaftsverlag, Berlin.
Schingoethe, D. J., Rook, J. A., and Ludens, F. (1977). *J. Dairy Sci.* **60**, 591.
Schmltz, J. (1976). M. S. Thesis, University of Nebraska, Lincoln.
Schroder, H. H. E. (1970). *J. Agric. Sci.* **75**, 231.
Schroder, H. H. E., and Gilchrist, F. M. C. (1969). *J. Agric. Sci.* **72**, 1.
Schroder, V. N. (1976). *Proc.—Soil Crop Sci. Soc. Fla.* **36**, 195.
Schulman, M. D., and Valentino, D. (1976). *J. Dairy Sci.* **59**, 1444.
Schwartz, C. C., Nagy, J. G., and Streeter, C. L. (1973). *J. Anim. Sci.* **37**, 821.
Scott, T. W., and Cook, L. J. (1975). *In* "Digestion and Metabolism in the Ruminant" (I. W. McDonald and A. C. I. Warner, eds.), p. 510. Univ. of New England Publishing Unit, Armidale, N.S.W., Australia.
Scott, T. W., and Hills, G. L. (1975). U.S. Patent 3,925,560.
Scott, T. W., Cook, L. J., Ferguson, K. A., McDonald, I. W., Buchanan, R. A., and Hills, G. L. (1970). *Aust. J. Agric. Sci.* **32**, 291.
Scott, T. W., Cook, L. J., and Mills, S. C. (1971). *J. Am. Oil Chem. Soc.* **48**, 358.
Sharma, H. R., Ingalls, J., and McKirdy, J. A. (1978). *J. Dairy Sci.* **61**, 574.
Shaw, G. H., Convey, E. M., Tucker, H. A., Reineke, E. P., Thomas, J. W., and Byrne, J. J. (1975). *J. Dairy Sci.* **58**, 703.
Shaw, J. C., Ensor, W. L., Tellechea, H. F., and Lee, S. D. (1960). *J. Nutr.* **71**, 203.
Sheehy, E. J., and Senior, B. J. (1942). *J.—Dep. Agric. (Irel.)* **39**, 245.
Sherrod, L. B., and Tillman, A. D. (1962). *J. Anim. Sci.* **21**, 901.
Shirley, B: (1980). *Beef* **17**(1), 34.
Shirley, B. (1982). *Beef* **18**(10), 19.
Shirley, R. L. (1970). *Proc. AFMA Nutr. Counc.,* **30**, 23.
Shirley, R. L. (1975a). *Feedstuffs* **47**(15), 20.
Shirley, R. L. (1975b). *BioScience* **25**, 789.

Shirley, R. L., Davis, G. K., and Neller, J. R. (1951). *J. Anim. Sci.* **10,** 335.
Shirley, R. L., Robertson, W. K., McCall, J. T., Neller, J. R., and Davis, G. K. (1957). *J. Fla. Acad. Sci.* **20,** 133.
Shirley, R. L., Peacock, F. M., Kirk, W. G., Easley, J. F., Palmer, A. Z., and Davis, G. K. (1963). *J. Fla. Acad. Sci.* **26**(1), 89.
Shoptaw, L., Espe, D. L., and Cannon, C. Y. (1937). *J. Dairy Sci.* **20,** 117.
Short, R. E., and Bellows, R. A. (1971). *J. Anim. Sci.* **32,** 127.
Shrivastava, V. C., and Talapatra, S. K. (1962). *Indian J. Dairy Sci.* **15,** 154.
Shultz, T. A., and Ralston, A. T. (1974). *J. Anim. Sci.* **39,** 926.
Shultz, T. A., Ralston, A. T., and Shultz, E. (1974). *J. Anim. Sci.* **39,** 920.
Singh, M., and Kamstra, L. D. (1981). *J. Anim. Sci.* **53,** 551.
Skelley, G. C., Stanford, W. C., and Edwards, R. L. (1973). *J. Anim. Sci.* **36,** 576.
Slyter, L. L. (1976). *J. Anim. Sci.* **43,** 910.
Slyter, L. L. (1981). *Proc., Annu. Meet.—Am. Soc. Anim. Sci., Raleigh, N. C.* Abstr. No. 743.
Slyter, L. L., and Weaver, J. M. (1972). *J. Anim. Sci.* **35,** 288 (abstr.).
Slyter, L. L., Oltjen, R. R., and Putnam, P. A. (1965). *J. Anim. Sci.* **24,** 1218.
Slyter, L. L., Oltjen, R. R., Kern, D. L., and Blank, F. C. (1970). *J. Anim. Sci.* **31,** 996.
Slyter, L. L., Chalupa, W. V., Oltjen, R. R., and Weaver, J. M. (1971). *J. Anim. Sci.* **33,** 300 (abstr.).
Slyter, L. L., Satter, L. D., and Dinius, D. A. (1979). *J. Anim. Sci.* **48,** 906.
Smith, G. E. (1971). *In* "Digestive Physiology and Nutrition of Ruminants." Nutrition, Vol. II. Chapter 23. (D. C. Church, ed.) O and B Books, Inc. Corvallis, Oregon.
Smith, A. M., and Reid, J. T. (1954). *J. Dairy Sci.* **15,** 515.
Smith, C. K., Brunner, J. R., Huffman, C. R., and Duncan, C. W. (1953). *J. Anim. Sci.* **12,** 932.
Smith, F. H. (1962). *In* "Procedings of the Cottonseed Quality Research Conference," p. 7. Natl. Cottonseed Prod. Assoc., Inc., Memphis, Tennessee.
Smith, G. S., Chambers, J. W., Neumann, A. L., Ray, E. E., and Nelson, A. B. (1974). *J. Anim. Sci.* **38,** 627.
Smith, G. S., Caddle, J. M., Walters, P., and Kiesling, H. E. (1976). *J. Anim. Sci.* **43,** 334 (abstr.).
Smith, L. W. (1973). *In* "Alternative Sources of Protein for Animal Production,," p. 146 Natl. Acad. Sci., Washington, D.C.
Smith, L. W., and Calvert, C. C. (1976). *J. Anim. Sci.* **43,** 1286.
Smith, L. W., and Lindahl, I. L. (1977). *J. Anim. Sci.* **44,** 152.
Smith, L. W., and Wheeler, W. E. (1979). *J. Anim. Sci.* **48,** 144.
Smith, L. W., Calvert, C. C., and Cross, H. R. (1979). *J. Anim. Sci.* **48,** 633.
Smith, M. F., Shipp, L. D., Songster, W. N., Wiltbank, J. N., and Carroll, L. H. (1980). *Theriogenology* **14,** 91.
Smith, N. E., Dunkley, W. L., and Franke, A. A. (1978). *J. Dairy Sci.* **61,** 747.
Smith, O. B., MacLeod, G. K., Mowat, D. N., Fox, C. A., and Moran, E. T., Jr. (1978). *J. Anim. Sci.* **47,** 833.
Smith, P. H., and Hungate, R. E. (1958). *J. Bacteriol.* **75,** 713.
Smith, P. H., Sweeney, H. C., Rooney, J. R., King, K. W., and Moore, W. E. C. (1956). *J. Dairy Sci.* **39,** 598.
Smith, R. H. (1969). *J. Dairy Res.* **36,** 313.
Smith, R. H., McAllan, A. B., Hewitt, D., and Lewis, P. E. (1978). *J. Agric. Sci.* **90,** 557.
Smith, V. R. (1941). *J. Dairy Sci.* **24,** 659.
Smuts, B. B. (1935). *J. Nutr.* **9,** 403.
Soberman, R., Brodie, B. B., Levy, B. B., Axelrod, J., Hollander, V., and Steele, J. M. (1949). *J. Biol. Chem.* **179,** 31.
Spears, J. W., and Harvey, R. W. (1984). *J. Anim. Sci.* **58,** 460.

# Bibliography

Spears, J. W., Ely, D. G., Bush, L. P., and Buchner, R. C. (1976). *J. Anim. Sci.* **43,** 513.
Sperber, I., and Heydén, S. (1952). *Nature (London)* **169,** 587.
Speth, C. F., Bohman, V. R., Melendy, H., and Wade, M. A. (1962). *J. Anim. Sci.* **21,** 444.
Stake, P. E., Owens, M. J., and Schingoethe, D. J. (1973). *J. Dairy Sci.* **56,** 783.
Stanhope, D. L., Hinman, D. D., Everson, D. O., and Bull, R. C. (1980). *J. Anim. Sci.* **51,** 202.
Stanley, R. W., and Morita, K. (1966). *Hawaii Agric. Exp. Stn., Tech. Prog. Rep.* **150.**
Stanton, T. L., Owens, F. N., Brusewitz, G. H., and Lusby, K. S. (1981). *J. Anim. Sci.* **53,** 473.
Starks, P. B., Hale, W. H., Garrigus, U. S., Forbes, R. M., and James, M. F. (1954). *J. Anim. Sci.* **13,** 249.
Stephens, E. L., Easley, J. F., Shirley, R. L., and Hentges, J. F., Jr. (1972). *Proc.—Soil Crop Sci. Soc. Fla.* **32,** 30.
Stern, M. D., and Hoover, W. H. (1979). *J. Anim. Sci.* **49,** 1590.
Stern, M. D., Hoover, W. H., Summers, R. G., Jr., and Rittenburg, J. H. (1977a). *J. Dairy Sci.* **60,** 902.
Stern, M. D., Hoover, W. H., and Leonard, J. B. (1977b). *J. Dairy Sci.* **60,** 911.
Stern, M. D., Hoover, W. H., Sniffen, C. J., Crooker, B. A., and Knowlton, P. H. (1978). *J. Anim. Sci.* **47,** 944.
Stiles, D. A., Bartley, E. E., Erhart, A. B., Meyer, R. M., and Boren, F. W. (1967). *J. Dairy Sci.* **50,** 1437.
Stiles, D. A., Bartley, E. E., Meyer, R. M., Deyoe, C. W., and Pfost, H. B. (1970). *J. Dairy Sci.* **53,** 1436.
Stirling, A. C. (1953). *Proc. Soc. Appl. Bacteriol.* **16,** 27.
Stokes, M. R., Clark, J. H., and Steinmetz, L. M. (1981). *J. Dairy Sci.* **64,** 1686.
Storry, J. E., Brumby, P. E., Hall, A. J., and Johnson, V. W. (1974). *J. Dairy Sci.* **57,** 61.
Stouthamer, A. H. (1969). *In* "Methods in Microbiology" (J. R. Norris and D. W. Ribbons, eds.). Vol. 1, p. 629. Academic Press, London.
Stouthamer, A. H., and Bettenhaussen, C. (1973). *Biochim. Biophys. Acta* **301.** 53.
Stouthamer, A. H., and Bettenhaussen, C. (1976). *Arch. Microbiol.* **111,** 21.
Stranks, D. W. (1959). *For. Prod. J.* **9,** 228.
Streeter, C. L., Oltjen, R. R., Slyter, L. L., and Fishbein, W. N. (1969). *J. Anim. Sci.* **29,** 88.
Stricker, J. A., Matches, A. G., Thompson, G. B., Jacobs, V. E., Martz, F. A., Wheaton, H. N., Currence, H. D., and Krause, C. F. (1979). *J. Anim. Sci.* **48,** 13.
Sudweeks, E. M. (1977). *J. Anim. Sci.* **44,** 694.
Sudweeks, E. M., Ely, L. O., and Sisk, L. R. (1979). *Proc. Ga. Nutr. Conf.* p. 80.
Sugden, B. (1953). *J. Gen. Microbiol.* **9,** 44.
Sullivan, J. T. (1973). *In* "Chemistry and Biochemistry of Herbage" (G. W. Butler and R. W. Bailey, eds.), Vol. 3, p. 1. Academic Press, London.
Sutton, J. D. (1971). *Proc. Nutr. Soc.* **30,** 243.
Swanson, E. W. (1960). *J. Dairy Sci.* **43,** 377.
Swanson, E. W., and Hinton, S. A. (1964). *J. Dairy Sci.* **47,** 267.
Swift, R. W., and French, C. E. (1954). "Energy Metabolism and Nutrition." Scarecrow Press, New Brunswick, New Jersey.
Swift, R. W., Bratzler, J. W., James, W. H., Tillman, A. D., and Meek, D. C. (1948). *J. Anim. Sci.* **7,** 475.
Swingle, R. S., and Waymack, L. B. (1977). *J. Anim. Sci.* **44,** 112.
Swingle, R. S., Roubicek, C. B., Wooten, R. A., Marchello, J. A., and Dryden, F. D. (1979). *J. Anim. Sci.* **48,** 913.
Tagari, H., Ascarelli, I., and Bondi, A. (1962). *Br. J. Nutr.* **16,** 237.
Tagari, H., Levy, D., Holzer, Z., and Ilan, D. (1976). *Anim. Prod.* **23,** 317.
Talapatra, S. K., Ray, S. C., and Sen, K. C. (1948). *J. Agric. Sci.* **38,** 163.

Tamminga, S. (1979). *J. Anim. Sci.* **49,** 1615.
Tamminga, S., and van Hellemond, K. K. (1977). *Protein and Non-Protein Nitrogen for Ruminants, Proc. Symp., 1977,* p. 9.
Tamminga, S., van der Koelen, C. J., and van Vuuren, A. M. (1979). *Lives. Prod. Sci.* **6,** 255.
Tarkow, H., and Feist, W. C. (1969). *In* "Cellulases and Their Applications" (R. F. Gould, ed.), p. 197 Am. Chem. Soc., Washington, D.C.
Tayler, J. C. (1959). *Nature (London)* **184,** 2021.
Teeter, R. G., Owens, F. N., and Sharp, M. W. (1981). *Proc., Annu. Meet.—Am. Soc. Anim. Sci., Raleigh, N. C.* Abstr. No. 756.
Templeton, J. A., Bucek, O. C., and Swart, R. W. (1970a). *J. Anim. Sci.* **31,** 255 (Abstr.).
Templeton, J. A., Swart, R. W., and Bucek, O. C. (1970b). *Proc., Annu. Meet.—Am. Soc. Anim. Sci., West. Sect.* **21,** 183.
Thiago, L. R. L. de S., and Kellaway, R. C. (1982). *Anim. Feed Sci. Technol.* **7,** 71.
Thomas, G. J. (1960). *J. Agric. Sci.* **54,** 360.
Thomas, E. E., Mason, C. R., and Schmidt, S. P. (1984). *J. Anim. Sci.* **58,** 1285.
Thomas, J. W., Yu, Y., and Hoefer, J. A. (1970). *J. Anim. Sci.* **31,** 255 (abstr.).
Thomas, O. O., and Langford, W. J. (1978). *Proc., Annu. Meet.—Am. Soc. Anim. Sci., West. Sect.* **29,** 454 (abstr.).
Thomas, P. C. (1973). *Proc. Nutr. Soc.* **32,** 85.
Thomas, P. C. (1977). *Proc. Int. Symp. Protein Metab. Nutr., 2nd, 1977,* p. 47.
Thomas, V. M., and Beeson, W. M. (1977). *J. Anim. Sci.* **46,** 819.
Thomas, V. M., Beeson, W. M., and Perry, T. W. (1975). *J. Anim. Sci.* **41,** 641.
Thomas, V. M., Glover, D. V., and Beeson, W. M. (1976a). *J. Anim. Sci.* **42,** 529.
Thomas, V. M., Beeson, W. M., Perry, T. W., and Mohler, M. T. (1976b). *J. Anim. Sci.* **43,** 850.
Thomas, W. E., Loosli, J. K., Williams, H. H., and Williams, L. A. (1951). *J. Nutr.* **43,** 515.
Thompson, G. E. (1973). *J. Dairy Res.* **40,** 441.
Thompson, L. H., Goode, L., Harvey, R. W., Meyers, R. M., and Linnerud, A. C. (1973). *J. Anim. Sci.* **37,** 399.
Thompson, W. R., and Riley, J. G. (1980). *J. Anim. Sci.* **50,** 563.
Thompson, W. R., Goodrich, R. D., Rust, J. R., Byers, F. M., and Meiske, J. C. (1981). *Proc., Annu. Meet.—Am. Soc. Anim. Sci., Raleigh. N. C.,* Abstr. No. 761.
Thonney, M. L., Duhaime, D. J., Moe, P. W., and Reid, J. T. (1979). *J. Anim. Sci.* **49,** 1112.
Thorbek, G. (1970). *Publ.—Eur. Assoc. Anim. Prod.* **13,** 129.
Thornton, J. H., and Owens, F. N. (1981). *J. Anim. Sci.* **52,** 628.
Thornton, R. F., and Yates, N. G. (1968). *Aust. J. Agric. Res.* **19,** 665.
Tilley, J. M. A., and Terry, R. A. (1963). *J. Br. Grassl. Soc.* **18,** 104.
Tillman, A. D., Singletary, C. B., Kidwell, J. F., and Bray, C. I. (1951). *J. Anim. Sci.* **10,** 939.
Timell, T. E. (1967). *Wood Sci. Technol.* **1,** 45.
Tiwari, A. D., Owens, F. N., and Garrigus, U. S. (1973). *J. Anim. Sci.* **37,** 1390.
Tollett, I. T., Swart, R. W., Ioset, R. M., and Templeton, J. A. (1969). *Proc., Annu. Meet.—Am. Soc. Anim. Sci., West. Sect.* **20,** 325.
Tomhave, A. E. (1932). *Del., Agric. Exp. Stn., Annu. Rep.* **179,** 23.
Tonroy, B. R., Perry, T. W., and Beeson, W. M. (1974). *J. Anim. Sci.* **39,** 931.
Torell, D. T. (1954). *J. Anim. Sci.* **13,** 878.
Trautmann, A., and Hill, H. (1949). *Pfluegers Arch. Gesamte Physiol. Menschen Tiere* **252,** 30.
Trei, J. E., and Scott, G. C. (1971). *J. Anim. Sci.* **33,** 301. (abstr.).
Trei, J. E., Hale, W. H., and Theurer, B. (1970a). *J. Anim. Sci.* **30,** 825.
Trei, J. E., Singh, Y. K., and Scott, G. C. (1970b). *J. Anim. Sci.* **31,** 256 (abstr.).
Tremere, A. W., Merrill, W. G., and Loosli, J. K. (1968). *J. Dairy Sci.* **51,** 1065.
Trenkle, A. (1978). *J. Dairy Sci.* **61,** 281.

Trenkle, A. (1981). *Fed. Proc., Fed. Am. Soc. Exp. Biol.* **40**, 2536.
Tritschler, J. P., Shirley, R. L., and Bertrand, J. E. (1983). *J. Anim. Sci.* **58**, 444.
Trowbridge, P. F., and Francis, C. K. (1910). *J. Ind. Eng. Chem.* **2**, 21.
Truman, E. J., Smithson, L., Pope, L. S., Renbarger, R. E., and Stephens, D. F. (1964). Effects of feed level before and after calving on the performance of heifers. *Okla. Agric. Exp. Stn., Livest. Feed. Day Rep.* **M-74**.
Turgeon, O. A., Jr., Brink, D. R., and Britton, R. A. (1981). *Proc., Annu. Meet.—Am. Soc. Anim. Sci., Raleigh, N.C.,* Abstr. No. 763.
Turner, H. A., Raleigh, R. J., and Young, D. C. (1977). *J. Anim. Sci.* **44**, 338.
Turner, H. A., Young, D. C., Raleigh, R. J., and ZoBell, D. (1980). *J. Anim. Sci.* **50**, 385.
Turner, K. W., and Robertson, A. M. (1979). *Appl. Environ. Microbiol.* **38**, 7.
Tyrrell, H. F., and Moe, P. W. (1972). *J. Dairy Sci.* **55**, 685. (abstr.).
Tyrrell, H. F., Moe, P. W., and Flatt, W. P. (1970). *Publ.—Eur. Assoc. Anim. Prod.* **13**, 69.
Tyrell, H. F., Moe, P. W., and Oltjen, R. R. (1971). *J. Anim. Sci.* **33**, 302. (Abstr.).
Uhart, B. A., and Carroll, F. D. (1967). *J. Anim. Sci.* **2**, 1195.
Ulyatt, M. J., MacRae, J. C., Clarke, R. T. J., and Pearce, P. D. (1975). *J. Agric. Sci.* **84**, 453.
Underwood, E. J. (1977). "Trace Elements in Human and Animal Nutrition," 4th ed. Academic Press, New York.
U.S. Environmental Protection Agency (1975). "Third Report to Congress - Resource Recovery and Waste Reduction, Publ. No. SW-161. USEPA, Washington, D.C.
Utley, P. R., and McCormick, W. C. (1972). *J. Anim. Sci.* **34**, 146.
Utley, P. R., and McCormick, W. C. (1975). *J. Anim. Sci.* **40**, 952.
Utley, P. R., and McCormick, W. C. (1980). *J. Anim. Sci.* **50**, 323.
Utley, P. R., Bradley, N. W., and Boling, J. A. (1970). *J. Nutr.* **100**, 551.
Utley, P. R., Lowrey, R. S., and McCormick, W. C. (1973). *J. Anim. Sci.* **36**, 423.
Utley, P. R., Chapman, H. D., Monson, W. G., Marchant, W. H., and McCormick, W. C. (1974). *J. Anim. Sci.* **38**, 940.
Utley, P. R., Newton, G. L., Wilson, D. M., and McCormick, W. C. (1977). *J. Anim. Sci.* **45**, 154.
Valentine, R. C., and Wolfe, R. S. (1963). *J. Bacteriol.* **85**, 1114.
Vance, R. D., Ockerman, H. W., Cahill, V. R., and Plimpton, R. F., Jr. (1971). *J. Anim. Sci.* **33**, 744.
Vance, R. D., Preston, R. L., Cahill, V. R., and Klosterman, E. W. (1972a). *J. Anim. Sci.* **34**, 851.
Vance, R. D., Preston, R. L., Klosterman, E. W., and Cahill, V. R. (1972b). *J. Anim. Sci.* **35**, 598.
Van Dyne, G. M., and Torell, D. T. (1964). *J. Range Manage.* **17**, 7.
Van Es, A. J. H. (1976). *In* "Feed Composition, Animal Nutrient Requirements and Computerization of Diets" (P. V. Fonnesbeck, L. E. Harris, and L. C. Kearl, eds.), p. 512. Utah State University, Logan.
Van Horn, H. H., and Bartley, E. E. (1961). *J. Anim. Sci.* **20**, 85.
Van Horn, H. H., Foreman, C. F., and Rodriguez, J. E. (1967). *J. Dairy Sci.* **50**, 709.
Van Horn, J. L., Thomas, O. O., Drummond, J., and Blackwell, R. L. (1969). *Proc., Annu. Meet.—Am. Soc. Anim. Sci., West. Sect.* **20**, 319.
Van Keulen, J., and Young, B. A. (1977). *J. Anim. Sci.* **44**, 282.
Van Nevel, C. J., and Demeyer, D. I. (1977). *Br. J. Nutr.* **38**, 101.
Van Soest, P. J. (1964). *J. Anim. Sci.* **23**, 838.
Van Soest, P. J. (1965). *J. Anim. Sci.* **24**, 834.
Van Soest, P. J. (1967a). *J. Anim. Sci.* **26**, 119.
Van Soest, P. J. (1967b). *J. Dairy Sci.* **50**, 989 (abstr.).
Van Soest, P. J. (1969). *In* "Cellulases and Their Applications" (R. F. Gould, ed.), p. 262. Am. Chem. Soc., Washington, D.C.
Van Soest, P. J. (1973). *Fed. Proc., Fed. Am. Soc. Exp. Biol.* **32**, 1804.

Van Soest, P. J. (1975). *In* "Digestion and Metabolism in the Ruminant" (I. W. McDonald and A. C. I. Warner, eds.) Univ. of New England Publishing Unit, Armidale, N.S.W., Australia.
Van Soest, P. J., and Mertens, D. R. (1974). *Fed. Proc., Fed. Am. Soc. Exp. Biol.* **33**, 1942.
Van Soest, P. J., and Robertson, J. B. (1974). *Proc.—Cornell Nutr. Conf. Feed Manuf., 1974,* p. 102.
Van Soest, P. J., Mertens, D. R., and Deinum, B. (1978). *J. Anim. Sci.* **47**, 712.
Van Walleghem, P. A., Ammerman, C. B., Chicco, C. F., Moore, J. E., and Arrington, L. R. (1964). *J. Anim. Sci.* **23**, 960.
Varner, L. W. (1974). *Mont., Agric. Exp. Stn., Res. Rep.* **52**, 34.
Varner, L. W., and Woods, W. (1972). *J. Anim. Sci.* **35**, 410.
Varner, L. W., and Woods, W. (1975). *J. Anim. Sci.* **41**, 900.
Vavra, M., Rice, R. W., and Bement, R. E. (1973). *J. Anim. Sci.* **36**, 411.
Veira, D. M., Macleod, G. K., Burton, J. H., and Stone, J. B. (1980). *J. Anim. Sci.* **50**, 937.
Verite, R., and Journet, M. (1977). *Ann. Zootech.* **26**, 183.
Vernlund, S. D., Harris, B., Jr., van Horn, H. H., and Wilcox, C. J. (1980). *J. Dairy Sci.* **63**, 2037.
Vijchulata, P., Henry, P. R., Ammerman, C. B., Potter, S. G., and Becker, H. N. (1980a). *J. Anim. Sci.* **50**, 41.
Vijchulata, P., Henry, P. R., Ammerman, C. B., Potter, S. G., Palmer, A. Z., and Becker, H. N. (1980b). *J. Anim. Sci.* **50**, 1022.
Virtanen, A. I. (1966). *Science* **153**, 1603.
Virtanen, A. I. (1967). *Agrochimica* **11**, 289.
Vohnout, K., and Jimenez, C. (1975). *ASA Spec. Publ.* **24**, 53.
Wada, S., Pallansch, M. J., and Liener, I. E. (1958). *J. Biol. Chem.* **233**, 395.
Wahlberg, M. L., and Cash, E. H. (1979). *J. Anim. Sci.* **49**, 1431.
Waldo, D. R. (1973). *J. Anim. Sci.* **37**, 1062.
Waldo, D. R., and Goering, H. K. (1979). *J. Anim. Sci.* **49**, 1560.
Waldo, D. R., and Jorgensen, N. A. (1981). *J. Dairy Sci.* **64**, 1207.
Waldo, D. R., and Schultz, L. H. (1956). *J. Dairy Sci.* **39**, 1453.
Waldo, D. R., Keys, J. F., Jr., and Gordon, C. H. (1971a). *J. Anim. Sci.* **33**, 305. (abstr.).
Waldo, D. R., Keys, J. E., Smith, L. W., and Gordon, C. H. (1971b). *J. Dairy Sci.* **54**, 77.
Waldo, D. R., Smith, L. W., and Cox, E. L. (1972). *J. Dairy Sci.* **55**, 125.
Walker, D. J., and Forrest, W. W. (1964). *Aust. J. Agric. Res.* **15**, 299.
Walker, D. J., and Nader, C. J. (1975). *Aust. J. Agric. Res.* **26**, 689.
Walker, D. J., Egan, A. R., Nader, C. J., Ulyatt, M. J., and Storer, G. B. (1975). *Aust. J. Agric. Res.* **26**, 699.
Walker, P. M., Weichenthal, B. A., and Cmarik, G. F. (1980a). *J. Anim. Sci.* **51**, 526.
Walker, P. M., Weichenthal, B. A., and Cmarik, G. F. (1980b). *J. Anim. Sci.* **51**, 532.
Waller, J. A., Black, J. R., Bergen, W. G., and Jackson, M. (1980). *Proc., Conf.—Distill. Feed Res. Counc.* **35**, 53.
Wallnofer, P., Baldwin, R. L., and Stagno, E. (1966). *Appl. Microbiol.* **14**, 1004.
Ward, G. M. (1966). *J. Dairy Sci.* **49**, 268.
Ward, G. M., Harbers, L. H., and Blaha, J. J. (1979). *J. Dairy Sci.* **62**, 715.
Ward, J. K. (1978). *J. Anim. Sci.* **46**, 831.
Ware, D. R., Self, H. L., and Hoffman, M. P. (1977). *J. Anim. Sci.* **44**, 722.
Warner, A. C. I. (1955). Ph.D. Dissertation, University of Aberdeen, Scotland.
Warner, A. C. I. (1956). *J. Gen. Microbiol.* **14**, 749.
Warner, A. C. I. (1964). *Nutr. Abstr. Rev.* **34**, 339.
Warner, R. G. (1970). *Proc., Conf.—Distill. Feed Res. Counc.* **25**, 11.
Warnick, A. C., Meade, J. H., and Koger, M. (1960). *Bull.—Fla., Agric. Exp. Stn.* **623**.

Warnick, A. C., Koger, M., Martinez, A., and Cunha, T. J. (1965). *Bull—Univ. Fla., Inst. Food. Agric. Sci.* **695.**
Watson, C. J., Kennedy, J. W., Davidson, W. M., Robinson, C. H., and Muir, G. W. (1949). *Sci. Agric.* **29,** 173.
Wayman, O., Iawanga, I. I., Henke, L. H., and Weeth, H. J. (1953). *Hawaii, Agric. Exp. Stn., Circ.* **43.**
Wayman, O., Johnson, H. D., Merilan, C. P., and Berry, I. L. (1962). *J. Dairy Sci.* **45,** 1472.
Webb, D. W., Bartley, E. E., and Meyer, R. M. (1972). *J. Anim. Sci.* **35,** 1263.
Webster, A. J. F. (1977). *Proc. Br. Nutr. Soc.* **36,** 53.
Weir, W. C., and Torell, D. T. (1959). *J. Anim. Sci.* **18,** 641.
Weiss, K. E. (1953). *Onderstepoort J. Vet. Res.* **26,** 241.
Welch, J. A., Anderson, G. C., McLaren, G. A., Campbell, C. D., and Smith, G. S. (1957). *J. Anim. Sci.* **16,** 1034 (abstr.).
Weller, R. A. (1957). *Aust. J. Biol. Sci.* **10,** 384.
Weller, R. A., Gray, F. V., and Pilgrim, A. F. (1958). *Br. J. Nutr.* **12,** 421.
Whanger, P. D., and Matrone, G. (1967). *Biochim. Biophys. Acta* **136,** 27.
Wheeler, W. E., and Noller, C. H. (1976). *J. Dairy Sci.* **59,** 1788.
Wheeler, W. E., and Noller, C. H. (1977). *J. Anim. Sci.* **44,** 131.
Wheeler, W. E., and Oltjen, R. R. (1977). *J. Anim. Sci.* **48,** 658.
Wheeler, W. E., Noller, C. H., and Coppock, C. E. (1975). *J. Dairy Sci.* **58,** 1902.
Wheldon, G. H., and MacDonald, R. E. (1962). *Nature (London)* **196,** 596.
Whetstone, H. B., Davis, C. L., and Bryant, M. P. (1981). *J. Anim. Sci.* **53,** 803.
White, T. W., and Hembry, F. G. (1978). *J. Anim. Sci.* **46,** 271.
White, T. W., Hembry, F. G., and Reynolds, W. L. (1972). *J. Anim. Sci.* **34,** 672.
White, T. W., Reynolds, W. L., and Hembry, F. G. (1973). *J. Anim. Sci.* **37,** 1428.
White, T. W., Hembry, F. G., and Reynolds, W. L. (1974). *J. Anim. Sci.* **38,** 844.
White, T. W., Reynolds, W. L., and Hembry, F. J. (1975). *J. Anim. Sci.* **40,** 1.
Wilgus, H. S., Jr., Norris, L. C., and Heuser, G. F. (1936). *Ind. Eng. Chem.* **28,** 586.
Wilkie, D. R. (1960). *Prog. Biophys. Biophys. Chem.* **10,** 259.
Wilkinson, J. I. D., Appleby, W. G. C., Shaw, C. J., Lebas, G., and Pflug, R. (1980). *Anim. Prod.* **31,** 159.
Wilkinson, J. M., and Penning, I. M. (1976). *Anim. Prod.* **23,** 181.
Willey, N. B., Riggs, J. K., Colby, R. W., Butler, O. D., and Reiser, R. (1952). *J. Anim. Sci.* **11,** 705.
Williams, D. B., Bradley, N. W., Little, C. O., Crowe, M. W., and Boling, J. A. (1970). *J. Anim. Sci.* **30,** 614.
Williams, J. E., McLaren, G. A., Smith, T. R., and Fahey, G. C., Jr. (1979). *J. Anim. Sci.* **49,** 163.
Williams, P. E. V., Day, D., Raven, A. M., and McLean, J. A. (1981). *Anim. Prod.* **32,** 133.
Williams, P. P., and Dinusson, W. E. (1972). *J. Anim. Sci.* **34,** 469.
Williams, P. P. and Dinusson, W. E. (1973). *J. Anim. Sci.* **36,** 151.
Williams, P. P., Davis, R. E., Doetsch, R. N., and Gutierrez, J. (1961). *Appl. Microbiol.* **9,** 405.
Williams, V. J., Nottle, M. C., Moir, R. J., and Underwood, E. J. (1953). *Aust. J. Agric. Res.* **10,** 865.
Willis, C. M., Boling, J. A., and Bradley, N. W. (1981). *Proc., Annu. Meet.—Am. Soc. Anim. Sci., Raleigh, N.C.,* Abstr. No. 775.
Wilson, A. D. (1970). *Aust. J. Agric. Res.* **21,** 273.
Wilson, G., Martz, F. A., Campbell, J. R., and Becker, B. A. (1975). *J. Anim. Sci.* **41,** 1431.
Wilson, J. R., Bartley, E. E., Anthony, H. D., Brent, B. E., Sapienza, D. A., Chapman, T. E., Dayton, A. D., Milleret, R. J., Frey, R. A., and Meyer, R. M. (1975). *J. Anim. Sci.* **41,** 1249.

Wilson, P. N., and Osbourn, D. F. (1960). *Biol. Rev. Cambridge Philos. Soc.* **34,** 324.
Wilson, R. K., and Pigden, W. J. (1964). *Can. J. Anim. Sci.* **44,** 122.
Wilson, T. R., Kromann, R. P., Warner, K. O., Hillers, J. K., and Martin, E. L. (1976). *Proc., Annu. Meet.—Am. Soc. Anim. Sci., West. Sect.* **27,** 292.
Wiltbank, J. N. (1973). *In* "Factors Affecting Calf Crop" (T. J. Cunha, A. C. Warnick, and M. Koger, eds.), p. 44. Univ. of Florida Press, Gainesville.
Wiltbank, J. N., Rowden, W. W., Ingalls, J. E., Gregory, K. E., and Koch, R. M. (1962). *J. Anim. Sci.* **21,** 219.
Wiltbank, J. N., Rowden, W. W., Ingalls, J. E., and Zimmerman, D. R. (1964). *J. Anim. Sci.* **23,** 1049.
Wiltbank, J. N., Bond, J., and Warwick, E. J. (1965). *U.S., Dep. Agric., Tech. Bull.* **1314.**
Winchester, C. F., and Howe, P. E. (1955). *U.S., Dep. Agric., Tech. Bull.* **1108.**
Winchester, C. F., and Morris, M. J. (1956). *J. Anim. Sci.* **15,** 722.
Winchester, C. F., Davis, R. E., and Hiner, R. L. (1967). *U.S., Dep. Agric., Tech. Bull.* **1374.**
Wing, J. M. (1974). *J. Dairy Sci.* **58,** 63.
Wing, J. M., and Becker, R. B. (1963). *Bull.—Fla., Agric. Exp. Stn.,* **655.**
Wise, M. B., Blumer, T. N., Matrone, G., and Barrick, E. R. (1961). *J. Anim. Sci.* **20,** 561.
Wise, M. B., Blumer, T. N., Craig, H. B., and Barrick, E. R. (1965). *J. Anim. Sci.* **24,** 83.
Wohlt, J. E., Sniffen, C. J., and Hoover, W. H. (1973). *J. Dairy Sci.* **56,** 1052.
Wohlt, J. E., Sniffen, C. J., Hoover, W. H., Johnson, L. L., and Walker, C. K. (1976). *J. Anim. Sci.* **42,** 1280.
Wohlt, J. E., Fiallo, J. F., and Miller, M. E. (1981). *Anim. Feed. Sci. Technol.* **6,** 115.
Wolff, J. E., and Bergman, E. N. (1972). *Am. J. Physiol.* **223,** 455.
Wolin, M. J. (1975). *In* "Digestion and Metabolism in the Ruminant" (I. W. McDonald and A. C. 'I. Warner, eds.), p. 134. Univ. of New England Publishing Unit, Armidale, N.S.W., Australia.
Wolin, M. J., Manning, G. B., and Nelson, W. O. (1959). *J. Bacteriol.* **78,** 147.
Wolin, M. J., Wolin, E. A., and Jacobs, N. J. (1961). *J. Bacteriol.* **31,** 911.
Wolstrup, J., and Jensen, K. (1978). *J. Appl. Bacteriol.* **45,** 49.
Woody, H. D., Fox, D. G., and Black, J. R. (1983). *J. Anim. Sci.* **57,** 710.
Woolfolk, P. G., Richards, C. R., Kaufman, R. W., Martin, C. M., and Reid, J. T. (1950). *J. Dairy Sci.* **33,** 385 (abstr.).
Wooten, R. A., Roubicek, C. B., Marchello, J. A., Dryden, F. D., and Swingle, R. S. (1979). *J. Anim. Sci.* **48,** 823.
Wordinger, R. J., Dickey, J. F., and Hill, J. R., Jr. (1972). *J. Anim. Sci.* **34,** 453.
Work, S. H., and Henke, L. A. (1940). *Proc. Am. Soc. Anim. Prod.* **39,** 404.
Wren, T. R., Bitman, J., and Sykes, J. F. (1961). *J. Dairy Sci.* **44,** 2077.
Wright, D. E. (1961). *N. Z. J. Agric. Res.* **4,** 216.
Wyatt, R. D., Johnson, R. R., and Clemens, E. T. (1975). *J. Anim. Sci.* **40,** 126.
Wyatt, R. D., Lusby, K. S., Whiteman, J. V., Gould, M. B., and Totusek, R. (1977a). *J. Anim. Sci.* **45,** 1120.
Wyatt, R. D., Gould, M. B., Whiteman, J. V., and Totusek, R. (1977b). *J. Anim. Sci.* **45,** 1138.
Wynn, R. M. (1967). "Cellular Biology of the Uterus." Appleton, New York.
Yadawa, I. S., and Bartley, E. E. (1964). *J. Dairy Sci.* **47,** 1352.
Yarns, D. A., and Putnam, P. A. (1962). *J. Anim. Sci.* **21,** 744.
Yeates, N. T. M. (1956). *Nature (London)* **178,** 702.
Yeck, R. G., Smith, L. W., and Calvert, C. C. (1975). Recovery of substances from animal waste—an overview of existing options and potentials for use as feed. *ASAE Publ.* **PROC-275,** 192.
Young, A. W. (1978). *J. Anim. Sci.* **46,** 505.
Young, A. W., and Kauffman, R. G. (1978). *J. Anim. Sci.* **46,** 41.

Young, A. W., Boling, J. A., and Bradley, W. N. (1973). *J. Anim. Sci.* **36,** 803.
Young, B. A. (1981). *J. Anim. Sci.* **52,** 154.
Young, B. A., and Berg, R. T. (1970). *Univ Alberta, 49th Feeders Day Rep.*, pp. 38–40.
Young, J. W. (1977). *J. Dairy Sci.* **60,** 1.
Young, J. W., Shrago, E., and Lardy, H. A. (1964). *Biochemistry* **3,** 1687.
Young, J. W., Trott, D. R., Berger, P. J., Schmidt, S. P., and Smith, J. A. (1974). *J. Nutr.* **104,** 1049.
Young, M. C., Theurer, B., Ogden, P. R., Nelson, G. W., and Hale, W. H. (1976). *J. Anim. Sci.* **43,** 1270.

# Index

## A

Acetate, 58, 79, 80, 87
  fermentative pathway, 95, 96
Acetic acid, 41, 124, 176
  acetate: propionate ratio, 95, 96, 202, 235, 238
Acetohydroxamic acid, 59
Acid detergent fiber, 87, 128, 205, 208, 215, 228, 231, 233, 259, 295, 301
  digestibility with tylosin, 87
Acid insoluble ash, 64, 66, 67
Acidosis, 44
  antibiotics and, 47, 48, 85
  beet pulp diets and, 45
  cattle on concentrate feed, 47
  corn–corn silage diets, 45
  endotoxins and, 49
  feedlot steers and, 49
  fermentable carbohydrates and, 44
  histamine and, 46
  rumen buffering and, 50
  wheat diets and, 47
Acid-utilizing bacteria, 18
  cytochromes and, 19
  formate and, 18
  lactate and, 18
Acrylate pathway, 96
Actinomycete counts, 176
Adenosinediphosphate, 91
Adenosinetriphosphate, 26, 43, 91, 96
Age, 175
Alanine, 144
Alcohol production by-products, 198–202
  by-passing the rumen, 199, 201
  condensed distillers solubles, 201
  dairy cattle and, 202
  sheep and, 201, 202
  wet distillers solubles, 201
  whole stillage, 201
Alfalfa, 107, 128, 151, 153, 159, 168, 170, 216, 217, 267
  silage of, 82
Alimentary tract, 102, 131–133
Alkali treatment of lignocellulose, 228–230
Alkaloids, 69
Amicloral, 59
  crude fiber digestibility with, 86
  feed efficiency and, 86
  organic matter digestibility and, 86
  propionic acid and, 86
Amino acids, *see also* specific amino acid
  degradation of, 24
  essential amino acids, 24
  inhibitors of degradation, 24, 25
Ammonia, 71, 72, 152, 159, 164–167, 176, 180, 181, 193, 199, 214
  ammonium acetate, 149
  ammonium bicarbonate, 149, 153
  ammonium hydroxide, 175
  ammonium lactate, 149
  ammonium sulfate, 154
  anhydrous ammonia and corn silage, 180, 181
  toxicity, 71, 72
Amylase, 134
Antibiotics, 113, 115
Apetite, 294
Apple pomace, 218, 219
Aquatic plants, 184, 215–217
Arginine, 24, 31, 128, 145
Ash, 102, 301

Aspartic acid, 144
Associative effects on net energy, 107–109
Avoparcin, 125, 129
  feed efficiency and, 86

**B**

Bacteria, 9, 33, 37, 144, *see also* Microbes
  interactions, 33, 34
  interrelationships with protozoa, 36–37
Bagasse, sugarcane, 214, 215
Barley grain, 33, 95, 107, 128, 134, 139, 170, 175, 224–240
  straw, 212–214
Basal metabolic rate, 290
Beef cattle, 176, 177, 195, 200, 201, 215, 218, 286, 294, *see also* specific type
Beet molasses, 170, 184, 185, 196, *see also* Molasses
Beet pulp, 107, 170
Bermudagrass, 128, 170, 205, 208, 209, 216, 245, 264
Biological value, 154
Biuret, 25, 149, 152–157, 160, 195, 206, 210, 211
  microbial adaptation to, 25
  toxicity, 72
Biuretolytic activity, 25
Bloat, 53
  animal factors, 55
  feed factors, 54
  feedlot animals, 56, 57
  microbial changes, 53, 54
  poloxalene in, 56
  polysaccharides in, 54
Blood meal, 128
Body composition, 275–284
  breed and, 276–279
  diet and, 276–280
  implants and, 281
  live animal determination of, 283
  pH of muscle and, 281
  predicting carcass composition, 281–283
  realimentation of cull cows and, 279, 280
  sex and, 276–279
Bolus, 37
Bone, 282
Breeds, 109, 111, 115, 122, 276, 277, 285
Brewers grains, 45, 81, 170, 200, *see also* Alcohol production by-products

Buffalo, 35
Buffers, 50, 51
  dairy cattle and, 51, 52
  drinking water, 51
  feed utilization with, 50, 51
Bulls, 82, 123, 126, 212, 250, 260, 271, 276–278
  calves, 179, 189, 214, 299
Butyrate, 58, 79, 80
Butyric acid, 41
  fermentation pathway, 95

**C**

$^{45}$Ca, 269
Cadmium, 265
Calcium, 216, 267, 268, 270
California Net Energy System (CNES), 2, 100–109
  feedlot phase, 102–106
  metabolism phase, 101, 102
  tables, 103, 105–107
Caloric efficiency, 246
Calorigenic effect, 96, 290
Calorimetry, 119
Calves, 131, 132, 154, 156, 160, 200, 205, 208, 220, 246, 247, 254, 255, *see also* specific type
Calving, 243, 244
Carbon dioxide, 53, 95, 289
Carcass, 102
  composition, 276–283
  density, *see* Specific gravity
  fat, 126, *see also* Fat
  protein, 126
Cattle, 100–107, 117, 134, 150, 151, 155, 159, 208, 209, 212, 223, 245, 271, 282, 291, 294, 298, *see also* specific type
  finishing, 252, *see also* Feedlots
Celery, 219, 227, 228
Cellulases, 6, 15–17, 80, 90
Cellulose, 15, 135, 146, 231, 265–267
  amorphous regions of, 15
  ash in, 16
  crystalline regions of, 15
  enzymatic degradation of, 15
  lignification of, 16
  solubility factors, 16
Cellulosic waste, 225, 226
Cell walls, 204–206, 215, 231, 293

**Index**

Central nervous system, 3
Cereal residues, 212–214, *see also* Crop residues
Chlorotetracycline, 87
Cholesterol, 137, 138
Choline, 139
Chromium, 265
  oxide, 64, 65
Citrus by-products, 188–191
  condensed molasses solubles, 190, 191
  dairy cattle and, 191
  meal, 188–190
  molasses, 170, 188, 189, 194
  pulp, 38, 82, 170, 184, 188–191, 211, 221, 258
Cobalt, 265, 270
Coccidia oocysts, 85
Cocoa bean, 227, 228
Coffee, 220, 221, 227, 228
Cold stress, 297–299
Coliform counts, 262
Comparative slaughter feedlot trial, 100, 137, 169, 173, 174, 179, 186, 213, 249, 276, 279, 282
Compensatory gains, 248–252
Composition of feeds, 301–304
  crude fiber, 302
  dry matter, 301
  energy, 302, 303
  minerals, 303
  neutral detergent fiber, 302
  protein, 301
  tables, 304
  vitamin A, 304
Compudose, 113
Conduction, 292
Convection, 292
Copper, 216, 265, 270
Corn, 159, 174–183, 206, 212
  gluten, 145, 170
  grain, 38, 45, 54, 108, 128, 129, 132, 133, 138, 170, 174–183, 193, 199, 235–237, 291
  ground, 33
  high moisture, 117, 175–177
  meal, 189
  molasses, 184, 185
  silage, 80, 108, 118, 128, 129, 160, 169, 170, 177–183, 193, 218, 223, 255
Cottonseed hulls, 107, 170

Cottonseed meal, 74, 107, 128, 150, 170, 191, 211, 240
Cows, 121–123, 135, 146, 150–152, 156, 158, 207, 270, 279, 280, 304, *see also* specific cattle type
Cow–calf production, 246, 247
Creep feeding, 247, 248
Crop residues, 206, 241, 242, *see also* Cereal residues
Crude fiber, 231, 232, 297, 302
Crude protein, *see* Energy and nitrogen utilization of feedstuffs
Cyanide, 67, 68
Cysteine, 154
Cystine, 144, 154

**D**

Dacron bags, 226
Dairy cattle, 191, 192, 201, 202, 256, 258, 269, 294, 296, 297
  cows, 167, 168, 217–219, 272, 287
  bulls, 181
Diaminopimelic acid, 26
Diarrhea, 270
Dicyanodiamide, 149
Diethylstilbestrol, 112
Digestible energy, 97, 120, 123, 126, 183, 212, 220, 245, 261, 302
  factors affecting availability of, 123–125
Digestible protein, 260, *see also* Protein
Digestion trials, *see* Metabolism trials
Dilution rates in rumen, 57
Dimethyl dialkyl quaternary ammonium compound, 57
Diphenyl iodonium chloride, 24, 59
Distillers grains, 145, *see also* Alcohol production by-products
Disulfide linkages, 23, 130
Dogs, 146
Dolomitic limestone, 61
Dry matter, 301
Dry matter intake, 113, 114
Dysprosium, 66

**E**

Electron microscopy, 26, 233
Embryos, 286
Endocrines, 285–287
  crossbreeding and, 285

Endocrines (cont.)
  diet and, 286
  milk production and, 286, 287
Energy and nitrogen utilization in feedstuffs, 173–202
  alcohol production by-products, 198–202
    condensed distillers solubles, 201
    rumen by-passing and, 199, 201
    wet distillers grains, 201
    whole stillage, 201
  citrus by-products, 188–191
    condensed molasses solubles, 190, 191
  corn, 174–183
    high moisture, 175–177
    new endosperm types, 183
    silage, 177–183
      anhydrous ammonia and, 180, 181
      formic acid and, 182
      microbial additives and, 181, 182
      monensin and, 182, 183
      urea and, 181
  feather meal, 193
  fermented ammoniated condensed whey, 192, 193
  grass silage, 183–185
    additives and, 184, 185
  hair meal, 193
  molasses, 194, 195
    liquid feeds and, 195–197
    Masonex, 197, 198
  sorghum grain, 185–188
    early-harvested, 188
    processing, 187, 188
    proximate analysis of, 186
    reconstituted, 188
    total digestible nutrients, 186
    variety and, 186
  sunflower meal, 191, 192
Ergs, 89, 90
Esophageal fistula, 211
Essential oil industry by-products, 226–228
Estradiol, 281
Estrogen, 286
Ethanol, 96
Ether extract, 102, 301
Evaporation, 292
Ewes, 126, 269
Excreta, animal, 257–264
  cattle, 257–262
    dehydration of, 259, 260
    ensiling of, 260, 261
    formaldehyde and, 260
    methane fermentation residue, 263
    nutrients in, 258
    poultry litter, 258, 259
    processing, 257, 258
    sewage sludge, 263, 264
    swine, 261
Exercise, 292

## F

Fat, 102, 107, 126, 135, 275–285
  diets containing, 135, 138
  encapsulation of, 136
  lipotropic factors, 139
  protein and, 138
Feather meal, 193, 194
Fecal energy, 92, 100, 101
Feed efficiency, 79–83, 86, 87, 146, 160, 174, 188, 190, 193, 196, 197, 200, 201, 224, 230, 235, 248, 259, 286, 291, 295–298
Feed intake, 5–7, 105, 144, 215, 232, 248, 270, 272, 291, 293–299
  acetate and, 5
  activity and, 5
  energy balance and, 5
  central nervous system and, 6
  digestion rate and, 6
  hypothalamus and, 5
  opiate peptides and, 6
  passage rate and, 6
  restriction of, 248
  water intake and, 6
Feeding frequency, 248, 249
Feedlot bloat, see Bloat
Feedlots, 100–109, 174–182, 252–255, see also Energy and nitrogen utilization
Fermentation, 44
  manipulation, 58
  methane inhibitors and, 59
  monensin and, 58
  pathways, 95
  pH, 44
Fertilizers, 208, 209
First Law of Thermodynamics, 89
Fish meal, 24, 128, 170, 196, 220
Fistula, 37, 39, 133–135, 143, 163, 164, 191, 193, 196, 218, 233, 238, 271, 295

Forages and roughages, 130, 156, 203–233, 241, 245
  alkali and, 22, 230
  apple pomace, 218, 219
  aquatic plants, 215–217
  cell walls, 204–206
  cereal residue, 212–214
  coffee grounds, 220, 221
  concentrate feeds and, 231–233
  corn residues, 206, 207
  dried celery tops, 219
  electron microscopy scanning and, 233
  essential oil industry by-products, 226–228
  grazing intensity, 208, 209
  infrared reflectance spectroscopy and, 233
  lignin and, 204–206
  mechanical processing, 214
  osyter shells and plastic polymers, 222
  paper, 221
  pecan hulls, 221, 222
  pineapple plant, 217
  potato residue, 220
  supplementation of, 209–212
  tomato pulp, 219, 220
  wood and wood by-products, 222–226
    pulp by-products, 224, 225
    sawdust, 222–224
    solid cellulose waste, 225, 226
Formaldehyde, 2, 22, 23, 136, 137, 139, 142, 185, 192, 260
Formate, 96
Formic acid, 41, 185
  corn silage and, 182
Frame size, 110
Fumarate, 45

## G

Gain projection, 109, 111
Gelatin, 31
Gelatinization, 33
Gestation, 292, *see also* Pregnancy
Ginger root, 227, 228
Glucanase, 15
Gluconeogenesis, 75–77
  amino acids and, 76, 77
  propionate and, 76
Glucose, 45, 75, 76, 77, 79, 84, 85
Glutamine, 149
Glycine, 128, 149

Goats, 35, 158
Gossypol, 74
Grain bloat, 53, *see also* Bloat
Grape marc, 190
Grass silage, 182–185
  additives and, 184, 185
Grazing, 208, 209
  intensity, 208, 209
  supplementation of, 209–212
Gross energy, 89, 100, 118
Growth promoters, 111, 112
  growth hormone, 285, 287

## H

Hair meal, 193, 194
Hay, 206
Heart rate, 295
Heat energy, 91, 92, 98
  dietary ME and, 98
  body tissue and, 98
Heat increment, 96, 101, 290, 294, 297
Heat stress, 290–296
Heifers, 105, 106, 117, 121, 122, 128, 150, 153, 154, 224, 244, 250, 252, 268, 269, 277, 286, 291, 304
  body composition of, 277
Hemaggluting agent, 73
Hemicellulose, 230
  composition of, 16
  digesters of, 16, 17
Histamine, 48
Histidine, 31, 145
Holsteins, 109, 111, 113, 143, 146, 202, 221, 266, 276, 286, 287
Hyacinths, 184
Hydrazine, 24
Hypophysis, 286

## I

Infrared reflectance spectroscopy, 233
Inorganic ions, 60
Inositol, 139
Insulin, 285, 287
Iodine, 265, 270
Iron, 265, 270
Isobutyrate, 40, 176
Isoleucine, 24, 145, 155
Isovalerate, 41, 79

## J

Joules, 89, 90

## K

Kiln dust, 52

## L

Lactate, 96
Lactic acid, 43, 45, 160, 182
　high grain diets and, 43
Lambs, 128, 185, 190, 192, 197, 212, 217, 219, 260
Lasalocid, 49, 79, 84–86
　coccidia oocysts and, 85
　feed efficiency and, 85, 86
　lactic acid and, 85
　methane production and, 85
Leucine, 24, 128, 145, 155
Lignin, 64–67, 204–206, 224, 230, 293, 302
Lignocellulose, 2
Limestone, 179, 267
Linseed meal, 24, 128, 202
Lipolytic bacteria, 18
　fermentative products, 18
Lipostasis, 5
Liquid feeds, 195, 197
Liver abscess, 78, 87, 222
Lysine, 24, 128, 142, 143, 145, 155, 183

## M

Magnesium, 216, 270
　carbonate, 51
Malic acid, 44
Malonic acid, 44
Maltase, 134
Manganese, 216, 265, 270
Marbling, 281
Markers, 64
　equations for, 64, 65
　external, 65, 66
　internal, 66, 67
Masonex molasses, 197, 198
Meal size, 5, 6
Meat scraps, 170
Meat and bone meal, 128
Melamine, 149, 151
Melengestrol acetate, 112, 286

Metabolic body rate, 290
Metabolic body size, 90, 102, 118, 121
Metabolism trials, 175, 178, 179, 184, 186, 194, 196, 205, 272, 273, 276
Metabolizable amino acids, 168–170
Metabolizable energy, 2, 93, 100–102, 117–120, 146, 174, 195, 188, 249, 261, 262, 296, 298
Metabolizable energy system (Blaxter), 117–119
Metabolizable protein, 165–171
Methane, 58–60, 96, 101, 102, 124, 263
Methanogenesis, 19, 85
Methionine, 24, 32, 142–146, 154, 155, 267
Methionine hydroxy analog, 32, 109, 136, 142–146, 186
Microbes of rumen, 37
　additives and, 181–183
　energy utilization by, 42
　protein synthesis by, 145
　relation to rumen functions, 37
Milk, 119, 138, 143, 146, 150, 158, 170, 193, 202, 215, 221, 225, 232, 268, 285–287, 292, 293, 297
Minerals, 265–270, 303, *see also* specific element
　calcium and, 267, 268
　nutrients in water and, 269, 270
　sulfur and nitrogen, 266, 267
　toxic elements in water and, 269, 270
　trace elements and, 268
Modeling, 61–63
Molasses, 194, 195, see specific types
　cows and calves, 195, 196
　finishing cattle, 196, 197
　lick wheels, 56
　Masonex, 197, 198
　salt blocks, 56
Molybdenum, 265
Monensin, 46, 58, 59, 79–87, 113, 179, 259
　acid detergent fiber digestion and, 87
　corn silage and, 182, 183
　feedlot cattle and, 79, 80, 113, 182, 183
　forages and, 82, 83
　reproduction and, 83, 84
Mud, 112

## N

$^{15}N$, 26, 27
Net energy, 97, 101, 102, 106–108, 113, 297,

303, *see also* Energy and nitrogen utilization
  associative effects on, 107–109
  breeds and, 111
  digestive stimulants and, 113
  dry matter intake and, 113, 115
  environment and, 111
  frame size and, 109–111
  gain and, 99, 101–108, 113, 142, 146, 174, 175, 186, 249, 262, 302, 303
  growth stimulants and, 111–113
  limitations of, 99, 100
  maintenance and, 98–102, 107–109, 142, 174, 175, 186, 249, 302, 303
  milk, 99, 100, 120
  previous nutritional treatments and, 113
  production and, 97, 98, 101
Net energy systems, 97–119
  Blaxter (British), 117–119
  California, 100–117
  Rostock (German), 99
Net protein, 114–117, 143
Neutral detergent fiber, 215, 231–233, 259, 295, 301, 302
Neutral detergent solubles, 295
Nickel, 265
Nitrate, 67–71
  adjustment to, 69, 70
  diet composition and, 70, 71
Nitrogen, 102, 120, 206, 208, 301, *see also* Energy and nitrogen utilization
  retention of, 163, 188, 190, 221, 260, 270
Nonprotein nitrogen, 149–172
  adaptation to, 151–154
  forages and, 156
  liquid supplements and, 159, 160
  maintenance and, 156
  milk production and, 158
  nitrogen sources and, 155, 156
  ruminal ammonia and, 161–164
  self-feeding supplements and, 159
  silage and, 160
  sulfur and, 154, 155
  weight gains and, 157
Nucleic acids, 166
Nutmeg hulls, 227, 228
Nylon bags, 204, 233

## O

Oats, 45, 170, 175
  hay, 107, 170
  hulls, 223
  pasture, 178
  straw, 213, 214
Oxalates, 68, 268
Oxalic acid, 45
Oxygen, 59, 60, 289
Oyster shells, 203

## P

$^{32}$P, 26, 269
Paper, 221
Particle size, 133
Pasture, 264
  bloat and, 53
Peanut meal, 170
Pecan hulls, 221, 222
Peppermint oil, 263
Pepsin–pancreatin digestion, 24
Pesticides, 63
pH, 44, 134, 281
Phosphorus, 27, 149, 195, 208, 216, 270
Plant chromagen, 64
Poloxalene, 56
Polyethylene glycol, 218
Polyethylene pellets, 203, 222
Potassium, 51, 60, 61, 149, 216, 270
Potatoes, 45
Poultry, 193
  litter and, 198, 258–261
  manure and, 82, 262
  waste, 207
Preconditioning, 253–255
Pregnancy, 120, 122, 150, 270, 286
Processing, 214, 235–242
  barley and, 239–240
  cereal straw and, 214
  cottonseed and, 240
  corn and, 235–237
  crop residues and, 241, 242
  forages and, 241
  sorghum grains, 237–239
  soybeans and, 240, 241
  wheat and, 240
Production practices, 243–255
  caloric efficiency of cow-calf, 246, 247
  compensatory gains, 248–252
  creep feeding, 247, 248
  feeding frequency, 248
  finishing cattle, 252
  forage grazing supplements, 245, 246
  growth and reproduction, 244

Production practices (*cont.*)
  preconditioning for feedlots, 253–255
  protein withdrawal, 252, 253
  spring versus fall calving, 243, 244
Progesterone, 281, 286
Propionate, 58, 79–86
Propionic acid, 41, 75, 76, 124, 133, 176
  fermentation pathways, 95, 96
Protein, 126, 127, 129, 138, 252, 253, 285, 286, 291
  degradation of, 21
  digestibility of, 238, 302
  labile, 147
  synthesis of, 26, 27, 145, 161, 197, 201, 224
    ammonia concentration and, 31
    low-roughage diets and, 31
    performed amino acids and, 31
    starchy cereal diets and, 31
    starea and, 33
    sulfur and, 26, 27
    TDN and NPN in, 26, 27
  solubility of, 129
  turnover of, 147
Proteolytic bacteria, 20, 21
  ATP formation by, 21
  degradation by, 21, 22
  disulfide linkages and, 23
  monensin on, 20
Protozoa, 34–36
  buffalo and goats, 35
  interrelation with bacteria, 35, 36
  transfaunation of, 34
Pulp, wood, 224, 225

## Q

$Q_m$ - Blaxter, 117, 119
$Q_f$ - Blaxter, 117, 119

## R

Ralgro, 112
Rams, 126, 212, 260
Rapeseed meal, 192
Rare earths, 64, 66
Realimentation, 250, 279
Reproduction, 244
Residues, plant, 206
Respiratory exchange, 119
Respiratory rate, 295

Ribonucleic acid, 26
Rice, 170
Roughages, *see* Forages and roughages
Roughage factor, 231
Rumen microbes, 9–60
  amino acids and, 12
  ammonia and, 12
  carbohydrates and, 11, 12
  cellulase and, 15
  cellulose and, 14–16
  fermentation and, 9
  glucanases, 15
  minerals and, 13
  nutrition of, 12
  peptide–nitrogen and, 12, 13
  protein degradation and, 11
  variation in numbers, 13, 14
  vitamins and, 13
Rumen–reticulum, 37
  mixing in, 38
Rumensin, *see* Monensin
Rumination, 39
  chewing and, 39
  hay intake and, 39
Rye grain, 170, 199
  pasture, 178
  straw, 170

## S

Saliva, 37, 40, 55
  buffers in, 40
  urea in, 40
Salt, 157, *see also* Sodium
*Salmonella*, 262
Sawdust, 222–224, 230, 264
Selenium, 185, 265, 268, 270
Serine, 144
Sewage sludge, 263, 264
Sex, 124, 276, 277
Shade, 294–296
Sheep, 35, 59, 134, 135, 153–158, 177, 188, 192, 197, 201, 205, 211, 216, 223, 245, 251, 257–260, 270–273, 281
*Shigella*, 262
Silage, *see* Corn
Sludge, clarifier, 225
Sodium, 60, 216
  arsenite, 24
  bentonite, 61
  chloride (salt), 269, 270

chlorite, 230
hydroxide, 175, 176, 223, 229, 230, 262
metabisulfite, 160
sulfite, 230
pyrophosphate, 281
Sorghum grain, 33, 45, 81, 142, 170, 174, 175, 185–188, 206, 212, 237–239
    early-harvested, 188
    processing of, 187, 188
    proximate analysis of, 186
    reconstituted, 188
    silage of, 178
    total digestible nutrients of, 186
    varieties of, 186
Soybean meal, 24, 38, 54, 82, 117, 132, 142–145, 149, 155, 158, 160, 169, 170, 174, 179, 191, 192, 200, 207, 214, 240, 241, 257, 267
Specific dynamic effect, 290
Specific gravity, 102, 282, 283
Spectral energy density, 145
Sprinkling, 296
Starch, 17, 18, 130–134
    alimentary tract and, 131–134
    enzymes and, 134
    fermentation products, 17, 18
    grain diets and, 130
    pH and, 134
    silage and, 130
Steers, 31, 38, 105, 106, 117, 128, 132, 136, 146, 148, 151–156, 161, 190, 193, 196, 205, 212–214, 219, 221–223, 226, 235, 238, 248, 255, 260–264, 268, 277–283, 289, 290, 295, 304
    body composition of, 276–279
Stress, *see* Cold stress; Heat stress
Succinic acid, 44, 96
Succinyl CoA, 41
Sudan grass, 205
Sugar, 18, 32, 214
Sugarcane, 214, 215
    bagasse from, 214
    molasses from, 170, 194–197
    silage of, 215
Sunflower meal, 32, 191, 192
Sweat, 268
Swine, 258, 264
    excreta, 261
Syncro-Mate, 84
Synovex-S, 6, 53, 108, 113, 281

## T

Tannins, 22, 205
Temperature, 289–299
    ambient
        diary cattle and, 279, 299
        diet and, 291, 292
        feedlot nutrition and, 294–298
        housing and management and, 299
        physiological responses and, 290, 291
        production problems and, 292, 294
        shade and, 296
        sprinkling and, 296
        water intake and, 294
    body
        calorigenic effect, 290
        fasting and, 289
        heat increment, 290, 296, 297
        heat stress and, 290–292
        metabolic body rate and, 290
Thiopeptin, 49, 85
Threonine, 24, 143, 145
Thymol, 24
Thyroprotein, 287
Thyroxin, 287, 288
Total digestible nutrients, 2, 120, 121, 123, 146, 161–163, 168–170, 186, 187, 195, 225, 236, 257, 262, 276, 302, 303
Toxic substances in rumen, 67–75
    alkaloids, 69
    cyanide, 67, 68
    gossypol, 74
    hemaggluting agent, 73, 74
    nitrates and nitrites, 69–71
    oxalates, 68, 69
    trypsin inhibitor, 72, 73
    urea, 71, 72
Tungstic acid precipitable nitrogen, 161, 163, 218
Tylosin, 86, 87
    acid detergent fiber and, 87
    feed efficiency and, 87
    liver abscess and, 87
Tyrosine, 144

## U

Urea, 2, 25, 31, 79, 117, 128, 129, 142, 146, 147, 149–160, 164, 169, 170, 174, 176, 180, 181, 189, 195, 196, 200–202, 206, 207, 210, 213, 214, 218, 267

# Index

Urea (cont.)
  corn silage and, 181
  methionine hydroxy analog and, 32
  soluble nitrogen factor, 25
  sulfur and, 33
  TDN and, 25
  toxicity of, 71, 72
  urea fermentation potential, 32, 196
Urease, 152
Urinary nitrogen, 101, 143, 148

## V

Valine, 24, 128, 143, 145, 155
Vanilla bean, 227, 228
Vitamin A, 159, 195, 280, 303, 304
Vitamin B, 44
Vitamin $B_{12}$, 139
Vitamin E, 159
Volatile fatty acids, 40, 41, 79, 124, 125, 130, 150, 193, 222, 294
  absorption of, 40
  ionophores and, 79, 80, 81, 83, 85

## W

Water, 51, 102, 269–273, 294
  absorption by omasum, 61
  buffers in, 51
  feed content of, 271, 272, 301
  feed utilization and, 270, 271
  intake by drinking, 272, 273
  nutrient elements in, 269, 270
  temperature and utilization of, 272, 294
  toxic elements in, 269, 270
Wet brewers grains, 145, 175
Wethers, 194
Wheat, 33, 35, 107, 170, 199, 240
  straw, 212, 213, 229
Whey, 192, 193, 226
Wood molasses, see Masonex

## X

Xylose, 205

## Y

Yeast, 170
Ytterbium, 66

## Z

Zeronal, 281
Zinc, 216, 265, 270